FINALIZATION IN SCIENCE

BOSTON STUDIES IN THE PHILOSOPHY OF SCIENCE

EDITED BY ROBERT S. COHEN AND MARX W. WARTOFSKY

VOLUME 77

GERNOT BÖHME, WOLFGANG VAN DEN DAELE,
RAINER HOHLFELD, WOLFGANG KROHN, WOLF SCHÄFER

FINALIZATION IN SCIENCE

The Social Orientation of Scientific Progress

Edited by

WOLF SCHÄFER

Johann Wolfgang Goethe-Universität, Frankfurt/Main

Translated by Pete Burgess

D. REIDEL PUBLISHING COMPANY

A MEMBER OF THE KLUWER ACADEMIC PUBLISHERS GROUP

DORDRECHT / BOSTON / LANCASTER

Library of Congress Cataloging in Publication Data

CIP

Gesellschaftliche Orientierung des wissenschaftlichen
Fortschritts. English.
Finalization in science.

(Boston studies in the philosophy of science ; v. 77)
Translation of: Die Gesellschaftliche Orientierung des wissenschaftlichen Fortschritts; pt. 3 added for English ed.
Includes bibliographies and index.
Contents: Authors' introduction / Gernot Böhme . . . et al. – Agricultural chemistry / Wolfgang Krohn and Wolf Schäfer – Autonomization and finalization / Gernot Böhme – [etc.]
1. Science—Social aspects. 2. Research—Social aspects.
I. Böhme, Gernot. II. Schäfer, Wolf, 1942– . III. Series.
[DNLM: 1. Science. 2. Research. 3. Sociology Q 175.5 G389f]
Q174.B67 vol. 77 [Q175.5] 501s [306'.45] 83–4443
ISBN 90–277–1549–1

Published by D. Reidel Publishing Company,
P.O. Box 17, 3300 AA Dordrecht, Holland.

Sold and distributed in the U.S.A. and Canada
by Kluwer Boston Inc.,
190 Old Derby Street, Hingham, MA 02043, U.S.A.

In all other countries, sold and distributed
by Kluwer Academic Publishers Group,
P.O. Box 322, 3300 AH Dordrecht, Holland.

Printed in The Netherlands

EDITORIAL PREFACE

These essays on *Finalization in Science – The Social Orientation of Scientific Progress* comprise a remarkable, problematic and controversial book. The authors propose a thesis about the social direction of scientific research which was the occasion of a lively and often bitter debate in Germany from 1976 to 1982. Their provocative thesis, briefly, is this: that modern science converges, historically, to the development of a number of 'closed theories', i.e. stable and relatively completed sciences, no longer to be improved by small changes but only by major changes in an entire theoretical structure. Further: that at such a stage of 'mature theory', the formerly viable norm of intra-scientific autonomy may *appropriately* be replaced by the social direction of further scientific research (within such a 'mature' field) for socially relevant or, we may bluntly say, 'task-oriented' purposes.

This is nothing less than a theory for the planning and social directing of science, under certain specific conditions. Understandably, it raised the sharp objections that such an approach would subordinate scientific inquiry as a free and untrammeled search for truth to the dictates of social relevance and dominant interests, even possibly to dictation and control for particularistic social and political interests.

The arguments and case-studies in this volume set out to show, in both theoretical and historical detail, why the authors think that this criticism is unwarranted and misplaced; but more, they provide a striking theory of the historical relations of theory and practice in science, and a model of the development of modern science that both rejects and goes beyond the models provided by Popper, Lakatos and Kuhn. At issue is the relation between 'internalist' and 'externalist' factors in the history of science. But this is no mere descriptivist history of the interaction of these factors. Rather, it is a critical-normative interpretation of the history of modern science, which ends with a program for a rational science policy which is derived from the authors' proposed theoretical periodization of scientific progress.

While not anticipating the formulations of the authors' thesis (and the revisions of their earlier formulations, which they also propose), it may still be useful to sketch the periodization here: the various sciences proceed through three stages, which are the *explorative* (pre-paradigmatic), *paradigmatic*, and *post-paradigmatic* (mature or 'closed' theory). During the first stage, external factors may and do play a significant role in orienting research. During the second, the typical 'autonomy' of internalist desiderata does (and should) dominate. However, in the *final* stage, it is appropriate that further theoretical development be oriented to the solution of problems which arise from the development of research in the service of social needs. 'Finalization', then, is not simply a description of a stage of scientific progress, nor even a hypothesis about such a stage, but at the same time a normative proposal for science policy, directed at the planning and funding of inquiry at this stage. We read that finalization "is a process through which external goals for science become the guideline of the development of scientific theory itself".

This is, of course, a bald statement of their thesis. In the last section of this volume, the prospects for such a 'normative finalization' are considered, and second thoughts and revisions of the original thesis are discussed. Moreover one of the authors (Schäfer) proposes that, in light of the research and conclusions thus far presented, an *alternative conception* of science needs to be formulated, in which the natural and social parameters of scientific *praxis* are synthesized in an ecologically-framed conception of a 'social natural science'.

But the body of this work is not schematic nor is it programmatic; it is a sustained elaboration of the thesis by way of the several historical case-studies, and of theoretical formulations and arguments which both guide the case-studies, and are tested by them. For historians and sociologists of science these studies may stand as a distinctive contribution, and also as provocative reconstructions of the patterns of scientific development. The case-studies are about agricultural chemistry in the nineteenth century in its relation to organic chemistry (Liebig *et al.*); fermentation research in relation to the development of microbiology, and the stimulating comparison of this field to that of fluid mechanics in the context of classical hydrodynamics; and cancer research (as a relevance-oriented field in its relation to, e.g. molecular biology, virology, immunology, etc.).

The project in these case-studies is not to impose some ready-made schema *on* the history but to see the uneven development of *different* models, at differentiated stages of development, in which the interplay of external and internal factors reveals the full complexities and varieties of the relations between theory and practice in science. The heuristic role of models from other disciplines is examined (e.g. in Liebig's use of the notion of reproductive cycles taken from economics and interpreted for the nitrogen, hydrogen and interdependent oxygen and carbon cycles in plant growth). So too, the distinction is drawn between 'applied' sciences, derived as 'transfer research' or 'technological development' from mature sciences, and *theoretical* development of a mature science in the search for specific applications which cannot be derived from the mature theory as it stands (e.g. in the theoretical developments of fluid mechanics for aircraft application, beyond the limits of classical hydrodynamics). The cancer research study is explored as a model of socially relevant 'task-oriented' research, directed to external goals, which involves a more complex relation to other disciplinary groupings. In short, the case-studies seek to give the body and detail required for an interpretation and a historical-empirical testing of the finalization thesis.

A host of interesting conceptual problems are posed here for the philosophy of science, and all the more for science policy studies: What is the criterion for an allegedly 'closed theory' (a concept, we must recall, introduced by Heisenberg and von Weizsäcker)? Is the progress of science finite? Does science converge to an 'end-state' in the future, in which a few wide-ranging 'closed theories' will have more or less exhausted what there is to be known in certain object-domains of scientific inquiry? Is the 'social steering' of theoretical knowledge a threat to the cognitive-theoretical search for truth, or is there some viable mediation of the two desiderata of truth and social relevance? Further – and crucially – *who* is to determine this social directedness of theoretical inquiry? How is the danger of particularistic power, or political interests, as the controller of inquiry to be avoided? (On this all-important question the authors remain, in our view, unfortunately programmatic while optimistic, and this surely requires much fuller critical discussion!)

The authors challenge the Popperian view of fallibilism in science. They challenge the unquestioned and unconditional criterion of autonomy in science. They challenge the ahistorical conception of the division between 'pure' and 'applied' science – and in one essay give a historical analysis of the origins of this division in the social and political contexts of the history of science itself. Also, they have prepared an account and a bibliography of the 'Finalization Debate' in Germany, so that we get the flavor and gist of the critical and polemical arguments against finalization.

We expect, and hope, that this volume will stimulate a lively discussion and debate among English-reading historians, philosophers and sociologists of science, medicine and technology, and within the science policy community as well. The unprecedented role of science, theoretical and applied;

of technology; and of science policy, public and private, in our contemporary society, form the background to this vigorous work. We are pleased to include it among the *Boston Studies in the Philosophy of Science.*

May 1983 ROBERT S. COHEN
Boston University MARX W. WARTOFSKY
Center for the Philosophy and History of Science

TABLE OF CONTENTS

EDITORIAL PREFACE v

FOREWORD xvii

AUTHORS' INTRODUCTION

GERNOT BÖHME, WOLFGANG VAN DEN DAELE, RAINER HOHLFELD, WOLFGANG KROHN AND WOLF SCHÄFER / Introduction 3

1. The Social Determinants of Knowledge 3
2. The Phases of Development in Scientific Disciplines 6
   2.1. Explorative phase 6
   2.2. Paradigmatic phase 7
   2.3. Post-paradigmatic phase 8
3. Science for Society as a Perspective 9

Notes 10
Bibliography 11

PART I. CASE STUDIES

SUMMARY OF CONTRIBUTIONS 15

WOLFGANG KROHN AND WOLF SCHÄFER / Agricultural Chemistry. The Origin and Structure of a Finalized Science 17

1. Introduction 17
2. The Demographic Context of Agricultural Chemistry 19
3. The Pre-history of Agricultural Chemistry 22
4. The Paradigm of Agricultural Chemistry 32
   4.1. Cycles of reproduction 32
   4.2. The constructive variation of cycles 36
   4.3. The progressive problemshift in agricultural chemistry 39

5. The Institutionalization of Agricultural Chemistry 43
6. Conclusion 46
Notes 47
Bibliography 50

GERNOT BÖHME / Autonomization and Finalization: A Comparison of Fermentation Research and Fluid Mechanics 53

1. Introduction 53
2. Fermentation Research – Fluid Mechanics: A Comparison of Two Theoretical Developments 54
2.1. Initial status and comparability 54
2.2. Historical overview 57
2.2.1. Fermentation research 57
2.2.2. Fluid mechanics 60
2.3. Analysis 64
2.3.1. Fermentation research 65
2.3.2. Fluid mechanics 71
3. Conclusion 83
Notes 86
Bibliography 88

RAINER HOHLFELD / Cancer Research. A Study of Praxis-Related Theoretical Developments in Chemistry, the Biosciences and Medicine 93

1. Introduction 93
2. The Development of Chemical and Biological Theories: Definitions and Assumptions 96
3. Cancer Approaches Based on Mature Theories 99
3.1. Classification of tumours in pathological anatomy 99
3.2. Chemical theories of carcinogenesis 100
3.3. The somatic mutation theory of tumour induction 101
3.4. The biochemical approach to cancer 103
3.5. Synopsis 105
4. Current Cancer Research in Cell and Developmental Biology 107
4.1. The molecular biology of higher cells 107
4.2. Experimental linkages between basic research in cell and developmental biology and cancer research 109
4.2.1. Tumour virology 110

4.2.2. Cell biology using systems of tumour cells 111
4.2.3. The system of blood-forming cells 112
4.2.4. A coincidence of tumour research and research into early developmental phases: the teratoma 113
4.3. Transfer research 113
4.4. Synopsis 114
5. Pure Empirical Cancer Research: Human Biology and Medicine 116
Notes 118
Bibliography 123

PART II. THEORETICAL CONSIDERATIONS

SUMMARY OF CONTRIBUTIONS 129

GERNOT BÖHME, WOLFGANG VAN DEN DAELE AND RAINER HOHLFELD / Finalization Revisited 131

I. *Problems in the Philosophy of Science*

1. Theoretical Maturity and Closed Theories 131
1.1. The phenomenon of the continuing validity of theories 131
1.2. The concept of closed theories 133
1.3. Systematic objections 135
1.4. Closed theories and the dynamics of theory 138
2. Fundamental Theories, Special Theories and the Problem of Applying theories 140
2.1. Theory-application as a problem in the philosophy of science 140
2.2. What does it mean to say ‘A theory is valid for an object, but cannot be applied to it’? 141
2.3. Principles of application 143
3. Theoretical Maturity as the Basis for Goal-orientation 145
3.1. “Theoretical maturity” in the case studies 145
3.2. Criteria for assessing the maturity of theories 147

II. *Science Policy Considerations*

1. Finalization and Direction 149

1.1. Orientation of science in the explorative phase 151
1.2. Orientation of science in the paradigmatic phase 152
1.3. Orientation of science in the post-paradigmatic phase 153
2. Theoretical vs Empirical Strategies for the Solution of Problems 157
2.1. The technical potential of theory 157
2.2. Orientation toward goals and criticism of goals 159
3. The Social Integration of Science 161
3.1. Finalization and autonomy 161
3.2. Disciplinary communities and "task communities" 164
Notes 166
Bibliography 170

GERNOT BÖHME, WOLFGANG VAN DEN DAELE AND WOLFGANG KROHN / The Scientification of Technology 173

1. Introduction 173
2. The Unification of Science and Technology in the Renaissance 174
2.1. The emergence of the science-technology relationship 174
2.2. Early technical and natural sciences 176
2.3. The knowledge of possible nature – the knowledge of the nature of the possible 177
2.4. The historical evolution of the interrelation of technology and science 179
3. The Separate Development of Natural Science and Technology 180
4. The Orientation of Science Toward Technology 183
4.1. Technological developments internal to science 183
4.2. The technology-orientation in the theory of mechanics and in the mechanistic world image 186
4.3. Scientific supply to production 188
5. The orientation of technology toward science 189
6. Theoretical technology 194
Notes 199
Bibliography 203

WOLF SCHÄFER / Normative Finalization 207

1. Introduction 207
2. The Theoretical Basis of Research in Finalization 207

2.1. Basic hypotheses reconsidered 207
2.2. Normative claims revised 212
2.3. The notion of a normative finalized science 213
3. Retraction of Basic Hypotheses and Suppression of Normative Claims 216
3.1. Theoretical retraction of basic hypotheses 216
3.2. Political suppression of normative claims 217
4. The Social Reconstruction of Science 219
4.1. A utopian perspective for science 219
4.2. Two strategies for research 220
4.3. Toward a revolution in the social paradigm of science 221
Notes 226
Bibliography 229

PART III. PROSPECTS

SUMMARY OF CONTRIBUTIONS 235

WOLFGANG VAN DEN DAELE AND WOLFGANG KROHN / Science in a Crisis of Legitimation 237

1. Introduction 237
2. Theoretical and Practical Prerequisites for the Legitimation of Science 237
3. First Paradox of Utility: from Experimenting for Human Beings to Experimenting on Human Beings 239
4. Second Paradox of Utility: the Pursuit of Technology for Its Own Sake 242
5. Control of Research and Change in Values: Between Dogmatism and Scientism 244
6. Accountability for the Social Consequences of Science 246
7. Perspectives 247
Notes 249
Bibliography 250

GERNOT BÖHME AND WOLF SCHÄFER / Towards a Social Science of Nature 251

1. The Project of the Modern Age – Recalled for Reform 251

2. Critical Science Studies Moved Forward: from Science and Society to the Politics of Nature 251
3. Science, Politics, Nature: Traditional and Contemporary Relations 252
3.1. Science and politics 252
3.2. Science and nature 254
3.3. Politics and nature 255
4. Digression: Going Natural, Going Luddite, and Going Technocrat. Three Ways of Going Astray 257
5. The Triangle of Science, Politics, and Nature Established 259
6. Normative Concepts of Nature in Social Reality 260
7. Normative Concepts of Nature in the Social Sciences 263
8. Research Prospects 266
Notes 268
Bibliography 268

APPENDIX

INTRODUCTORY NOTE 273

WOLF SCHÄFER / The Finalization Debate: A Reply to our Critics. With a Bibliography of the Finalization Discussion and Debate 275

1. Introduction 275
2. The First Proscription: It is Forbidden to Study the Progressive Goal Orientation of Science 277
2.1. The humus theory of basic research – an anti-finalist position 277
2.2. The dissolution of the coincidence of the social and cognitive autonomy of basic research – a finalist hypothesis 281
3. The Second Proscription: It is Forbidden to Advocate the Emancipatory Partisanship of Scientists 286
4. The Third Proscription: It is Forbidden to Believe in the Evolutionary Convergence of Theoretical and Practical Discourse 288
4.1. The disjunction of theory and practice – an anti-finalist position 288

4.2. The convergence of theoretical and practical discourse – a finalist hypothesis 290
5. Conclusion 293
Notes and References 294

BIBLIOGRAPHY OF THE FINALIZATION DISCUSSION AND DEBATE 301

I. The anti-finalization campaign and debate in the media 301
II. Reviews of *Starnberger Studien 1* 303
III. Contributions to the academic finalization discussion and debate 303

NOTES ON AUTHORS 307

INDEX OF NAMES 309

FOREWORD

The studies and essays contained in this volume incorporate that part of the historical and theoretical work of the research group "Alternatives in Science" which was directed at investigating the possibilities for the social orientation of science.

The Group, whose work is presented here in its collective context for the first time in an English edition, was formed at the Max Planck Institute for the Study of the Conditions of Life in the Scientific-Technical World, Starnberg, West Germany in the early 1970s: it initially consisted of three members (Gernot Böhme, Wolfgang van den Daele and Wolfgang Krohn), later expanded to six (Wolf Schäfer, Rainer Hohlfeld and Tilman Spengler), and united not merely a diverse range of temperaments but also wide-ranging specialisms – physics, law, philosophy, history, biology and sinology. The Institute, which had been established by Carl Friedrich von Weizsäcker in 1970, and jointly headed by himself and the sociologist Jürgen Habermas was closed in 1981 after considerable controversy: this decision removed the institutional basis for the work of the group within a *single* team. Nevertheless, the former members of the group continue to work in this area in a number of different universities and institutes in West Germany (Bielefeld, Darmstadt, Erlangen and Frankfurt).

The fundamental theoretical framework for the contributions in this volume was originally outlined in 1973 by Gernot Böhme, Wolfgang van den Daele, and Wolfgang Krohn under the title 'Die Finalisierung der Wissenschaft' in *Zeitschrift für Soziologie* (Vol. 2, pp. 128–144). Most of the papers in the present volume – Parts I and II – were published in Germany in 1978 in Volume I of the *Starnberger Studien* series (Böhme, van den Daele, Hohlfeld, Krohn, Schäfer, Spengler, *Die gesellschaftliche Orientierung des wissenschaftlichen Fortschritts*). The two essays in Part III were written especially for this volume with the aim of clarifying both for ourselves and for the reader how our emphases have changed and how we envisage the next steps in our current work: by setting out the beginnings and the perspectives for new research projects they serve to conclude the earlier research of the group.

The working procedure of the group and the structure of this volume can be seen to follow a common logic of development. An initial hypothesis (*cf.* the English version of Böhme *et al.*, 1973: 'Finalization in Science', *Social Science Information* **15**, 1976, pp. 307–330), was followed by the empirical testing of its basic theoretical premisses through a number of selected case-studies (Part I); in turn, the findings of the case-studies inspired a reflection on and revision of the original position (Part II); the reconstruction of the theory, together with the growing impact – or more acute perception – of a number of urgent contemporary problems (the crisis of science, ecological crisis) finally demanded the formulation of a number of new hypotheses (Part III). – The essay on the Finalization Debate in the Appendix gives some insight into a different, and rather more negative, challenge which became associated with our previous theoretical politics.

I would like to thank all those friends and colleagues – Seyla Benhabib, Gernot Böhme, Everett Mendelsohn and Marx Wartofsky – who stimulated and supported me in the preparation of this volume. My thanks also go to Debra Nails for her conscientious administrative assistance and to Pete Burgess, with whom it was a special pleasure to work. Particular thanks are reserved for Robert S. Cohen, who not only granted me the hospitality of the international Center for Philosophy and History of Science, but who also donated his friendship and advice.

Frankfurt, August 1982 WOLF SCHÄFER

AUTHORS' INTRODUCTION

GERNOT BÖHME, WOLFGANG VAN DEN DAELE, RAINER HOHLFELD, WOLFGANG KROHN AND WOLF SCHÄFER

INTRODUCTION

1. THE SOCIAL DETERMINANTS OF KNOWLEDGE

Systems of knowledge, including systems of knowledge about the natural world, originate within the framework of patterns of social activity. They are dependent on the structures through which people have ordered their economic, political, social, cultural and religious relations. Any process leading to the acquisition of knowledge is not only rooted in needs and interests, but is also built upon the images and definitions which emerge from this social environment, and which offer themselves as models for the classification and explanation of natural phenomena. Although empirical knowledge represents a form of analysis of an independently conceived nature, it is also a practical activity carried out in a social context at a particular point in time. And although knowledge leads to solutions to problems, the validity of which can be confirmed through the degree of success they encounter in practical application, the specification of what is problematic, and the interpretation of these solutions, takes place through conceptual schemes which originate in social reality.

It is no longer disputed that this situation characterised the empirical systems of knowledge of the *archaic cultures*. We encounter them as intimately bound up with magic rituals, the symbolism of language, numerology, witchcraft, astrology and religious myths. Although natural knowledge was formulated in social categories it evidently worked in practice.

In the case of *Aristotelian natural philosophy* it seems plausible to assume that the forces and purposes attributed to nature have been modelled on the labour of the skilled artisan.[1] And in *scholastic natural philosophy* we find the influence both of Christian religion and hierarchical conceptions of feudal society.[2]

Such influences can also be detected in the *sciences of the modern period*, whose development began in the seventeenth century. The interpretation of nature as an ordered system of laws represents a probable analogy to the absolutist state and the juridical character of bourgeois society.[3] Darwin acknowledged that Malthus' theory of population had served as a model for biological evolution.[4] And Liebig based his principles of agricultural

G. Böhme et al., Finalization in Science, ed. by W. Schäfer, 3–11.

chemistry on the notion of the cycle as advanced in bourgeois political economy.[5]

Despite the existence of such examples, a number of objections have been made to the drawing of such parallels between early natural knowledge and modern science. There are two main counter-arguments. The first contends that the full development of the hypothetical approach and experimental method in the seventeenth century created a set of tools for the acquisition of knowledge which, by its nature, has eliminated the influence of social factors in determining the character and substance of objective knowledge. And the second holds that with quantification as an ideal of analysis, linguistic analogies and metaphorical associations become inconsequential in science. These two arguments have led a number of philosophers of science to locate any social influence solely within the context of discovery, and deny it any role in affecting the character of knowledge once it is established. The dominant view amongst both natural scientists and philosophers of natural science is that the methods and the research programme of modern natural science have established an "internalism" which rules out "external", social factors – and in particular, which excludes any subordination of science to social problems or social purposes.[6]

There is no simple decision which can resolve whether the "internalist" or "externalist" position is the correct one. Rather than attempting an abstract resolution of this dichotomy, we argue here that the question must be viewed in concrete historical terms. The answer will depend on the particular epoch of science under consideration, the discipline involved, and the different demands placed on science by society. It is equally important to distinguish the different *types* of the social determination of science. Unlike the systems of knowledge of earlier cultures, twentieth century sciences are no longer tied to magical or mythical beliefs; and in contrast to the sciences of the seventeenth century, they are no longer oriented towards comprehensive world-views. There are barely any "axiomatic beliefs" in contemporary science derived from philosophical or religious principles. However, the neutrality of science in relation to these systems of thought does not necessarily mean that no other factors may have taken their place.

The conventional model for a consistent "internalist" conception of the history of science is provided by physics. The passage from Newtonian mechanics to relativity and quantum theory took place largely independently of economic, religious or political determination.[7] But if we direct our attention away from one particular case – although it may be of philosophical interest – and look at the broad sweep of current scientific development, such an internalist position can no longer be viewed as tenable: the bulk of public research funds, together with facilities in the private sector, are not supplied with the aim of augmenting scientific knowledge in general, but function to tackle specific industrial, military and civil projects. Science has become a social resource, which is mobilised to solve economic, political and social problems.[8] This social function of science in turn feeds back into its own development – a fact manifested both in the organizational form of research and the increasing importance of technical areas of research within the classical disciplines. The indisputable fact that research is financially and organizationally externally steered is usually reconciled with the internalist position by distinguishing between the *development* of theories, frequently termed basic research – which continues to be regarded as an autonomous enterprise, accessible to promotion but not direction – and *planning* and *direction*, which can only occur once theoretical results have been made available for external objectives – customarily termed applied research. The studies which make up this volume oppose this claim, and argue that it is possible for *theoretical* developments themselves to be subject to social influences and even political planning, and further, that this phenomenon of orientability is to be found at the *current* level of scientific attainment. However, the manner in which the social forming of theories is exercised has shifted from linguistic patterns and metaphors to the perception of what constitutes a scientific problem, and the establishment of criteria for solving them. Thus, the social orientation of present-day sciences can be seen to be operative in a number of different areas:

– the *selection* of research problems; in its simplest form the ordering of priorities in accordance with military, economic, welfare, or other, interests;

– the *definition* of problems; for example, the establishment of concepts of illness, and objectives in combating it – in general terms, the translation of social goals and standards into concrete research problems. Standards (such as profitability, utility, environmental acceptability) are matched with technical norms (increasing efficiency, durability, reduction in noise levels), which in turn serve to define particular research problems (in thermodynamics, fracture mechanics, fluid mechanics);

– the specification of *explanatory ideals* which are to be regarded as appropriate for a given field (micro-theoretical explanation, macro-theoretical explanation, technical forecasts, computer simulation etc.);

– the *limiting conditions* of research – that is, the boundaries beyond which further analysis of the field is not held to be of interest. The conditions follow from the specification of the explanatory ideals for the field.

These influences do not mean that the sciences are impoverished or reduced to the status of adjuncts of ideologies. In fact the opposite is the case; the susceptibility of the substance of theories to forming influences originating in society is a sign that empirical knowledge has been pursued successfully. In other words: the methods of the modern natural sciences render more and more disciplines open to the possibility of social direction. The research dynamic of the modern sciences contains within itself the potential for an increase in those conditions which favour the social steering of theoretical knowledge. The following is a summary of the model we used to investigate the interaction between scientific progress and social goals.

2. THE PHASES OF DEVELOPMENT IN SCIENTIFIC DISCIPLINES

Our discussion of the sciences primarily focuses on the disciplines of the *natural* sciences. Although differing in terms of the specific conditions of their initial emergence, the various object-fields of these disciplines can be characterized as "simple" or "homogeneous".[9] Here these disciplines are considered historically given starting points for the analysis of the dynamics of current science. A preliminary comparative glance at their histories shows that each has passed through similar phases in development. These changes can be made explicit in the form of a three-phase model. Following Kuhn's terminology we distinguish an "explorative", a "paradigmatic" and a "post-paradigmatic" phase.[10]

2.1. *Explorative Phase*

The explorative phase of a discipline embraces the period of its development prior to the emergence of theories which serve to organise the scientific field. Methods are predominantly inductive in character, and research is determined by strategies aimed at classification and experiment. The dynamics of the field are characterised more by discovery than explanation. The fine structure of the object of study remains largely unknown, and is handled in a manner closely paralleling cybernetics' famous "black box". The scientist knows the relevant inputs and outputs – but what goes on between remains a mystery. Kuhn's "pre-paradigmatic" sciences such as pre-Lavoisier chemistry, or the theory of electricity prior to Franklin represent historical instances of this phase: modern examples can be found in the fields of elementary particle physics and molecular developmental biology.

2.2. *Paradigmatic Phase*

The onset of the paradigmatic phase is marked by the emergence of a theoretical approach which is able to organise the field. The introduction and elaboration of this approach represents a theoretical development with a definitive end. Following Kuhn, we term this the establishment of a paradigm. The acquisition of a paradigm initiates a type of scientific dynamic in which the progress of research is determined by the problems of theory. Although scientists still have certain options as far as raising questions and deciding on strategic priorities are concerned, development nevertheless takes place in accordance with a reasonably clearly identifiable internal logic. Diverse approaches converge onto a limited number of fundamental – and often hierarchically structured – problems. The formulation of the "central dogma" of molecular genetics since 1953 represents one recent example of theoretical development during the paradigmatic phase. Crick and Watson's "dogma" contained a number of basic suppositions about the functions of DNA for the replication of genetic information and the synthesis of proteins. The direction of the research programme in which these suppositions have been elaborated and tested has embraced the experimental verification of central hypotheses, the reformulation of the classical concept of the gene, the formulation of the genetic code, and elucidation of how protein synthesis is quantitatively regulated.[11]

The theoretical dynamic of the paradigmatic phase is evidently one which can come to a conclusion – that is, can lead to mature theories which contain a fundamental, and in certain respects a conclusive, understanding of the discipline's research object. This, at any rate, is the interpretation offered by scientists themselves who operate on the assumption that the problems of motion, heat, light, electricity, the make-up of matter under terrestrial conditions, heredity, and other fundamental problems have been essentially resolved by current theories. The history of science also reveals that solutions of this type are not simply rendered obsolete, even if the field is shaken by a theoretical revolution *à la* Kuhn, in which a new paradigm is established. With a few notable exceptions – such as the phlogiston theory of combustion – superseded paradigms remain both as instruments of research and parts of the accumulated body of scientific knowledge. In fact, the success and limitations of such paradigms can often only be explained by a succeeding theory. Heisenberg's notion of *closed theories* represents a first step in providing an explanation of the relative stability of theories. The further development of those fields which possess dominant mature theories has three aspects. Firstly, existing theories become grounded in more general theories. Secondly, the scope of the paradigm is extended to embrace objects not previously covered – the application of physics and chemistry to heredity being a case in point. And thirdly, the paradigm is specifically applied to groups of phenomena which are already in principle established as lying within the scope of the paradigm.

Of these three aspects, the third is the most significant for the post-paradigmatic development of individual disciplines, and fits most closely with Kuhn's definition of normal science. Instead of aiming to advance beyond the paradigm's basic propositions, this type of research attempts to exploit the existing framework to the full. This naturally presupposes that the paradigmatic phase has already yielded explanations of the field's phenomena which can be treated as basically stable or mature – that is, not open to further improvement.[12]

2.3. *Post-paradigmatic Phase*

Where the organising theories of scientific disciplines are clearly formulated and comprehensive, the possibilities of revolutionary changes or spectacular generalizations of their basic principles are commensurably reduced. Instead, the dynamics of theoretical development will be determined by the application of paradigmatic theories for the explanation of complex systems which can be subsumed within them.[13]

The application of basic theoretical principles is not a process in which results are simply obtained from pre-existing theories through logical deduction. The situation is rather that of the development of special theories, and by implication, the continuance of basic research within a field. However – and this is the central contention of this study – this research can be oriented and planned in line with external goals. In contrast to the paradigmatic phase, post-paradigmatic development in a discipline is not characterised by an internal logic which determines the direction of progress and selects problems for research. The attainment of theoretical maturity means that – even where no reference is made to external objectives – it is possible to define an increasing number of problems which are open to theoretical treatment. However, there is no clear hierarchy of results and research fronts. Science is therefore open to orientation in accordance with goals which do not originate within science. Research fronts can be planned – not necessarily in the form of an exhaustive step-by-step specification of what has to be done, but rather through the institutionalization of specific areas of basic research with certain goals in mind. Further theoretical development within the discipline can then proceed broadly along the path indicated by such external goals. We term such a process of theoretical development "finalization".

3. SCIENCE FOR SOCIETY AS A PERSPECTIVE

The potential for an orientation to external objectives increases during the course of scientific development. The success of science is not only evidenced by the accumulation of technically exploitable individual findings, but also through the fact that more and more disciplines arrive at the stage of post-paradigmatic research, at which point it is possible to observe a convergence between the development of science and technology at the level of the formation of theory.

Although the unity of natural and technical knowledge was already implicit in the operational concept of early modern scientific knowledge, which took as its reference construction and prognosis, theories from the disciplines of natural science did not assume the function of special technical theories until the twentieth century. With the finalization of theoretical development, the methodological and sociological separation of science and technology ceases to be tenable.[14]

The political consequence of the transition to post-paradigmatic development is the disintegration of science's traditional claim to autonomy. Understood as a particular form of organization, autonomy was achieved for the disciplines during the phase of paradigmatic theoretical development, in which the course of research is governed by an internal logic and where the concretization and execution of research priorities must be left to the scientists, free from any intervention dictated by criteria external to science. To this extent the institutional autonomy of science can be legitimated by appeal to the cognitive logic of science. However, once scientific areas attain theoretical maturity and enter a phase of post-paradigmatic development, autonomy can no longer be justified. External criteria can not only be invoked to determine the use of the results of mature sciences, but can also influence the actual process of theoretical development. In the case of such sciences it is in principle both feasible and justifiable to call for the social orientation of scientific progress.

Daniel Bell has forecast that "theoretical knowledge will be the strategic resource of post-industrial society".[15] If our diagnosis of the dynamics of modern science is correct, fewer and fewer fields of science will be characterised by a relationship between science and society in which society is the passive partner and increasingly by one in which society takes an active and guiding role.

The paradigmatic phase of scientific development, often presented as the "Golden Age" of science, must eventually come to a close in all disciplines. This does not, however, mean the termination of theoretical developments within those fields; rather, it signifies the beginning of the possibility of theoretical development in pursuit of social goals.[16] The realisation of this possibility, which is inextricably related to the task of responsibly determining the aims of the social development of science (normative finalization), requires political processes which can neither be left to state or industrial bureaucracies nor to a "republic of science".

NOTES

[1] Moscovici (1968), pp. 78 f.

[2] Nitschke (1967); Stürner (1975), especially pp. 185 ff.

[3] Zilsel (1976), pp. 93 ff.

[4] Darwin (1892), pp. 42 f. *Cf.* too the letter from Wallace, 2nd December 1887 (*ibid.*), p. 200.

[5] See 'Agricultural Chemistry. The origin and structure of a finalized science', in this volume.

[6] See Lakatos (1971), in particular p. 105.

[7] For a reconstruction of this inner logic see, for example, Mach (1889); Reichenbach (1942).

[8] *Bundesforschungsbericht V*, pp. 80–81.

[9] Of significance are: *states of matter* – mechanics, hydrodynamics, theory of gases, plasma physics; *structure of matter* – elementary particle physics, nuclear physics, quantum mechanics, solid-state physics; *fields of phenomena* – theory of electricity, optics, acoustics; *configurations or structures* – astrophysics, chemistry, genetics, neurology.

[10] If sub-fields or levels with independent theoretical characteristics can be distinguished within one discipline, then the three-phase model can also occur within each of these sub-fields. The most important examples of this are disciplines in which research at the phenomenological or macrotheoretical level takes place independently of the microtheoretical, such as molecular and macro-genetics, statistical and phenomenological thermodynamics, electrodynamics and physical optics.

[11] See Stent (1969), pp. 49 ff.

[12] See 'Finalization Revisited' below for a precise specification of this idea.

[13] If we consider contemporary physics, as presented in the American Academy of Science's report *Physics in Perspective*, out of eight sub-fields of physics only "elementary particles" and "astrophysics and relativity" (and to a lesser extent nuclear physics) were accorded a clear "potential for the discovery of fundamental laws." The five other

areas (acoustics, optics, plasma and fluids, solid-state, atomic physics) are developing through the application of stable general theories, in particular quantum mechanics. These areas were felt to have some "potential for the discovery of generalisations with wide scientific applicability," and they rate highly in their "potential contribution to technology" (see *Physics in Perspective 1972*, p. 404). In post-quantum mechanics physics, the remaining fundamental problems themselves have become esoteric special domains. The field is dominated by post-paradigmatic sciences. In 1969 40% of all published research papers in physics were in solid-states physics. 25.5% of doctoral graduates work in this field (see *Physics in Perspective* 1972, p. 926).

14 See 'The Scientification of Technology' below.

15 Bell (1976), p. 49.

16 This was expressed by Franklin Long on the occasion of the bi-centennial celebration for chemistry in 1974: chemistry is a mature discipline with stable paradigms, whose prospects for development in the next century are to found in a strengthened committment to the "service of humanity" – however that may be interpreted in specific detail. The direct demand that science should solve social problems will be the interesting challenge of the future. In the year 2074 "science for society" will have its "golden age" (*Science* 1975, p. 4154).

BIBLIOGRAPHY

Bell, D.: 1976, 'Welcome to the post-industrial society', *Physics Today* **29**, No. 2, pp. 46–49.

Darwin, C.: 1958, *The Autobiography of Charles Darwin and Selected Letters*, Dover, New York.

Federal German Ministry of Research and Technology: 1975, *Forschungsbericht V der Bundesregierung*, Bonn.

Lakatos, I.: 1971, 'History of science and its rational reconstruction', in *Boston Studies in the Philosophy of Science,* **VIII**, pp. 91–136.

Long, Franklin quoted in Hammond, A. L. 1974, 'Chemistry and science: the next hundred years', *Science* **185**, No. 4154, p. 847.

Mach, E.: 1889, *Die Mechanik in ihrer Entwicklung*, Brockhaus, Leipzig (9th edition, 1933).

Moscovici, S.: 1968, *Essai sur l'histoire humaine de la nature*, Flammarion, Paris.

National Academy of Sciences: 1972, *Physics in Perspective*, Vol. I, Washington D.C.

Nitschke, A.: 1967, *Naturerkenntnis und politisches Handeln im Mittelalter* (*Stuttgarter Beiträge zur Geschichte und Politik*, Vol. 2), Klett, Stuttgart.

Reichenbach, H.: 1927, *Von Kopernikus bis Einstein*, Berlin.

Stent, G. S.: 1969, *The Coming of the Golden Age*, Natural History, New York.

Stürner, W.: 1975, *Natur und Gesellschaft im Denken des Hoch- und Spätmittelalters.* (*Stuttgarter Beiträge*, Vol. 7), Klett, Stuttgart.

Zilsel, E.: 1976, *Die sozialen Ursprünge der neuzeitlichen Wissenschaft*, W. Krohn (ed.), Suhrkamp, Frankfurt.

PART I

CASE STUDIES

SUMMARY OF CONTRIBUTIONS

The case studies of *Agricultural Chemistry* and *Cancer Research* together with the comparison of *Fermentation Research* and *Fluid Mechanics* form our analyses of the sciences' capacity for orientation.

Agricultural Chemistry – which emerged around 1840 – is an instance of finalization, a scientific special development, made possible through the methodological maturity of organic chemistry; however, Agricultural Chemistry was not a product of problems internal to the research programme of organic chemistry, but rather an upshot of the external necessity to feed a growing industrial population. As such, it did not merely represent the application of chemistry to certain specific problems, but rather demanded the development of a special theory, that of the cycles of substances within the chemical interrelations of soil, air, animals and plants.

Fluid Mechanics represents a second case of finalization, in which the growing significance of aircraft construction prompted a special development of hydrodynamics. It represents an instance in which equations developed within the original science – hydrodynamics – could not be applied directly, necessitating the creation of a theory for the application of basic theories – the boundary-layer theory. The same case study compares *Fluid Mechanics* with *Fermentation Research*. Although there is no material relation between the two areas, they can be compared structurally: both are underpinned by certain industrial problems (the production of beer and vinegar on the one hand, aircraft production on the other), and in both cases a scientific solution to these problems was both expected and pursued. However, this search was only successful in the case of Fluid Mechanics. Fermentation Research lost its practical reference for some time, and became submerged in the fundamental problems of microbiology – autonomization and not finalization took place.

Cancer Research tackles a similar theme: the problem of this illness touches on a number of disciplines (molecular biology, virology, immunology, cell biology, tumour virology, experimental medicine, epidemiology), all of which in principle might be expected to produce a solution. Each reacts differently, depending on their stage of theoretical development. The study classifies the variations in their relation to practical application and shows

G. Böhme et al., Finalization in Science, ed. by W. Schäfer, 15–16.

that even in "sub-finalization" conditions, goal-oriented research relevant for the development of special forms of therapy can be carried out. At the same time, the paper also suggests that a fundamental understanding of the mechanisms through which cancer develops are most likely to emerge from an elucidation of the molecular processes regulating the growth and differentiation of cells; and that such a theory would constitute the scientific basis for a causally-based form of cancer-therapy.

WOLFGANG KROHN AND WOLF SCHÄFER

AGRICULTURAL CHEMISTRY. THE ORIGIN AND STRUCTURE OF A FINALIZED SCIENCE

1. INTRODUCTION

This case-study of agricultural chemistry illustrates a number of those interconnections between social needs, cognitive patterns in science and strategies for the institutionalization of science which are relevant to a cognitive sociology of the development of science and technology. In particular, it allows a consideration of the interaction between, (i) the existence of a social problem, (ii) the perception of this problem, (iii) the limited ability of science to offer a solution to the problem at a particular stage of development, (iv) the development of experimental techniques and models, (v) the institutionalization of a cognitive variant in science. The specific structuring of these relationships within agricultural chemistry enables it to be classified within that category of science dynamics termed by us "finalized science".[1] Finalized science occupies a place between those types of scientific activity which – at least conventionally – are regarded as the investigation of the "construction of reality", carried out independently of any social influences, and those whose main concern is the application of scientific findings to special technical problems using the theoretical tools already at hand. In contrast, agricultural chemistry involves both the development of fundamental theories *and* the consideration of the social goals which prompt and shape these theories.

External influence on science can take a number of forms. Science can be stimulated directly through the allocation of resources or the creation of a particular institutional framework – for example, the promotion of research into magnetism (declination of compasses) aimed at solving the problem of finding position at sea. Such stimuli do not, however, impinge on the cognitive structure of a discipline. External influence can also act on the content of theories. This is normally confined to the semantics of a theory when patterns of thinking which originate in the social belief-system help to articulate new scientific knowledge. Political and economic liberalism, for example, gave its particular stamp to the formulation of Darwinian theory. Although such patterns may play a heuristic role in determining the direction of research, no *particular* pattern is necessary. Finally, external

G. Böhme et al., Finalization in Science, ed. by W. Schäfer, 17–52.

influence can operate upon the cognitive content of a science via the process through which the objects of the research field are constituted. This occurs when concepts within the field imply objectives which cannot be formulated in terms of the natural characteristics of a pre-given reality. Agricultural chemistry represents exactly such a case.

Agricultural chemistry is concerned with the possibilities for the selection and construction of levels of biological reproduction in accordance with social objectives. The emergence of agricultural chemistry allows us to observe the process by which social goals act as constitutive elements in the formation of theories. "Agricultural chemistry" does not mean merely the growth of cultivated plants, but embraces all those attempts to specify and manipulate existing, or possible, chemical interrelations established by the mechanisms of respiration, nutrition, absorption and putrefaction between the atmosphere, the soil, plants and human beings. Its object consists of the chemical conditions for the primary reproduction cycles of cultivated plants. The particular circumstances of the period enabled it to play a key role in the social developments of the nineteenth century.

Naturally, social norms are not the only forces which act to structure scientific theories. The sciences themselves must possess the appropriate cognitive and institutional features before they can respond to social problems.[2] Prior to the emergence of agricultural chemistry proper, chemistry was unable to make any positive impact on agricultural problems, despite their seriousness, merely offering *ad hoc* solutions not markedly superior to technical and economic approaches.

The development of agricultural chemistry can be outlined in terms of six main points – which we also use as the structure for our study.

(1) Needs existed which were external to science: these needs occasion the emergence of agricultural chemistry and influence its content.

(2) Following Wallerius (1764), agricultural chemistry became an academic area, and was included in the research programme of "rational husbandry" (Thaer) from around 1800; however, as long as its theoretical basis was made up of phlogiston chemistry or early agrarian science, real advance was impossible.

(3) Agricultural chemistry was not a necessary step in the development of chemistry, and nor was it an indispensable part of the research programme of nineteenth century chemistry. In fact, it represented a special development of chemistry.

(4) Nevertheless, agricultural chemistry did build on the foundations established by the relative *theoretical* maturity of inorganic chemistry,

embodied in the formulation of certain fundamental laws, and the *methodological* maturity of organic chemistry.

(5) However, agricultural chemistry was not the mere application of the results of other "more fundamental disciplines", but an identifiable theoretical development in its own right, generating its own specific heuristic procedures and explanatory models.

(6) Agricultural chemistry became formally institutionalized. Its establishment and consolidation cannot be attributed simply to theoretical or experimental success, but was also promoted through the planning of research and science policy.

2. THE DEMOGRAPHIC CONTEXT OF AGRICULTURAL CHEMISTRY

Liebig's "application of chemistry to agriculture and physiology"[3] was not undertaken with the aim of solving the scientific question of the nutrition of plants, but with overcoming the social problem of the nutrition of people. A number of pieces of evidence are required to show that this objective not merely prompted the development of agricultural chemistry but also influenced the structure of its theory. We therefore have to demonstrate, (i) that the period in which Liebig worked witnessed an exceptional growth in population, (ii) that there was evidence of an impending subsistence crisis, (iii) that Liebig was aware of this crisis, and (iv) that it is possible to discern the orientation of agricultural chemistry to external goals in the structural features of its theories.

The "context of discovery" for agricultural chemistry is constituted by two circumstances: on the one hand the interdependence of population growth and subsistence crisis, and on the other, Liebig's consciousness of this situation. The population explosion of the eighteenth and nineteenth centuries which followed in the train of the Industrial Revolution necessitated a new relationship with the material world, in which natural resources had to be more intensively exploited for agricultural purposes as the yields from traditional agriculture were simply insufficient to guarantee an adequate supply of food to the growing industrial population. In such a context, the development of agricultural chemistry was a progressive step, not simply because it enabled the industrial population to be fed, consequently releasing subsequent social development from its dependence on pre-industrial agriculture, but also because it established the theoretical foundations for a new relationship with the material world.

The establishment of a new relationship to the material world can take a number of forms: new sources of power can be developed – as in the Industrial Revolution – or, as was the case with Liebig's agricultural chemistry, an existing relationship to nature can be *reconstructed*. In fact, ever since the onset of the industrial and demographic revolution in the mid-eighteenth century agriculture had been expected to make progress and to some extent it really did but was not able in its "pre-1840" state[4] to understand and prevent the growing danger of a rapacious exploitation of the soil. The practical man of agriculture was unable to grasp that the intensification of arable-farming and stock-production rested on the dubious art of draining the fertility of the soil at an ever-accelerating pace; this "practical man" characterized by Liebig as the "sworn enemy of all 'theory'" had "theorized that the fertility of his soil was inexhaustible . . . and had acted in full accord with the *theory* that the sources required to restore the yields of his fields were inexhaustible".[5]

Doubts began to be raised about this optimistic scenario around the beginning of the nineteenth century. Malthus' *Essay on the Principle of Population* (1798) represented a systematic attempt to calculate the disparity between the revolutionary population explosion and the conservative production of food, based on the now famous principle that, "population when unchecked goes on doubling itself every twenty five years, or increases in a geometrical ratio"[6] whereas, "the means of subsistence under circumstances the most favourable to human industry could not possibly be made to increase faster than in an arithmetical ratio".[7]

Although Malthus was incorrect to represent the movement of population as a natural law,[8] his perception of the historical situation in the immediate wake of the Industrial Revolution was acute, and the impact of the thesis considerable. By contending that traditional agriculture was powerless in the face of demographic realities Malthus gave a new edge to the crisis mentality of the modern period, and introduced the vision of a *secular subsistence crisis* into the nineteenth-century world view.

> When acre has been added to acre till all the fertile land is occupied, the yearly increase of food must depend upon the melioration of land already in possession. This is a fund, which, from the nature of all soils, instead of increasing, must be gradually diminishing. But population . . . would go on with unexhausted vigour; and the increase of one period would furnish the power of a greater increase the next, and this without any limit.[9]

The new situation of a widening gap between the growth in population and the increase in means of subsistence in which the industrialised countries found themselves was therefore already recognized by the beginning of the nineteenth century. Two solutions were open: either drastically reduce the growth in population, or increase the production of foodstuffs far beyond the yields obtained from the traditional agriculture of the "practical man". The alternative was a "natural" regulation of the gap between the surplus population and the available food supply; in other words, a dramatic increase in mortality.

The problems created by the unprecedented development of population in the early phases of industrialization were immense. World population statistics recorded an increase of more than 250 million between 1800 and 1850 (from 906 to 1171); the population of Europe almost doubled within a century – from 140 million in 1750 to 266 million by 1850; Great Britain's population tripled between 1750 and 1860 (from 7 million to 23 million).[10] The fact of this extraordinary increase in population, together with Malthus' warning that "population must always be kept down to the level of the means of subsistence"[11] served as the stimulus for the development of agricultural chemistry.

Liebig was aware of the threat of a secular subsistence crisis, and its exacerbation through the fall in the share of the population working on the land. Unlike Malthus, however, he recoiled from the prospect of "preventive checks to population" and instead looked for an alternative to "present husbandry". For Liebig the "social question" of his time did not pose itself as a question of morals or of population policy but as a *scientific and technical problem*. By tackling the population problem as a problem of nutrition, he confronted chemistry with the social question of establishing the conditions under which "large and continuously increasing harvests" could be obtained "in perpetuity".[12] This recasting of Malthus' gloomy vision consequently became the starting-point, and the *leitmotiv*, for a scientific project. Liebig himself clearly believed that had certain scientific breakthroughs not been made, a crisis would have been inevitable.

A combination of circumstances has brought about an increase in the population of all the European countries in proportions which do not correspond to the productive capacities of these countries, and which are therefore unnatural. Population has reached such a level that it can only be sustained with present techniques of husbandry under two conditions. One, if a Divine miracle intervenes to restore the fields to the degree of productivity stripped from them by folly and ignorance; and two, if deposits of manure or guano are discovered in volumes approximating to those of the English coalfields. No one with any knowledge or reason would expect either of these conditions to be fulfilled. Within a few years the guano deposits will be exhausted, and no scientific or, if one wishes, theoretical debates will be required to prove the existence of that law of nature which obliges humanity to attend to the conditions which maintain life ... Self-preservation will compel nations to slaughter and annihilate each other in never-ending wars in order to restore an equilibrium, and, God forbid, if two years of famine such as 1816 and 1817 succeed each other again, those who survive will look on as hundreds of thousands perish in the streets. Add a war, and as in the Thirty Years' War, mothers will drag home the bodies of the slain enemy to still the hunger of their children with their flesh; and as in Silesia in 1847, they will exhume the bodies of diseased animals to prolong their death agony with the carcass. These are not vague and obscure prophesies, images of a sick imagination; for science does not indulge in prophecy but calculates – not *if* but *when* is what is uncertain.[13]

We are now better able to appreciate the contradiction between the industrial-demographic revolution and the accompanying intensification of traditional agricultural practices. Despite external appearances, far from being a revolution which heralded stable improvements, this agrarian revolution was a manifestation of the latent depradation of natural resources which had been the hallmark of agriculture since its Neolithic origins, a process expressively captured in the German term *Raubbau* (literallly robbing the soil). In Liebig's view the "future of the countries of Europe was balanced on the tip of a needle."[14] Agricultural chemistry was set the task of extending this precarious basis – and successfully met the challenge. By "explaining the chemical process of the nutrition of plants,"[15] and consequently exposing "modern farming as a practice of depradation"[16] it became possible to develop a progressive agricultural practice whose manifest characteristic was the use of artificial fertilisers, and whose significance for the human history of nature was – and still is – the concept of a non-exploitative relation to nature "like a wave motion within a cycle."[17]

This idea of a balanced relationship between nature and humanity remains unrealised. Although increases in yields have been achieved which would have been scarcely credible in the years before 1840, new ecological problems have arisen to jeopardize the "maintenance of the conditions for life."[18] However, the programme which is currently required to overcome these dangers is not one which contradicts the premisses of the paradigm of agricultural chemistry; advances in physical, chemical and biological understanding as to the causes of ecological disequilibria in agriculture, together with the knowledge needed to establish stable cycles, will emerge through scientific and technical procedures which rest on Liebig's paradigm.

3. THE PRE-HISTORY OF AGRICULTURAL CHEMISTRY

In common with every other special development in chemistry, Liebig's agricultural chemistry was based on the foundations laid down by inorganic chemistry. These comprised the definitions of the chemical elements and chemical bonding into compounds, the fundamental laws established by Lavoisier (conservation of weight), Proust (constant proportions), Dalton (multiple proportions), Gay-Lussac and Humboldt (simple volume proportions), together with the theory of the atom (Dalton), a usable theory of constitution, at least in taxonomic terms (Berzelius' binary theory of compounds), and finally, the precise determination of atomic weights (Berzelius).

The disciplinary matrix of inorganic chemistry, which Liebig encountered around 1820 was at the following stage.

> Elements were clearly distinguished from compounds; chemical combination was explained by the supposition that it was due to the close approximation of atoms of opposite kinds, and the union of atoms to form a chemical compound was attributed to the attraction caused by charges of electricity of opposite nature which were supposed to be resident on the atoms.[19]

This disciplinary matrix was put into a consistent form principally through the empirical work carried out by Berzelius, whose prime concern was the isolation of the chemical elements and the determination of their atomic weights, through which the preconditions for a quantitative analysis of chemical compounds could be established.

> In order to obtain recognition that the law of fixed proportions contained in *Dalton's* theory represented an incontrovertible foundation for all subsequent analyses of the quantitative composition of chemical compounds, it was necessary to show that wherever the composition of compounds could be accurately determined, the law could be seen to be valid via induction. Of all Berzelius's achievements, this could well represent the most significant.[20]

Dalton's work had not succeeded in determining the atomic weights of the chemical elements through chemical analysis; in fact, the quantitative data was so plentiful and imprecise that it could have been used just as well to oppose the law of simple and multiple proportions as to support it (the dispute between Proust and Berthollet). Berzelius published his results in the period between 1810 and 1818,[21] a decade in which the system of quantitative chemistry achieved both *theoretical consistency* and *empirical efficacy*.

The fundamental findings of inorganic chemistry did not provide the only basis for agricultural chemistry; a number of successes in the field of organic chemistry also made a contribution. The main experimental and theoretical task of inorganic chemistry was the isolation through analysis of chemical elements – their discovery in fact – together with the description of their elementary physical and chemical properties, However, since the number of elements involved in all the organic compounds is only quite small,[22] organic chemistry is more concerned with the analysis of compounds and groups of compounds (the radicals), than with examining the fundamental properties of the constituent elements. In addition, the discovery of isomers necessitated that attention be given to chemical *processes* besides chemical products.[23]

Although inorganic chemistry could claim title to an established disciplinary matrix by around 1820, organic chemistry remained virtually unorganized during this period. Set against existing knowledge of inorganic compounds[24] and the enormous number of possible organic compounds, the analytic repertoire of organic chemistry was minuscule.

> The composition and nature of 'organic' compounds was practically unknown. A few such substances had been isolated, e.g., milk sugar and grape sugar were known as distinct substances, and were differentiated from common sugar. Alcohol, nearly pure, had been known in the form of spirit of wine from early times. Acetic acid was known, as well as several acids found in vegetable tissues such as oxalic, formic, malic, tartaric and benzoic acids. There were, however, no means of determining their composition and although Lavoisier had devised an apparatus in which organic compounds could be burned in oxygen, and the water and carbon-dioxide thus formed could be collected, the process was both cumbrous and incapable of yielding exact results.[25]

The successful accomplishment of these tasks in the 1820s and 1830s – the isolation and quantitative analysis of substances – engaged the activities of a large number of researchers, including Liebig and his laboratory in Gießen. Although space prohibits a full description of the development of organic chemistry in the years preceding 1840, we can indicate those main features which characterised overall developments, and which provided the basis for the creation of agricultural chemistry. Firstly, *methodological maturity*: in the pre-1840 period, methods were developed which permitted the isolation and quantitative determination of the elements of organic compounds. These procedures were sufficiently accurate to yield the empirical formulae for all the compounds of high molecular weight, once they had been isolated. In particular, Liebig's method of elementary analysis, published in 1831, combined precision, ease of operation and application over a broad field. Whereas Berzelius, "was obliged to repeat (his) analysis a large number of times using different methods to obtain the method leading to the right result,"[26] Liebig's procedure, "was the invention and perfection of a method for analyzing organic compounds which in its essential features is still practised everywhere."[27]

Liebig spent virtually seven years working on the routinisation of organic elementary analysis. Although his own individual innovations did not constitute a breakthrough in the field of scientific and technical instruments, taken as whole they represented a widely applicable method, with a low error tolerance which imposed minimal demands on the individual skill of the analyst: "in Gießen organic analysis became such a matter of routine that Liebig was able to boast 'now any fool can be a chemist'."[28]

Secondly, *theoretical clarity*: the theory of electrochemical affinity was only of limited value for compounds of high molecular weight, and was, moreover, rendered suspect by the discovery of anomalous isomers. Dumas, Wöhler, Liebig and others attempted to develop systems of classification which would apply specifically to the isolation and the reactions of organic compounds. The motive behind these endeavours was the desire to specify the so-called *radicals*, and classify the various compounds of a group as additives to these radicals. Radicals are not elements but core compounds which possess properties akin to elements; in particular, they retain their molecular structure throughout certain chemical reactions. Gay-Lussac suspected in 1815 that cyanogen (CN) might represent such a radical. Later, in their 1832 study on bitter almond oil, Liebig and Wöhler were able to show that the benzoyl group (C_7H_5O) remained constant in all the oil's derivates. Further studies identified other groups of compounds.[29]

Although the theory of radicals began to wane in influence in the 1840s, having served merely as linguistic convenience and convention for many of its advocates, it did, despite the controversy surrounding it, nevertheless represent the first theory to have been worked through both cooperatively and purposively – and to have penetrated the gloom of the "jungle" of organic chemistry, a "jungle populated by the most remarkable things, an incredible undergrowth, with no exit or end, into which no one dared to venture."[30] By introducing a sense of structure, the theory of radicals allowed and encouraged the systematic search for compounds, whose existence was at first simply conjectured.

The conjunction of a structuring theory and a set of experimental techniques formed the essential precondition for the development of agricultural chemistry, paving the way for the huge number of experiments required to establish the nature and properties of the multiplicity of phenomena involved – plants and their parts, soils and air. Although the concept of cycles of reproduction may have been conceivably arrived at without these preconditions, they were however vital in the attempt to undertake and understand the processes required to *reconstruct* such cycles, either wholly or in part. Liebig constantly pushed ahead with the carrying out of long series of investigations and with further routinisation, with its associated simplification of analytical procedures, even at the cost of sacrificing improvements in precision.[31] By 1840 Liebig considered that chemistry had reached the required state of maturity.

Chemistry had become among the sciences so independent in its structure that it could participate in the development of other areas; and in that the efforts of the chemists were turned to studying the living conditions of plants and animals, they touched upon agriculture . . . Chemistry began to study according to its rigorous methods all the parts of plants with as great a degree of precision as possible; it examined what was in leaves, stems, roots and in fruit; it pursued the processes of the animals' nutrition and what was to become of the nutrition in their bodies; it analyzed arable lands in different regions of the world.[32]

This concept of maturity nevertheless requires more accurate formulation. Originally the finalization theory claimed that theoretical special developments presuppose that basic theories are *theoretically mature*. Theoretical maturity is consequently regarded as the conclusion of a process of theoretical development within a field, or more cautiously, the dwindling away of problems associated with explaining the structures and properties of the objects which constituted the field. However, the state of chemistry in Liebig's time bore no resemblance to such a concept. Although quantitative chemistry had established a theoretical matrix via experiment, the hypotheses which were used to make sense of experimental phenomena remained unexplained (atomic theory, non-transmutability of elements). And, as was acknowledged by their initiators, the theory of radicals and other methods of classification were of mere transitory importance.

It is therefore more appropriate to regard the chemistry of *c*. 1840 as being mature *methodologically*,[33] meaning that when required it could produce a quantitative determination of any given complex reaction, and in particular of those reactions involved in nutrition; that is, supply the empirical formula, an indication of its chemical synthesis, and the process of its chemical breakdown. It was this type of maturity which in fact allowed chemistry to sustain special theoretical developments, and by extension, also endowed agricultural chemistry with a stability not available to previous approaches. The contrast between Davy and Liebig illustrates this difference quite well: Davy's motives in turning to agricultural chemistry at the beginning of the nineteenth century were very similar to Liebig's – although, unlike Liebig, he was not rewarded by any noteworthy theoretical or practical successes. In 1801 he was appointed lecturer and demonstrator in chemistry at the Royal Institution, established by Count Rumford in 1799 as a weapon of science against hunger and poverty.[34] Davy tried to carry out his mandate by giving his public lectures a practical bent – such as the chemistry of tanning – while continuing his own research into electrical phenomena. He also started to experiment in the field of agriculture in this period. In 1802 he was appointed Professor of Chemistry at the Board of Agriculture, where he regularly lectured on agricultural chemistry. His *Elements of Agricultural Chemistry* (1815) was a collection of these lectures;[35] as a scientific work it is highly eclectic, and as a contribution to economics, euphoric. Davy himself noted in the Foreword:

> Agrirultural chemistry has not yet received a regular and systematic form. It has been pursued by competent experimenters for a short time only: the doctrines have not as yet been collected into any elementary treatise . . . I know, however, that your candour will induce you not to expect anything like a finished work upon a science as yet in its infancy.[36]

Despite such *caveats* Davy goes on to cite examples of the benefits of agriculture conducted in accordance with the teachings of agricultural chemistry: apart from its importance as a "theory of the art" of husbandry, chemical studies are also, "useful in affording simple and easy experiments for directing his (*i.e., the farmer's*) labours and for enabling him to pursue a certain and systematic plan of improvement." Davy cites the following examples: in trying to improve unproductive land one "sure method" is to establish the *cause* of the sterility which had to consist in a deficiency in the soil's composition, "easily discovered through chemical analysis". Rich soils may also be infertile, however; here chemical tests are also appropriate since the soil must contain a "noxious principle" which "may be easily discovered and probably easily destroyed."[37] After several more examples, such as the chemical analysis of the usefulness of different limestones as manure, and the chemical explanation of optimal manuring techniques, Davy concludes:

> It would be easy to adduce a multitude of other instances of the kind; but sufficient, I trust, has been said to prove that the connection of chemistry and agriculture is not found on mere vague speculations, but that it offers principles which ought to be understood and followed, and which in their progression and application can hardly fail to be highly beneficial to the community.[38]

In terms of sheer propaganda Liebig had nothing to add to this when he turned to agricultural chemistry forty years later. Liebig was also able to draw on a number of Davy's own studies which retained their usefulness.

Nevertheless, considered as a whole Davy's approach was eclectic and theoretically unorganised. At the end of his eloquent Introduction he noted:

Discoveries in the cultivation of the earth are not merely for the time and country in which they were developed, but they may be considered as extending to future ages, and as ultimately tending to benefit the whole human race.[39]

In saying this Davy glossed over two problems which demanded a theoretical solution: generalisation in time and generalisation in space. The absence of such a theory meant that neither knowledge of the long-term exhaustion of the soil, nor of the "similarity" or "dissimilarity" of sets of initial conditions was available at the beginning of the nineteenth century.

In the absence of a paradigm, agricultural chemistry could only be practically utilised in localised situations; and since it lacked a corpus of generalizable experimental findings, it simply became part of the general stream of those disciplines with aspirations of utility in agriculture (the other contenders being animal and plant physiology, and "rational husbandry"). To discover why it was chemistry which eventually emerged as the leader in terms of its theoretical contribution to agriculture, it is necessary to take a closer look at the state of these disciplines at the beginning of the nineteenth century. Organic chemistry and physiology were not really separable disciplines during the eighteenth century.[40] Differences of content and method only began to appear when quantitative and elementary chemical analysis became defined. Organic chemistry departed from the analysis of the purpose and function of bodily organs, and instead concentrated on the substances which constituted and sustained them. Physiology was displaced into the role of a heuristic for problems originating within chemistry. Comparative anatomy and microscopy were able to provide insights into transport within plants and animals, and to the structures of organs, as well as specifying the physiological conditions within which chemical processes took place – they could not, however, "explain" how plants and animals actually grew.

A parallel shift from an equivalent to a superior status can also be observed in the relation between agricultural chemistry and husbandry: husbandry had differentiated itself from the preceding "oeconomia ruralis et domestica" at the beginning of the seventeenth century, it evolved into a pragmatic "experimental economy" in the latter half of the eighteenth century, and culminated in the early-nineteenth century "rational agriculture" of the Thaer school. In his study of English agricultural developments, written between 1798 and 1804, Thaer lists the achievements of eighteenth century agriculture as, (i) techniques of soil melioration, (ii) land reallocation and consolidation and enclosures, (iii) improvements of the three-field system, (iv) introduction of crop-rotation, (v) improvement in animal husbandry and new methods of stock-breeding, (vi) introduction of new plants, such as potatoes, maize and sunflowers, and (vii) improvements in farm implements and buildings.[41]

The experimental institutes of "rational agriculture" concentrated their efforts on obtaining higher profits through the adaptation of modern, and in particular mechanical, techniques to the specific conditions prevailing on individual farms. Higher yields could be attained either through improved technology or more rationalized farm management: in turn, these higher yields produced the sought-after higher profits. Under "rational agriculture" the economic aspects of improved management took precedence over the technical.

Two theories dominated the work of the experimental institutes: the humus theory which considered the humus to be the main food of plants, and the economic theory of farm location.[42] Although the progressive isolation of all the variables relevant to agriculture created a disciplinary background for the emergence of agricultural chemistry, the approach was marred by a number of major theoretical deficiencies: the economistic approach neglected the ecological problems associated with *Raubbau* agriculture, and the humus theory of plant nutrition, with its vitalist foundations and commitment to phlogiston chemistry, lagged behind developments in chemical theory. As a result, experimental findings could not be applied or generalized beyond the immediate geographical setting in which they were made.

During the first four decades of the nineteenth century none of the approaches then under consideration – the technical-mechanical, the economic-management, the chemical and the physiological – had taken a decisive lead: although each had enjoyed particular successes, all were constrained by limitations. Davy's theories were not perceptibly superior to Thaer's or Thünen's.

The publication of Liebig's *Organic Chemistry in Its Application to Agriculture and Physiology* in 1840 changed this situation radically: chemistry emerged as the science able not only to *explain* the processes of agriculture, but also *shape* them in accordance with human purposes. It was therefore able to integrate technical and economic knowledge into a wider theory, with each exercising its appropriate functional role within the whole. These, anyway, were the claims made by Liebig himself – meeting with both strong and unexpected assent. The year 1840 therefore suggests itself as the year in which agricultural chemistry acquired a paradigm. Although such exercises always have an element of arbitrariness, this particular year has made its own mark in the historiography of agricultural chemistry: although Liebig's successors may have opposed him on specific questions, they did so on the foundations of the new agricultural chemistry which he established.

There are two reasons why Liebig can be singled out as being of particular importance in agricultural chemistry – although any more comprehensive treatment of the subject would require that this one-sided focus be balanced out with the contributions of others in the field. Firstly, and this is a crucial factor in the sociology of science, he knew how to locate himself, his laboratory, and his publications at the centre of scientific and public interest: he succeeded in strategically institutionalizing agricultural chemistry. Secondly, he knew how to formulate both the available scientific findings, and the most urgent gaps in knowledge, in a comprehensive and consistent way. He gave a new discipline an essentially reliable paradigm, and although this may have been more of a reflective than a theoretical achievement, it was of decisive importance in forming the new direction for research. Liebig himself assessed his own role in much the same terms.

> In my agricultural chemistry I have simply tried to put a light into a dark room. All the furniture was there, even tools and objects of comfort and enjoyment; all these objects were, however, not clearly visible to the society which used this room for its welfare and utility. Groping haphazardly one person found a chair, another a table, the third a bed in which he made himself as comfortable as possible; yet most remained blind to the harmony and interrelationship of the furnishings. If an object received a little, if weak, illumination, many began to complain that the light had not wrought any significant change in the room; one person had already known this, another had known and used something else, between them they had all felt and touched everything that was present in the room. Chemistry, the light of knowledge, can never however be removed from this room without harmful consequences. *This* goal has been fully achieved.[43]

The peculiar pre-history of the *Agricultural Chemistry* is itself an indicator of the book's status within the development of chemistry. Whilst on a visit to England in 1837 Liebig was given the "honourable task of preparing a Report on the state of Organic Chemistry" for the British Association for the Advancement of Science in Liverpool.[44] Why did Liebig choose to meet this request by writing a book on agricultural chemistry?

Berzelius had coined the term "organic chemistry" as a limiting concept to mineral chemistry, based on the (incorrect) supposition that one large group of compounds could be distinguished by the fact that they could only be synthesized by living organisms *in vivo.*[45] Organic chemistry was consequently defined as the science of the chemical processes of organic substances, or the science of the chemical conditions of life. Well into the nineteenth century, these chemical conditions were accompanied by an additional, non-chemical condition – the *vital force*, which was thought to be responsible for organising chemical processes into particular organic forms. As far as chemists were concerned, the idea of the vital force was predominantly employed as a metaphysical device, delimiting certain questions which they felt it unproductive or distracting to pursue. Liebig's intention to "apply the principles of natural science . . . to the development and nutrition of plants"[46] therefore fitted exactly with the initial programme of organic chemistry – the attempt to produce a scientific explanation of the organism's ability to carry out chemical synthesis. On the other hand, the years between 1820 and 1840 had witnessed a rapid development of organic chemistry in another direction, namely, the weakening of its separate identity from inorganic chemistry. Following the synthesis of urea by Wöhler and Liebig in 1828, it became generally accepted that all organic materials could, in principle, be produced artificially. Organic chemistry detached itself from its original programmatic aim, and evolved into the chemistry of – in modern terms – carbon compounds, whose specific properties lay not in their organic origin, but in the chemical characteristics of carbon.

The development of organic chemistry into carbon chemistry meant that, on cognitive grounds, a whole set of exogenously determined goals became irrelevant to the direction traced by theory. The theory of radicals[47] – the first structuring theory in organic chemistry – marked the beginning of a controversy around the structure of carbon compounds and chemical bonding forces. Liebig's adherence to the aim of explaining the specific chemistry of living things therefore became, both theoretically and experimentally, a special development within the wider context of the re-orientation of organic chemistry.

The emerging agricultural chemistry of the mid-nineteenth century is not easily classified amongst the scientific activities of that period. In comparison to organic chemistry's tendency to develop into a separate chemistry of hydrocarbons, agricultural chemistry represented a special scientific development. It integrated elements of physics, agronomy and physiology, without at the same time being exhausted by these individual disciplines.[48] If such a confluence of disciplines was to avoid eclecticism, it required the coherence of an integrating principle, which could link the separate elements in accordance with their function for the structure of the discipline as a whole. Liebig's Introduction, cited above, does not refer to the application of chemistry, but rather to the application of the principles of science, in the context of which "application" implies not that the results of one discipline are technically exploited in another, but that a particular domain – the manipulation of the nutrition of cultivated plants – becomes "scientificated", as distinct from hit-or-miss, localised and narrowly empiricist treatments. The emergence of such an innovation for the explanation of, and intervention into, a particular field makes it legitimate to speak in this context of a special theoretical development. The next two sections explore the content of this theory.

4. THE PARADIGM OF AGRICULTURAL CHEMISTRY

4.1. *Cycles of Reproduction*

The fundamental concept in the paradigm which organised agricultural chemistry was that of the *cycle of processes* constitutive for the reproduction of organic structures.

> Our present research in natural history rests on the conviction that laws of interaction not only exist between two or three, but between all the phenomena of the animal, vegetable and mineral spheres which determine life on the surface of the earth. Thus, none of them is separate but are always linked to one, or to several others, and these in turn to yet further, all ultimately linked together, without beginning or end. The sequence of these phenomena, their origins and their departures can be compared to a wave motion within a cycle.[49]

The notion of the cycle had a number of different origins and meanings.[50] As found in political economy and biology it had two distinct components: firstly, the concept of the re-establishment of an initial position after a succession of metamorphoses; and secondly, the functional maintenance of an equilibrium through control mechanisms. Of these, it was the first which was basic for agriculture: plant and animal life, together with meteorological processes, jointly circulate certain "substances"; apart from the irreversible transformation of energy into heat, living processes do not "use up" nature, but reproduce the conditions for their continued existence.

Since this cycle consists of a large number of independent cycles of varying duration, which ensure the maintenance of individual plants and animals, species and biotopes, the second aspect of the concept of the cycle also plays a role: feedback mechanisms function to reestablish an original position. Throughout the history of settled agriculture, the activity of the farmer has consisted in the making of small adjustments to maintain an ordered pattern which does not exist in an uncultivated nature, oblivious to the existence of "harvests" or "weeds". The phenomena of biology and that of the procession of the seasons therefore allude to a general notion of cycles.

Nevertheless, a number of factors suggest that Liebig was more influenced by the concept of the cycle as elaborated in political economy. The central agricultural notion of a yield or surplus has a greater affinity with economics than with biology. Liebig's reproach to the agronomists was directed at the inability of the theory of marginal revenue to differentiate between those practical achievements which used up the working capital supplied by nature, and those which conserved it, with added interest.[51] In a cycle intended to remain in long-term equilibrium, "capital" should circulate, not diminish. The alternative would not only block an expansion in production, but might ultimately heighten the probability of a rapid and final breakdown of the cycle.

Agricultural production does not differ fundamentally from the customary industrial enterprise. The factory-owner and the manufacturer are only too well aware that their investment and working capital must not steadily diminish if their business is to continue. In the same way, the rational agricultural enterprise requires a farmer seeking higher yields to increase the number of active things in the soil with which he obtains his product.[52]

In the same vein, Liebig criticised Smith, and subsequent political economists, for the fact that although they treated agriculture as the basis of an industrial economy, "they nonetheless consider it self-evident that a field which has supplied crops can continuously and repeatedly be returned to its original condition through human labour and certain forms of management, and that, therefore, once the soil has brought forth its effect (produced crops), no portion of it is actually consumed."[53]

The analogy between the concept of the cycle in agriculture and that of the cycle, or circuit, of capital was productive because it led Liebig to draw a distinction between *natural cycles* and cycles based on *culturally determined norms*. The analogy between the turnover of capital and a natural cycle and between the yield on capital and agricultural earnings required a standard of measurement in agriculture which could serve as a criterion for determining the amount of necessary investments. Is an increase in year *X* a real net gain, or a hidden consumption of capital? Does mechanical aeration of the soil produce higher yields, or increased despoliation? "*Practice poses these questions*: but practice cannot resolve them."[54]

The concept of the cycle therefore determined Liebig's theoretical approach. The first step was to give the concept a firm theoretical foundation. Liebig raised the question: If the nutrition of plants and animals constantly consumes particular resources found in the soil and air, why do these nevertheless remain available in the same amounts over any given period of time? Three answers were possible: either the resources were infinite, or they could be generated spontaneously, or they were regularly reproduced. Existing chemical knowledge enabled the first two solutions to be rejected out of hand. If the quantitative consumption of oxygen by animals can be worked out, and the total volume of the atmosphere is known, a simple calculation will reveal the length of time before all the oxygen in the atmosphere is exhausted.[55] Moreover, as far as the second suggested answer was concerned, it was a fundamental part of the paradigm of chemistry that elements were non-transmutable – although this message had not, at that time, reached every physiologist. The idea of natural cycles consequently represented the only possible schema for conceptualizing the chemical conditions for life-processes reconcilable with the broader disciplinary matrix of chemistry.

Organic nature *had* to follow cycles if it was to produce stable forms over time passing through chains of chemical reactions. The concept of cycles was, therefore, a postulate, not the result of induction. A cycle presupposes the determination of an initial state, in relation to which the completion of the cycle can be defined. If several relatively independent cycles interact, these initial conditions can only be stated approximately, and it becomes necessary to possess the means to distinguish between desired, or "ideal" values, and actual values. In agriculture, the desired values can be expressed as the sum of the factors which make up the fertility of the soil. This fertility must be reproduced after the completion of one agricultural cycle: since such a value can only be approximately specified, a scale has to be constructed against which possible deviations can be detected, and means need to be available for intervening and correcting such deviations. Such correcting mechanisms are already built into natural cycles, as feedback systems which operate to control departures from equilibrium. In cycles involving crops, it is the farmer who has to take on the role of regulator. It was with this in mind that Liebig specified the cycle "on the surface of the earth" as an "unceasing exchange, a repeated departure from and return to equilibrium." He posed this question directly in terms of its relevancy for technical application: "Does the art of agriculture rest on anything other than the reestablishment of balance?"[56] Liebig's concept of the cycle therefore embraces both reproduction and self-regulation, or – in the case of agriculture – regulation through technical intervention.

Explaining the existence and stability of cycles placed new demands on chemical causal analysis. What had to be pinned down was the guiding thread which, via the complex interplay of composition and decomposition in the atmosphere, soil and physiology of plants, eventually led back to the re-establishment of the initial conditions. Such a task was entirely new, both for organic and inorganic chemistry: the path was indicated heuristically through the concept of the cycle, and could base itself on work already carried out in organic chemistry and physiology on the concepts of balance and metamorphosis. These concepts were precursors of the concept of the cycle, and were analogous to the notion of "balancing" in public and commercial financial practices – the keeping and balancing of accounts for assessing and recording the inward and outward movement of values. Drawing up such a balance presupposes that the postings can be qualitatively distinguished, which in the case of a commercial transaction is a relatively trivial matter. However, in chemistry and physiology it represented the central problem. What substance in each product (that is, the plant considered as the product of the soil and the air, or the animal as the product of its fodder and the air) corresponded to an equivalent substance in the factors of production? The transformations involved, termed metamorphoses, were the object of chemical physiology and ash-analysis. Boussingault, for example, performed such balancings during the 1830s and called for a type of "running account to be kept between the inorganic components of harvests and those present in fertilizers."[57]

Ash-analysis was not sufficient, however, to establish reliable balancings, and it was also improbable that long-term soil melioration could be obtained through processes of trial-and-error. It was essential, therefore, to pinpoint the specific chemical reactions in animals and plants which were integral to the cycles. Further, in order to make controlled interventions into these cycles, it was also important to identify *all* the necessary and *only* the necessary conditions in the chains of chemical processes. This could not be accomplished solely through laboratory analysis. The physiology of food assimilation required *in vivo* research. While tackling the practically significant question of whether carbohydrates could be converted into fats Liebig took the opportunity to rebuke the practice of animal physiologists.

Without any knowledge of the conditions for the building up of fats, or any inquiry into whether such conditions exist, they begin by excluding everything which might make it possible to answer the question. Animals are put into a state of artificially induced disease, and deprived of all nourishment; great care is taken to exclude all those factors involved in the formation of the blood and in the sustenance of the vital functions which effect the formation of fats. They then believe that such miserable and cruel experiments will be able to demonstrate that sugar, a non-nitrogenous substance, cannot be transformed into fat, another non-nitrogenous substance! These experiments serve only to prove the ignorance and total incapacity of these experimenters to offer a solution to such questions – and apart from this, they prove nothing else at all.[58]

Liebig directed similar criticisms at the plant physiologists, who he accused of "conducting experiments of no value at all for resolving any question," since they possessed neither theories nor proper methods for controlling the structure of their experiments.

They conduct experiments without any acquaintance with the circumstances necessary for life, the qualities of the proper nutriment of the plant, and the nature and components of its organs: they consider such experiments as providing convincing proofs, whereas in fact they merely awaken pity.[59]

Liebig's preoccupation in making such criticisms was to hammer home the message that experiment had to be preceded by theory: "an experiment is the expression of an idea."[60] The posing of an experimental question required first the stating of the conditions required to answer it – otherwise no experiment could ever actually falsify a hypothesis or produce an unexpected result. In agricultural chemistry these conditions were defined in advance by the concept of the cycle of the elementary components of plants, animals, soils and the atmosphere. Practical analysis of the metamorphoses and balancing of the elements in life-processes is therefore methodologically secondary to the principle of the cycle.

4.2. *The Constructive Variation of Cycles*

The principle of cycles of reproduction, and its theoretical implications, form only one half of the paradigm of agricultural chemistry. As such no specific reference is included to the normative "culture" of agricultural chemistry, since the principle of cycles applies to the "conditions necessary for the life of *all* plants."[61] Agricultural chemistry is not, however, merely the science of natural cycles but a technology for the variation of such cycles in accordance with human purposes – the very reason why we have designated agricultural chemistry a finalized science.

Whereas nature's intrinsic finality consists in stably reproducing itself over time, the main concern of the "practical man" is to "cut as many shoes as possible from the inexhaustible stock of leather of the soil; and the best teacher will be the one who appears most successful at this type of intensive cultivation."[62] However, once it is established that such a depradation of the soil will eventually prove fatal to society's survival, the interest of society and the interest of the "practical man" will cease to coincide – society's interest is not short-term profit, but "large and increasing harvests in perpetuity."[63] Liebig's view was that the task of accomplishing this demanding feat fell upon agricultural chemistry: by accepting the challenge, agricultural chemistry left the domain of the natural sciences and became a technical science, whose purpose was the realisation of specific social objectives. Instead of merely scientifically investigating the reproduction of plant life, it took on the role of constructing purposive natural cycles, which used nature-given means to achieve purposes not intrinsic to nature itself. This relationship between analysis and construction, in which the specification of objectives influenced the structure of the theory, can be examined in more specific detail by considering how the goal of permanently rising harvests was actually operationalized.

Clearly, the idea of "increase", meaning an *additional* yield of a collection of plants, is a concept which has no place in nature: "In their free and uncultivated state all the parts of a plant develop in accordance with the conditions of the nutrition supplied by their location; they grow equally as perfectly on meagre, infertile soils as on the most fertile and luxuriant."[64] It would be meaningless to contend that a dwarf-pine is in some sense less perfect than a forest-pine simply because it has less branches, or is smaller: a dwarf-pine "increased" in dimensions would not be able to sustain itself. As the science of the long-term extraction of socially-desired use-values from nature, agricultural chemistry is not concerned with the laws of "natural" nature, but with those potentialities in nature which can be identified as possessing desirable characteristics, and which therefore function as "ideal values" for the reconstruction of natural cycles. The "laws" in nature which are of interest to agricultural chemistry are associated with norms superimposed on natural qualities, in part as the qualities of plants, part as the qualities of soils, and part as the qualities of cycles. According to Liebig these were: (i) norms relating to the abnormal development of parts of plants and vegetable substances, such as fine, soft, pliable straw for making straw-hats, strength and rigidity of the stem, maximum number of ears, maximum amount of nitrogen in the ears, and so on;[65] (ii) norms relating to soils, in particular the various forms of fertility which depend on the sum of the nutrients which can be absorbed, and duration of fertility, which depends on the sum of nutrients actually available;[66] (iii) norms which relate to cycles, which emerge from the selection of crops, and the succession of crops.[67]

Irrespective of how one might try to structure these goals in a systematic schema, no such structuring could ever correspond to any scaling appropriate to natural states. The aim of agricultural chemistry is not to define a state of normality, and correct any observed deviations from it, on the premiss that such deviations can be measured. Rather, agricultural chemistry is engaged in a strategy for maximising the supply of desired social use-values through natural means: unlike departures from and returns to an accepted state of "health", such use-values are not pre-given. The selection of goals determines the pattern of description, and the methods, which eventually lead to the formulation of a theory.

Finalization in the field of agricultural chemistry means that the construction of suitable soils, plants and fertilisers is not simply the application of available theory, but actually represents an integral part of the process of theory formation itself. The theoretical notion of "suitability" cannot be deduced from nature, but serves to structure possible states of nature. This does not mean, however, that the results obtained are not objective natural laws: they are, both in the sense that once discovered, they can be seen to be derivable from more general, antecedent, laws of chemistry, physics and physiology, and in the sense that the experimental, and ultimately technical, practice which they point to entails objective experimental and technical procedures. The concept of finality does not relate to the form of these laws, but to the possibility of social goals being conceptualised directly in scientific terms.

The central concept of agricultural chemistry is not, therefore, the concept of the cycle but rather that of the desired goal – increasing harvests in perpetuity. This may appear inconsistent at first sight, since if all processes of animal, plant, soil, and atmospheric chemistry are regarded as being linked in cycles on the one hand, and it is proposed to remove increasing amounts of materials from these cycles on the other, by what means can the long-run stability of the cycles be guaranteed? The answer which agricultural chemistry provided was that the conception of cycles opened the way for their individual variation in accordance with social objectives, *whilst maintaining overall natural equilibrium*. It also follows, by definition, that the removal of a substance from one cycle implies its addition to other chemical processes. In an economy based on *Raubbau* agriculture, many of these processes are not subsequently reintegrated, leading to the exhaustion of resources at one pole, and the accumulation of vegetable waste at the other. However, if these processes are controlled so that any removal of material is carried out in association with a *reorganization* of the cycles directed at ensuring that the initial conditions are re-established, rising harvests in perpetuity are no longer a utopian dream, but the end-purpose of a logically consistent research programme.

4.3. *The Progressive Problemshift in Agricultural Chemistry*

The subsequent development of theoretical research in agricultural chemistry was, and is, characterised by the incorporation of problematics originating in microbiology, genetics, and isotope physics. However, we do not propose to deal with these areas here, since what was crucial for the progressive problemshift in agricultural chemistry was whether the paradigm outlined by Liebig was theoretically convincing, and able to generate empirical research. We consider this aspect in the following six points:

(1) The concept of the cycle, the first constituent of the paradigm, led to the discovery of fundamental chemical processes which could not have been arrived at via chemistry's then characteristically linear approach to phenomena. These were, the discovery of the nitrogen cycle, the hydrogen cycle and the interdependent oxygen and carbon cycles. One direct spin-off of these discoveries was the ability to explain the long-term constancy of the concentrations of the elements in the soil and air. It also allowed inferences to be drawn about the reduction in the amount of carbon in the air prior to the emergence of animal life, and the equivalence of this mass of carbon to deposits of coal and oil. The heuristic relevance of the category of the cycle can scarcely be exaggerated: it influenced not only animal physiology, provided a link between plant and animal physiology, but also opened the way for those fields currently grouped together under the term "recycling".

(2) The concept of metamorphosis, originally developed in organic chemistry, acquired a key role in agricultural chemistry. Berzelius noted in 1839:

> In our studies of organic bodies our attention was directed principally at a particular product, with only the barest consideration to the by-products. We gradually became aware of a number of these, and eventually of all of them. In fact it was *Liebig* and *Wöhler* who began studies in this field . . . and who shed light on an entirely new field in the study of organic chemistry . . . that of the chemical *metamorphoses* of *organic oxides*. Their excellent work on the transformation of uric acid came just at the moment when it was best able to serve as a guide in these involved questions . . . The time has now come to pay the closest attention to these metamorphoses.[68]

In agricultural chemistry, the concept of metamorphosis guided the analyses of fermentation, decay and putrefaction, together with the processes of nutrition and respiration. A knowledge of all the products of transformation involved in these processes was fundamental both for an understanding of the chemically self-contained character of the reproduction of plant and animal life, and, accordingly, for an assessment of the effects of substances introduced into nature.[69] The chemistry of organic decomposition was itself drawn into a theoretical framework through the systematic structure it had gained via agricultural chemistry.

(3) Agricultural chemistry allowed the first scientific assessments of the experiments carried out by traditional agriculture by making it possible to formulate them as scientific hypotheses expressing causal or law-like relationships, which could be subject to testing. "What Thaer found good and proper in his fields at Möglin came to be held as good and proper for all German soils, and what Lawes discovered on the small strip of land at Rothampstead came to be held as axioms applying to all English soils."[70] Liebig was not, of course, advocating inductionism: on the contrary, he realised the need for a theoretical standard against which the range of technical capabilities could be explained. Only with such a theoretical backdrop could a "trial" become an "experiment". "An experiment which is not preceded by a theory, that is, an idea, bears much the same relationship to scientific research as the shaking of a baby's rattle does to music."[71] As far as Liebig was concerned, trial-and-error methods in which the object of investigation figures as a "black box" were devoid of theoretical value since they did not allow the relevant factors within trial conditions to be distinguished from the merely contingent.

> The farmer who begins to carry out experiments without the guidance of a proper scientific principle is unlikely to be rewarded by success. Thousands of agriculturalists carry out similar experiments in a variety of directions, the results of which eventually embrace a number of practical experiences, which taken together constitute a method of culture, leading to the attainment of a given goal for a given area. However, the farmer next door will fail using these methods. Imagine the amount of capital and the effort wasted in *these* experiments. And consider how much more safe the different path traced by science is: once we take to this path we are spared the hazards of failure and guaranteed the certainty of success.[72]

The new theoretical background of agricultural chemistry opened the way for a number of experimental techniques, such as laboratory experiments with synthetic nutrient solutions,[73] analyses of physiological metamorphoses, and experimental sequences on the chemical constitution of plants.[74]

(4) Although the development of successful artificial fertilizers took around twenty years, the failures which occurred on the way were the result of inadequate empirical knowledge, and not any deficiency in the basic paradigm. The problems could be broken down into two categories. In the

first, the composition of the fertiliser suggested by theoretical reflection failed to match up to the empirical conditions under which it was required to function; Liebig, for example, thought that water-soluble fertilizers would be washed away by rain, and he therefore put insoluble mineral salts into his artificial fertilizer – with the result that plants could not absorb the alkalis. It was a further sixteen years before Liebig fully understood the underlying reasons, and gave due attention to the physical characteristics of the soil. The second problem arose through Liebig's underestimation of the role of nitrogen. He initially believed that plants could obtain sufficient nitrogen from the ammonia in the atmosphere,[75] and it took some time before he conceded his error on this point. The necessity for nitrogen fertilizers would have blurred the distinction between the traditional humus theory and his own theory of minerals – two theories which for strategic and institutional reasons he wished to keep distinct.[76]

(5) The construction of an artificial fertilizer was based on two fundamental insights, which expressed as the "law of the minimum" came to constitute a type of technical natural law. They were: (a) that plant growth is determined not by the organic but by the mineral components of the soil; and (b) that the elements contained in the nutrient mineral salts are not mutually interchangeable.

Plant growth was therefore seen to depend on the availability of the rarest mineral substance. Sprengel had already formulated the law of the minimum prior to Liebig; however, Liebig did not think very highly of Sprengel's contribution, since although Sprengel had recognised the importance of the mineral constituents of the soil for plant growth, he was not able to distinguish between the actual contents and the amounts theoretically needed for growth, or their possible substitutes. Plants will take up all the soluble salts present in the soil.

> Because of this capacity, no conclusion can be drawn as to the necessity of the presence of particular salts in plants, or whether they are there merely by chance. It is not possible to state that all the inorganic constituents found in the ashes are necessary without further investigation.[77]

It is this distinction between the actual and the necessary constituents which absolves the law of the minimum from the charge of triviality; the establishment of the minimum for individual species required not merely ash-analysis but theory and experiment too. As a law rooted in experimental science, it possessed both a technical and a strategic significance; absent substances could be identified and added, or appropriate substitutes found, and the amount of the necessary constituents determined.

(6) One good indicator for the stability of a theoretical approach is its capacity to outlive the errors of the sub-theories which it generates. The failure of Liebig's first artificial manure, which caused him to suffer a number of institutional setbacks, was in fact explained by means of *his own theory*. Faulty suppositions, such as the belief that plant cells did not need to respire oxygen, or that micro-organisms were unimportant, did not hinder the long-term development of agricultural chemistry, but through the process of their correction, actually assisted it.

The progressive problemshift in the paradigm of agricultural chemistry indicates that the discipline did not simply represent the successful application of accumulated chemical knowledge, but actually possessed a theoretical dynamic of its own. Nevertheless, there are differences between a discipline such as agricultural chemistry, and disciplines such as classical mechanics and inorganic chemistry. Agricultural chemistry lacks the fundamental axiomatic structure of mechanics, or the basic laws of chemistry. The characteristic feature of finalized disciplines lies in their roots in a set of social problems and interests which exist independently of theoretical considerations. Finalized disciplines naturally require a theoretical basis from which standards can be derived for the evaluation of progress, and solutions specified for subsequent layers of problems. In the absence of such paradigmatic autonomy the corresponding field of research will often prove to be nothing more than an assembly point for the findings of other fields. Even without Liebig's paradigm agricultural chemistry would have existed as an eclectic union of findings from chemistry, physiology, botany and physics. However, in this form it would not have displayed a progressive problemshift sustained by its own paradigmatic structure. Finalized disciplines emerge in the intermediate zone between interdisciplinary eclectism, whose sole *raison d'être* is the urgency of social problems, and theoretical contexts generated exclusively by scientific interests.

Thus far we have considered agricultural chemistry as a special theoretical development based on chemistry, whose pre-history is located within the history of early-nineteenth century organic and inorganic chemistry. In the post-1840 period (to take the publication of Liebig's book as a marker) the methodological maturity of chemistry facilitated developments which allowed the scientification of fields which were not fundamental to the exploration of chemistry's basic problems. Agricultural chemistry represents such a development: it established a new explanatory model which although fully in consonance with the basic laws of chemistry, and its structuring principles, cannot be derived from them simply by deductive reasoning.

Agricultural chemistry required the development of concepts and methods of research which had no particular utility within chemistry itself – such as the concept of cycles of reproduction. We have termed such theories "principles of application" (*Anwendungsgrundlagen*).[78] This indicates that the application of basic knowledge to spheres not originally covered by it requires a *theoretical concept* of application, and not simply the ability to use known formulae under different sets of initial conditions.

Looked in terms of their structure, attempts to construct theories for the application of basic theories are not markedly different from the construction of general theories. Both represent attempts to establish a paradigm able to serve as the basis for new special developments. This approach appears plausible, for example, if the development of agricultural chemistry is seen as the paradigmatic conclusion to earlier initiatives within agricultural science, rather than as a special development of chemistry. In relation to these initiatives, agricultural chemistry represents a second phase of development. Humus theorists (Thaer), economic agronomists (Thünen), physiologists (de Saussure) and chemists (Davy) all established competing approaches none of which could rule out the others on theoretical grounds. This is typical of the first phase of disciplinary development. After Liebig, agricultural chemistry acquired the means for deciding between these competing theories, and evaluating their individual contributions. It initiated a phase in which the detailed working out of the paradigm of cycles of reproduction could begin. Depending on the point of reference, therefore, agricultural chemistry is a third phase of chemistry or a second phase of agricultural science.

5. THE INSTITUTIONALIZATION OF AGRICULTURAL CHEMISTRY

So far we have concentrated our attention on the *cognitive* structure of agricultural chemistry as a finalized science. Does such a cognitive structure correspond to any particular type of institutional structure? There are good grounds for thinking this might be the case, since a discipline with a strategic orientation to external objectives could be expected to be accompanied by an equally strategic institutionalization, or at least, identifiable attempts to create such an institutionalization.

We should preface this discussion by noting that the institutionalization of agricultural chemistry and the institutionalization of chemistry in general were inseparable in a number of respects. A general training in chemistry was a precondition for practising agricultural chemistry, and consequently a degree of overlap between the institutionalization of agricultural chemistry and the institutionalization of chemistry as a scientific discipline within universities was inevitable.

For Liebig the development of agricultural chemistry was marked by its relation to social concerns from the outset. His "insuperable disgust and aversion to the current practice of chemistry"[79] was in large measure a product of the fact that he considered it to be "asocial", producing useful applications neither in "medicine, physiology or industry."[80] In contrast, he was moved to show that unless "physiologists and agronomists use the culture of chemistry, no permanent and valuable progress can be expected in physiology and agriculture."[81] Specifically, Liebig was impelled by the conviction that unless chemistry turned toward agriculture it would be impossible to avoid a catastrophe of historical dimensions – hardly the position of a detached scholar who felt it sufficient to let the truth speak for itself. In order to promote and push this view forward, Liebig resorted to an array of strategies, tactics, diplomacy and polemics, for which he was more than once reproached – either as an expression of a flaw in his character, or as an attempt by him to use an illegitimate style of "power play".[82]

Liebig made frequent written reference to the problems encountered in establishing a new discipline. His open antagonism to the humus theorists and physiologists owed as much to his actual disagreement with these theories, as to his acute sense of how to mount an effective public opposition.

> A glance at the history of science reveals that if a new doctrine takes the place of a prevalent one, the new doctrine will not be a further development but in fact the very opposite of the old one.[83]

Liebig consequently advocated opposition and polemic.

> For a long time I believed that in agriculture it was enough to teach the truth for it to become disseminated, as is common in science, and to pay no further attention to errors; I have finally realised that this was wrong. In order to place the truth on a firm foundation it is necessary to shatter the altars of lies.[84]

Liebig acted accordingly: his conviction in the correctness of his position in the years after 1840 was so great that his belief in the efficacy of his, theoretically deduced, artificial manure amounted to nothing less than blind faith. In consort with the manufacturer of his patent fertilizer, he used the English press, in particular, to wage a promotional canpaign urging the rapid adoption of his fertilizer. Unfortunately for Liebig, and the progress of the institutionalization of agricultural chemistry, the insolubility of the mineral salts in the fertilizer rendered it highly ineffective in practice. "Liebig's teachings seemed so completely discredited that the *Journal of the Agricultural Society* refused to publish Liebig's reply to Lawes' and Gilbert's reports."[85] It was a further sixteen years before Liebig could provide a theoretical and experimental explanation of the disaster; his battle with rational husbandry correspondingly took the same amount of time. This episode provides proof that internal recognition is not, in itself, sufficient to secure the institutionalization of a finalized science: the results also have to work.

The training of students represents one of the most important means for the institutionalization of sciences. Liebig's establishment of a laboratory in which students received regular experimental instruction, enabled him to promote agricultural chemistry as a special field of education.[86] Once having created a cadre of chemists trained in experimental techniques, Liebig set about dividing up the complex problems of agricultural chemistry into discrete tasks, which in turn entailed the standardization of methods and procedures and the rapid execution of a large number of analyses. Liebig consequently strove to simplify the design and construction of experimental apparatus, although this could not always be achieved without some loss of precision.

Students could only be trained if they had the prospect of subsequent employment. Liebig produced a number of reports attacking the level of competence of teachers of rational husbandry, with the aim of displacing them from their posts in favour of his own students.[87] Reflecting on the success of his strategy in Austria, he noted that, "for many years no candidate who had not received training in Giessen or completed studies there was able to obtain a teaching post."[88] His efforts were equally successful in Hesse, Saxony, Bavaria and later in Prussia.

Liebig's tight grip over the policy of the leading German chemical journal, the *Annalen der Chemie und der Pharmacie* enabled him to dispense with the need to found his own journal: the *Annalen* was at that time also the leading international journal for organic chemistry. In order to allow some of the success of organic chemistry to rub off on agricultural chemistry, in 1840 Liebig restructured the format of the *Annalen*, with more prominence for contributions on agricultural chemistry, which had previously appeared under "Miscellaneous Notes". He subsequently used the *Annalen* as the vehicle for all relevant work in the field, including his own. Liebig's considerable involvement in the preparation of handbooks and textbooks also enabled him to ensure that agricultural chemistry was given the attention he thought it deserved.

The final period of institutionalization took place after 1860: in 1862 the first chair of agricultural chemistry was established at Halle, followed by Leipzig (1869), Giessen (1871), Göttingen (1872), Munich (1874), Königsberg and Kiel (each 1876), and Breslau and Berlin (1880). During the same period the experimental institutes, the bastions of rational husbandry, were closed down: Regenwalde (1859), Möglin (1862), Waldau (1868), Tharandt (1869), Hofgeismar (1871), Eldena (1877) and Proskau (1880). Others were taken over by students of Liebig.

6. CONCLUSION

Agricultural chemistry marked a new type of theory-formation in the natural sciences. We have termed this type of scientific activity "finalized science", a concept intended to indicate that social needs and interests can encroach quite explicitly into the process through which a new science constitutes its relevant field. Such social needs and interests form the external guidelines for theory-formation. The science of agriculture is a theory about the possibilities of the rational construction of nature, with the purpose of satisfying human nutritional needs. Set against the previous (ir)rationality towards nature which characterized the methods through which these needs were met before, it simultaneously represented a *critical* science.

Such a science clearly differs from the classical disciplines of physics, chemistry and biology. Although research in these sciences yields utilizable knowledge, and although scientific thought is influenced by its cultural environment, the research object remains a reality separate from human interests. No such distinction can be drawn between agricultural chemistry and the so-called applied sciences of engineering or medicine, which also take one segment of nature and approach it with the objective of satisfying human needs. Nevertheless, our aim here is not to introduce a distinction between agricultural chemistry as a finalized science and applied science, but rather to offer a more precise meaning for the vague notion of "applied science". The term "applied science" gives the misleading impression that goal-oriented science simply involves the application of an existing science, rather than the creation of a new *theoretical* development. This in turn feeds the misconception that pure science is superior to applied science. Although examples exist of where an applied science or technology has emerged without any additional theoretical labour, the production of applicable science frequently involves a process of theory-formation which proceeds independently of the mother-discipline. This occurs, for example, where the application under consideration requires a degree of specialization in which the fine detail is of no relevance to the mother-discipline. Pure and applied research also involve two differing types of objectivity: on the one hand, the investigation of nature independently of any particular objective, and on the other, the purposive construction of nature with specific goals in mind.

Considered historically, agricultural chemistry represents a very early example of a successful, goal-oriented science which, in contrast to preceding goal-oriented sciences, not merely formulated models and analogies for its chosen phenomena but produced a *theory* which was deductive, comprehensive and heuristic, and able to meet the requirements of a paradigm for the application of basic principles.

This type of applied science tends to be overlooked by the philosophy of science, with the consequence that the particular status of its theories remains unrecognized. This may be attributable to the fact that natural philosophy has placed a higher value on the universal qualities of nature, and their investigation, rather than the study of diseases, machines, fertilizers and so on. In contrast to this, we consider "human" nature, that is an artificially created reality directed at a particular aim, to be no less interesting than "natural" nature.

It is clearly not possible to discuss the question of the role of pure science in the genesis of the applied and technical sciences on the basis of a few case-studies. The intention of this study has merely been to show that under certain circumstances an applied science must develop its own theoretical model, and that such a model cannot be directly or solely deduced from the foundations of the pure sciences. Traditional philosophy of science only reveals *one* aspect of the development of modern science – its reductionist programme. The *other* aspect is captured through the idea of finalization, which denotes the human construction of reality through scientific theory-formation in the pursuit of social objectives and interests.

NOTES

[1] *Cf.* Böhme *et al.* (1973).

[2] *Cf.* van den Daele and Weingart (1976).

[3] The seventh (1862) edition of Liebig's *Agricultural Chemistry* was entitled *Chemistry in its Application to Agriculture and Physiology*.

[4] Liebig used 1840, the year in which the first edition of his *Agricultural Chemistry* appeared, to periodize the history of agriculture.

[5] Liebig (1862), I, Introduction, p. 9.

[6] Malthus (1971), p. 4.
[7] Malthus (1971), p. 6.
[8] The fall in the birth-rate towards the end of the nineteenth century contradicted the thesis that generative behaviour is historically and sociologically invariant.
[9] Malthus (1971), p. 4.
[10] *Cf.* Köllmann (1965), Vol. 4, pp. 1 and 6.
[11] Malthus (1971), p. VI.
[12] Liebig (1862), I, Introduction, p. 12.
[13] Liebig (1862), I, Introduction, pp. 125 ff.
[14] Liebig (1862), I, Introduction, p. 155.
[15] The title of the first volume of the 1862 edition of the *Agricultural Chemistry.*
[16] Liebig (1862), I, p. XV.
[17] Liebig (1862), I, Introduction, p. 87.
[18] Liebig (1862), I, Introduction, p. 125.
[19] Tilden (1921), p. 190.
[20] Kopp (1873), p. 311.
[21] Kopp (1873), pp. 319 ff.
[22] Principally carbon, nitrogen, oxygen; additionally, in particular, sulphur, halogen, etc.
[23] Isomers are chemical compounds with the same empirical formula, but with different structures and consequently different properties.
[24] Berzelius alone dealt with 2000; cf. Szabadváry (1966), pp. 142 ff.
[25] Tilden (1921), p. 190.
[26] Bunge (1929), Vol. 1, p. 434.
[27] Tilden (1921), p. 194.
[28] Hiebert (1964), p. 322.
[29] In 1834 Dumas and Peligot discovered the methyl radical ($C_2H_3CCH_3$), with other discoveries following in subsequent years. In 1839 Bunsen was able to show that the cacodyl ($C_4H_{12}As_2$) is the radical of arsenic methyl compounds.
[30] Wöhler to Berzelius, 1835; *cf.* Bunge (1930), Vol. 2, p. 14.
[31] *Cf.* Volhard (1909), I, p. 231 and 239.
[32] Liebig (1862), I, Introduction, p. 97.
[33] The expression employed by Liebig at the beginning of the *Agricultural Chemistry* gives an indication of this perspective: "Chemistry began to study according to its rigorous methods . . . ".
[34] The Royal Institution was established as "Public Institution for diffusing the knowledge and facilitating the introduction of useful mechanical inventions and improvements, and for teaching by courses of philosophical lectures and experiments the application of science to the common purposes of life . . . " (quoted in Tilden *op. cit.*, p. 152).
[35] *Cf.* Paris (1831), p. 105.
[36] Davy (1815), p. 177.
[37] Davy (1815), p. 178.
[38] Davy (1815), p. 180.
[39] Davy (1815), p. 199.
[40] The pre-history of scientific agriculture, especially in the eighteenth century, contains many chemists who undertook work in physiology, and physiologists who worked with chemical methods; for example, Priestley (discovery of oxygen production of

green plants in sunlight), Ingen-Housz (carbon-dioxide production at night), Senebier (photosynthesis); *cf.* Browne (1944), Chap. V.

41 Thaer (1798–1804).

42 Thünen (1826).

43 Volhard (1909), II, p. 32.

44 Liebig (1840*a*), Dedication to Humboldt.

45 Berzelius (1847), pp. 1–7.

46 Liebig (1840*a*), Dedication.

47 Hjelt (1916), p. 62.

48 "In order to provide a complete solution to these questions, we call upon the knowledge and powers of the plant physiologists, the agronomists and the chemists" (Liebig, 1862, I, p. 159). Physics was competent to understand the processes of mechanical corrosion which arise in the soil, for the mechanical working of the soil and climatic research; *cf.* Liebig (1862), II, pp. 65 ff.

49 Liebig (1862), I, Introduction, p. 87.

50 For example, in astronomy, the revolutions of celestial bodies; in medicine, the circulation of the blood; in economics, the circulation of money; in technology, feedback systems; in bourgeois economics, market mechanisms; *cf.* Mayr (1971) and Schmidtgall (1973).

51 *Cf.* Liebig (1862), I, Introduction, p. 143.

52 Liebig (1862), I, Introduction, p. 147.

53 Liebig (1862), I, Introduction, p. 136. In fact according to Liebig Smith refers to the "special generative power" of the earth: *cf.* Liebig (1862), I, Introduction, p. 137.

54 Liebig (1844*a*), p. 290.

55 According to Liebig's estimate, "in 800,000 years the atmosphere would no longer contain any trace of oxygen, and sometime before that it would have become totally unfit for processes of respiration and combustion; a reduction of its oxygen content to 8 per cent would have proved fatal to animal life". As Liebig was well aware, this was an optimistic assessment (Liebig, 1862, I, p. 18 note).

56 Liebig (1844*a*), p. 279.

57 *Cf.* Browne (1944), p. 246. Boussingault established such balancings after five-year experimental series: in 1839 he employed this method to try and resolve the dispute as to whether cows obtain nitrogen from the atmosphere: his balancings suggested that they obtained all their nitrogen from grass; *cf.* Browne (1944), p. 249.

58 Liebig (1844*b*), p. 41.

59 Liebig (1840*a*), p. 41.

60 Liebig (1840*a*), p. 36.

61 Liebig (1862), I, p. 137.

62 Liebig (1862), I, Introduction, p. 5.

63 Liebig (1862), I, Introduction, p. 12.

64 Liebig (1862), I. 157.

65 *Cf.* Liebig (1862), I, p. 159.

66 *Cf.* Liebig (1862), II, pp. 172 ff. and pp. 252 ff.

67 *Cf.* Liebig (1862), II, p. 257.

68 Berzelius (1839), pp. 20 ff.

69 *Cf.* Liebig (1840*a*), pp. 199 ff.

70 Liebig (1862), I, p. XIII.

71 Liebig (1863), p. 49.
72 Liebig (1862), I, p. 180.
73 *Cf.* Liebig (1843), p. 73.
74 Kellner and Immendorff (1913), pp. 415 and 464.
75 *Cf.* Liebig (1840*a*), p. 84.
76 For the history of the artificial manure see Schmitt and Ertel (1958).
77 Liebig (1841), p. 36; *cf.* Sprengel (1837).
78 See Böhme, 'Autonomization and Finalization: A comparison of fermentation research and fluid mechanics', in this volume, p. 72.
79 Letter to Berzelius, 26 April 1840; Carrière (1893), p. 210.
80 *Ibid.*, p. 211.
81 Letter to Berzelius, 3 September 1840; Carrière (1893), p. 215.
82 It is not necessary to refute these allegations. Browne (1944), one of the most well-informed historians of agricultural chemistry, suggests that Boussingault's contribution should be rated more highly than Liebig's, since he was more circumspect, more open and as a consequence made less mistakes. Nevertheless, one should set against these aspects, which we have not dealt with in our study, the fact that in a science which evolves in relation to external objectives and through external factors, the social introduction of its findings and working methods (that is, its institutionalization) forms a significant element in the historical process.
83 Liebig (1862), I, Introduction, p. 13.
84 Liebig (1862), I, pp. XIV ff.
85 Volhard (1909), II, p. 37.
86 Kellner and Immendorf (1913), p. 465.
87 See Liebig (1838), (1840*b*), and (1855). – Peter Borscheid brought to our attention the long drawn out war which Liebig waged against the Akademie Hohenheim (up until the mid-1860s) in the most widely circulated German daily newspaper of that time, the *Augsburger Allgemeine Zeitung*: Liebig's aim was to reach the important individuals who could play a key role in the institutionalization of agricultural chemistry – in this instance the members of the Württemberg *Land* Parliament – and mould their opinions in his favour. Borscheid summarized the result of this campaign in the following terms: "The view of various members was that 'If a man like Liebig, with a record of success in England, has got something critical to say about Hohenheim, then it is the Government's duty to take such criticisms seriously, follow them up and implement the necessary changes so that Hohenheim can continue as the leading agricultural academy in Germany!'." (Letter from P. Borscheid to the authors, 25 April 1977); *cf.* also Borscheid (1976), pp. 28 ff.
88 Cited in Volhard (1909), I, p. 369.

BIBLIOGRAPHY

Berzelius, J. J.: 1839, 'Über einige Fragen des Tages in der organischen Chemie', *Annalen der Pharmacie*, Vol. XXXI.

Berzelius, J. J.: 1847, *Lehrbuch der Chemie*, Vol. 4, 5th edition, Dresden and Leipzig.

Böhme, G., van den Daele, W., and Krohn, W.: 1973, 'Die Finalisierung der Wissenschaft', *Zeitschrift für Soziologie*, Vol. 2, 128–144.

Borscheid, P.: 1976, *Naturwissenschaft, Staat und Industrie in Baden (1848–1914)*, Klett, Stuttgart.

Browne, C. A.: 1944, *A Source Book of Agricultural Chemistry*, Chronica Botanica Company, Waltham, Mass.

Bunge, G.: 1929/30, *Das Buch der großen Chemiker*, 2 Vols., Chemie Verlag, Berlin.

Carrière, J.: 1893, *Berzelius und Liebig. Ihre Briefe von 1831–1845*, J. F. Lehmann, Munich and Stuttgart.

van den Daele, W. and Weingart, P.: 1976, 'Resistance and receptivity of science to external direction', in *Perspectives on the Emergence of Scientific Disciplines*, Mouton, The Hague, Paris, pp. 247–275.

Davy, H.: 1815, *Elements of Agricultural Chemistry*, Vols. 6/7 of *Collected Works of Sir Humphry Davy*, Smith, Elder & Co., London, 1840.

Hiebert, E. N.: 1964, 'The problem of organic analysis', in *L'aventure de la science. Mélanges Alexandre Koyré*, Hermann, Paris, pp. 303–325.

Hjelt, E. von: 1916, *Geschichte der organischen Chemie*, Vieweg, Brunswick.

Kellner, O. and Immendorff, H.: 1913, 'Beziehungen der Chemie zum Ackerbau', in Meyer, E. von: 1913, *Chemie. Allgemeine Kristallographie und Mineralogie*, Teubner, Liepzig and Berlin.

Köllmann, W.: 1965, *Bevölkerung und Raum in neuerer und neuester Zeit*, Ploetz, Würzburg.

Kopp, H.: 1873, *Die Entwicklung der Chemie in der neueren Zeit*, Oldenbourg, Munich.

Liebig, J.: 1838, 'Der Zustand der Chemie in Österreich', *Annalen der Pharmacie*, Vol. XXV, pp. 339–347.

Liebig, J.: 1840*a*, *Die organische Chemie in ihrer Anwendung auf Agricultur und Physiologie*, Vieweg, Brunswick. Published in 1840 in English, edited by Lyon Playfair, under the title *Organic Chemistry in its Applications to Agriculture and Physiology*.

Liebig, J.: 1840*b*, *Über das Studium der Naturwissenschaften and über den Zustand der Chemie in Preußen*, Vieweg, Brunswick.

Liebig, J.: 1841, *Die organische Chemie in ihren Beziehungen zu den Herren Dr. Gruber in Wien und Dr. Sprengel*, Winter, Heidelberg.

Liebig, J.: 1843, 'Die Wechselwirtschaft', *Annalen der Chemie und Pharmacie*, Vol. XLVI, pp. 58–97.

Liebig, J.: 1844*a*, *Chemische Briefe*, Winter, Heidelberg; first published in English in August 1843 under the title *Familiar Letters of Chemistry*, Taylor & Walton, London.

Liebig, J.: 1844*b*, *Bemerkungen über das Verhältnis der Thier-Chemie zur Thier-Physiologie*, Winter, Heidelberg.

Liebig, J.: 1855, *Herr Dr. Emil Wolff in Hohenheim und die Agricultur-Chemie*, Vieweg, Brunswick.

Liebig, J.: 1862, *Die Chemie in ihrer Anwendung auf Agricultur und Physiologie*, 7th edition in 2 volumes, with an Introduction, Vieweg, Brunswick, no English edition.

Liebig, J.: 1863, *Über Francis Bacon und die Methode der Naturforschung*, Literarisch-artistische Anstalt, Munich.

Malthus, T. R.: 1971, *An Essay on the Principle of Population*, Seventh Edition, 1872, reprint, Kelley, New York.

Mayr, O.: 1971, 'Adam Smith and the concept of the feedback system', *Technology and Culture*, Vol. 12, pp. 1–22.

Paris, J. A.: 1831, *The Life of Sir Humphry Davy*, Bentley, London.
Partington, J. R.: 1972, *A History of Chemistry*, Vol. 4, Macmillan, London.
Schmidtgall, H.: 1973, 'Zur Rezeption von Harveys Blutkreislaufmodell in der englischen Wirtschaftstheorie des 17. Jahrhunderts', *Sudhoffs Archiv*, Vol. 57, pp. 416–430.
Schmitt, L. and Ertel, H.: 1958, *100 Jahre erfolgreiche Düngerwirtschaft*, J. D. Sauerländer, Frankfurt.
Sprengel, C.: 1837, *Die Bodenkunde oder die Lehre vom Boden*, Leipzig.
Szabadváry, F.: 1966, *History of Analytical Chemistry*, Pergamon Press, Oxford.
Thaer, A. D.: 1798–1804, *Einleitung zur Kenntniß der englischen Landwirthschaft*, 3 Volumes, Hahn, Hannover.
Thaer, A. D.: 1809–1812, *Grundsätze der rationellen Landwirthschaft*, 4 Volumes, Realschulbuchhandlung, Berlin.
Thünen, J. H.: 1826, *Der isolirte Staat in Beziehung auf Landwirthschaft und Nationalökonomie*, Hamburg.
Tilden, W. A.: 1921, *Famous Chemists. The Men and their Work*, Routledge, London.
Volhard, J.: 1909, *Justus von Liebig*, 2 Volumes, J. A. Barth, Leipzig.
Wallerius, J. G.: 1764, *Chymische Grundsätze des Ackerbaus*, A. Wever, Berlin.

GERNOT BÖHME

AUTONOMIZATION AND FINALIZATION: A COMPARISON OF FERMENTATION RESEARCH AND FLUID MECHANICS

1. INTRODUCTION

The aim of this study is to examine the distinction between two types of scientific development – autonomization and finalization – through a consideration of two characteristic instances, fermentation research and fluid mechanics. The title 'Autonomization *and* Finalization' is intended to emphasise that these two tendencies are not mutually exclusive in the scientific developments considered here. No one type of development ever completely dominates a field; rather, under certain conditions a field of research will display a tendency towards autonomy, and under others a tendency towards finalization, in which the development of the field is influenced by extrinsic goals. Such conditions may be the product of the social and economic situation in which a science develops, with the actual nature of the science itself playing no essential role.[1] Alternatively, these conditions may be located in the cognitive state of a science, with differences existing between individual sciences at any given point in time. There is no necessary reason why the internal conditions for autonomization and finalization, and the external conditions, should be in unison. A science which does not meet the internal conditions (such as theoretical maturity) may pursue finalization at any early stage in its history – a process we have termed "functionalization".[2] Alternatively, a mature science can retreat from the expectations placed on it by society and collapse into academic sterility. This study is concerned with the *cognitive* conditions for autonomization and finalization; these are responsible for the extent to which, and the how successfully, individual sciences have been able to match the demand of the global conditions presented by society in the nineteenth and twentieth centuries.

What are the relevant questions as far as examining and explaining the cognitive developments of sciences are concerned? As far as autonomization – the detachment of a science from its social origins – goes, one of the most fundamental must be: "Why doesn't science 'whither away' under such conditions of isolation?" The answer to this requires identifying a mechanism for the internal generation of scientific problems. Moreover, does the cognitive

G. Böhme et al., Finalization in Science, ed. by W. Schäfer, 53–91.

status of a particular science, or science in general, contain the basis for tendency towards autonomization and isolation?

Some explanation of the tendency towards isolation may be found in the consideration of the development of specialist terminologies, and of the tendency towards autonomization through the differentiation of scientific discourse.[3]

Explaining the cognitive foundations of the tendency towards finalization requires, as a first step, some elucidation of the "secret" harmony between natural science and technology: the sociology of knowledge and the history of science, in particular that of the Renaissance period, contain a number of approaches to this problem. What are the cognitive presuppositions which allow a science to respond to external direction at all: when can a science transform technical problems into scientific ones, and on what type of cognitive basis can it organise a solution – for example, in the form of a direct orientation to the technical problem, possibly even with such organisational aspects as a specific division of scientific labour and a programmatic schedule? Finally, does the integration of such goals mean that such a science will undergo a modification in terms of its cognitive apparatus, theoretical development and methodology?

We can only tackle *some* of these issues through the format of a case study. The main areas considered are (i) the mechanisms through which problems are generated and (ii) the mode of evaluation. A precondition for approaching these questions is some understanding of the notion of a "research front" and a "scientific breakthrough", both for individual sciences and for science in general. We also consider the type of relation between scientific knowledge and its application. In the second case-study, this question is posed as one of the relations between theories, with a fundamental theory on the one hand, and application-oriented special theories on the other. Finally, we attempt to determine more closely the relationship between the development of problems and technological use. Each of the case-studies exhibits its own characteristic structures of theoretical development, which can be linked through the types of consensus and dissensus within each specialist community, and the mechanisms by which problems are generated.

2. FERMENTATION RESEARCH – FLUID MECHANICS: A COMPARISON OF TWO THEORETICAL DEVELOPMENTS

2.1. *Initial Status and Comparability*

Fermentation is currently widely regarded as simply the process by which alcohol is obtained from sugar through the activity of yeast – essentially, one phenomenon. In contrast, flow phenomena are both numerous and have a large number of practical implications in such fields as wind and water mills, canal and ship construction, ballistics, steam-engines, furnaces and heating installations, aviation and space-travel. A counterposition of nineteenth-century fermentation research – with its meagre resources in terms of personnel, and its sporadic and non-institutionalized character – and twentieth century research into flow phenomena – with its enormous state and private research facilities – clearly requires the establishment of some foundation on which comparison can take place. At the beginning of the nineteenth century "fermentation" still denoted a very extensive class of phenomena, ranging over alcoholic fermentation, all processes of souring and decaying, most processes of nutrition and digestion, and often even inorganic processes such as erosion. This diffuse class of phenomena shared the the common feature that in each case a self-actuating agent, the ferment, brought about certain chemical changes. Fermentation therefore covered that class of chemical changes which occurred without the intervention of an external agency such as heat or mechanical activity or electricity. The concept of fermentation was progressively narrowed down during the nineteenth century, and it was not until almost at the close of the century that it regained its former scope, although in a changed form, through its subsumption within the class of enzyme processes.

As far as the range and possible application of fermentation phenomena were concerned, fermentation research had the potential for becoming an independent and institutionalized research field; in fact, enzymology has developed into precisely such a field in the twentieth century. However, a difference does exist between the way in which the usefulness of nineteenth century fermentation research was approached, and the approach to the practical application of twentieth century fluid mechanics.

It is evident from the mere definition of fermentation that such processes are of considerable practical value: organic life, and consequently human existence itself, it strongly determined by the nature of such processes. Nevertheless, at the beginning of the nineteenth century virtually all the industrially exploitable chemical technologies were inorganic in nature, embracing areas such as salt-extraction, metallurgy, soda-making and the production of sulphuric acid. Only those technologies with their roots in the kitchen were organic; beer-brewing, wine and vinegar making, and baking. The latter were just beginning to assume industrial dimensions and methods of production at the start of the nineteenth century. This marked the first time that scientific research into fermentation could be expected to yield practical consequences, and expectations in this direction were indeed formulated. Nevertheless, the impulse behind the emergence of fermentation research in the nineteenth century did not come from such expectations. Although Pasteur is widely acknowledged to have been consulted on a number of problems in industrial fermentation by an alcohol-manufacturer shortly before beginning his studies of fermentation, it is unlikely that this was the factor which prompted his work in the field.[4] Pasteur had already given a course on the chemistry of alcohol production as a part of his duties in the training of technicians at the University of Lille; moreover, Pasteur himself explained his interest in fermentation as stemming from the optical activity of amyl alcohol.[5] Since amyl alcohol was a product of fermentation – in Pasteur's opinion, therefore, a product of the life-activity of yeast – this raised the question as to the existence of a relationship between optical activity and life. Nevertheless, although practical considerations may have conceivably prompted ***Pasteur*** to take up the study of fermentation, this argument cannot be extended to research in this field in general in the nineteenth century: Liebig, for example, encountered the problem of fermentation via theory, Mitscherlich was interested in the analogy between fermentation and "catalytic" processes, and Cagniard de la Tour developed his physiological theory of fermentation as a result of discovering the vegetable nature of yeast through the microscope.

The relation between scientific research in fluid mechanics and technical application in industry was quite different to that which prevailed in fermentation research. Scientists had always been interested in the area to some extent, and after the eighteenth century the technical questions related to flow phenomena were treated more and more explicitly; the main problems were located in the fields of hydraulic engineering, the reduction of resistance in various contexts, and ballistics. The offering of a prize for an explanation of the relation between resistance and the ratio of the cross-section of a canal to the cross-sectional area of a vessel by the French Académie des Sciences in 1770 in connection with contemporary French canal building projects represents a characteristic instance.[6] Up until the twentieth century scientists were only able to respond to practical demands with applications of procedures obtained from experimental rather than theoretical innovation, with the exception of the theory of the turbine formulated by Euler in 1754. The fact that the science of flow phenomena, hydrodynamics, was only able to respond to the demands of practice through the experimental treatment of practical problems is all the more remarkable since flow phenomena had been subsumed within a fundamental theory by 1850: the significance of this is discussed more fully below. It is sufficient to note here that it was the successful attainment of powered-flight which provided the "take-off" for research in flow phenomena.

Whereas research into flow phenomena in the twentieth century was able to develop on the basis of a closed, fundamental theory, nineteenth century fermentation research did not have access to a comparable theory. Although following Lavoisier basic chemical techniques had been established – that is, the identification of elements, establishment of conversion formulae, and quantitative methods – the characterization of chemical changes lacked the means to establish an energy-balance, let alone to consider such changes in structural-chemical terms, which were methods which developed in parallel with fermenation research. No comprehensive, and crucially, no universally acknowledged chemical theory existed for chemical bonding or chemical reactions. In contrast, the experimental investigation of fermentation processes had access to a large number of recognised *experimental* techniques. Standard indicators were known for a large number of substances, and Liebig had even drafted a general schema for the analysis of organic substances.[7]

Although flow and fermentation phenomena shared the feature of having a wide scope – both in terms of the phenomena embraced and the breadth of practical applications – the starting position of fermentation research in the early nineteenth century was very different from that of fluid mechanics in the early twentieth. Fluid mechanics possessed a closed theory, and was driven forwards by a wide variety of practical demands; fermentation research lacked a fully developed foundation in the underlying chemistry, and was not propelled by quite such specific practical interests.

2.2. *Historical Overview*

2.2.1. *Fermentation research*. If it were necessary to choose a date to mark the beginning of fermentation research, then according to the criterion that the "beginning" of a research field coincides with the emergence of a corpus of relevant scientific literature, the 1830s would have the strongest claims. Of course, scientists had long been interested in fermentation and had accumulated some experience with it. However, the 1830s witnessed the emergence of three different theories of fermentation, or expressed more cautiously, three theoretical approaches towards the explanation of fermentation phenomena which subsequently succeeded in integrating all further scientific work in this field into *one* principal framework. The pre-history of fermentation research took place in three distinct areas, where experience was accumulated, and which later influenced the various theoretical approaches. The first of these areas was the process of fermentation itself, and in particular the centuries of experience accumulated in the brewing and wine-making trades. The chemical process of alcoholic fermentation was described in terms of its raw-material inputs and its end-products by Lavoisier, and given quantitative specification by Gay-Lussac, Thénard and de Saussure by *c*. 1800.

The second area was the discovery of the living character of yeast, largely as a result of advances in microscopy. Leeuwenhoek had observed yeast globules through the microscope as early as 1680. In 1787 Fabroni used chemical criteria to establish that yeast was a living substance by observing that it gave off ammonia during dry distillation. The third area was the result of the early-nineteenth century discovery of a number of chemical processes later termed "catalytic": these were processes which took place in the presence of certain substances, or more correctly, because certain substances were present, without these substances entering into the process, at least as far as could be discerned from the initial and final products of these processes. G. S. C. Kirchhof, for example, discovered that starch could be broken down into sugar using diluted acids. Tenard used alkalis, silver, platinum, gold and fibrin to break down hydrogen peroxide. Humphry Davy demonstrated the burning of alcohol and ether over platinum, and Döbereiner the combustibility of hydrogen gas in air through platinum sponge at room temperature. In 1834 it was realized that starch could be broken down through diastase.

The three theories of fermentation were, (i) the physiological theory, (ii) the catalytic theory, and (iii) Liebig's theory – termed the sympathetic or mechanical theory in this study. The physiological theory was formulated by Cagniard de la Tour between 1835 and 1837; alcoholic fermentation was regarded as a physiological process of the yeast organism. The catalytic theory originated with Mitscherlich (1834) and Berzelius (1835)[8] and was based upon two theoretical steps: firstly, the supposition that all those instances in which substances influenced the course of chemical reactions without entering into them were analogous, and secondly, the hypothesis that one particular cause underlay this class of events. Mitscherlich took both steps in his 1834 work on ether,[9] initially choosing the term "contact" to describe the nature of the cause. In 1835 Berzelius christened such processes "catalyses", and ascribed the "catalysts" with a specific force, the catalytic force. Berzelius was also the first to include fermentation in the class of catalyses. Liebig's theory of fermentation first appeared in his 1839 paper 'Über die Erscheinungen der Gärung, Fäulnis und Verwesung und ihre Ursache'[10] published in *Annalen der Pharmacie*. According to Liebig, organic molecules displayed two types of bonding or affinity; those between carbon, hydrogen, oxygen and nitrogen, and those between these elements and the radicals. Within a complex molecule both types of bonding could be found in a relation of tension, which Liebig termed the chemical difference. The causes of chemical change in organic materials were identified as those factors tending to increase chemical difference. Liebig cited three possibilities: (1) a change in the state of cohesion of the material through the effects of heat, (2) contact with a third body, which does not enter into any resultant compound, and (3) the addition of the elements of water. "Contact" in this context was not regarded as a cause in its own right, in contrast to the catalytic theory, but rather as the mechanism through which one body transmitted its breakdown to another. In fermentation, Liebig considered the ferment to be in a continuous process of decomposition into which those substances undergoing fermentation (sugar in the case of alcoholic fermentation) were drawn. We term this theory "sympathetic" since the substances which become fermented enter the process in sympathy with the decomposition of the ferment. Liebig could be regarded as looking for a mechanical cause; in his view, in situations where chemical affinity was weak a mechanical cause was required to overcome the inertia inherent in complex molecules. For fermentation this implied that,

> . . . the movement in which its atoms are engaged (*that is, the atoms of the decomposing ferment: author's note*) is communicated to the atoms of the elements of the sugar; they cease to persist in the state in which they constitute sugar and order themselves in accordance with their specific characteristic relations.[11]

The three approaches co-existed separately, but in constant interaction, up until 1896. All three concentrated their attention on alcoholic fermention, although this phenomenon was classified differently depending on the theoretical approach involved. A number of significant scientific steps were made within this overall development: different types of fermentation were identified and the corresponding yeasts discovered. For Pasteur this became a general research programme for the systematic study of the connection between chemical processes and micro-organisms, culminating in the identification of micro-organisms as the causes of certain diseases. The metabolism of the yeast cells and their life-cycle was subjected to detailed empirical scrutiny. The advocates of the catalyst theorem unearthed and investigated a steady stream of new catalytic processes and catalysts; of particular importance was the collection of organic catalysts – soluble ferments, of which around a dozen were known by the end of the nineteenth century.[12] Other advances made within the overall context of fermentation research were the refutation of spontaneous generation, and the development of processes for sterilizing and conserving organic materials. This period of research was brought to a close by the publication of Eduard Buchner's *Alkoholische Gärung ohne Hefezellen*,[13] in which Buchner described his isolation of a substance from yeast cells – the enzyme zymase – which was able to induce fermentation without any living yeast being present. Buchner's work marked the unification of the various directions pursued within fermentation research, and initiated the science of enzymology. In fact, Willi Kühne had already presaged this unification, at least terminologically, by coining the expression "enzyme" in 1978;[14] Kühne had contended that catalytically operating soluble ferments in the living cell corresponded to substances which were responsible for the fermenting effect of micro-organisms.

The development of fermenation research between 1830 and 1890 can be summarised as follows: the field was defined by the continued pursuit of three competing theoretical approaches, and dominated by a number scientific controversies. The main outcome of this development was the subjection of fermentation phenomena to experimental study, and finally, the convergence of the three theoretical approaches in enzymology. Any practical consequences followed more as a side-effect of *theoretical* development, than from a direct orientation to practical applications. Fermentation research was not socially institutionalized; there were no specific journals, no professorial chairs, scientific associations or research institutes.

2.2.2. *Fluid mechanics*. In contrast to fermentation research, twentieth century fluid mechanics research has been determined both by practical requirements and social institutionalization. The leading role in the field was occupied by aerodynamics.[15] Three particular events can be noted: in 1902 Kutta succeeded in calculating the lift on an aerofoil, in 1903 the Wright brothers achieved powered flight, and in 1904 Prandtl advanced his theory of the boundary-layer.[16] These developments occurred against a background of an essentially closed and mathematically established theory of hydrodynamics. However, prior to Kutta and Prandtl, the theory could not be used to resolve practical problems.

By the end of the nineteenth century fluid mechanics was divided into two more or less completely separate fields. *Hydrodynamics*, which rested on Euler's equations for the motion of a frictionless fluid, had reached a high degree of perfection and completeness. However, this had little significance for practical applications since the results of this so-called hydrodynamics were flatly contradicted by experience under numerous circumstances. As a consequence, engineers prompted by the important problems raised by rapidly developing technology developed their own, distinctly empirical, science of *hydraulics*, based on a large number of experimental results, and differing substantially from the aims and methods of theoretical hydrodynamics.[17]

It was the concepts of lift and circulation on the one hand, and the theory of the boundary-layer on the other, which allowed the general theory to be linked to the resolution of practical problems. This linkage in turn spawned a large number of special areas of research into flow phenomena, each with corresponding special theories. To cite just a few: the theory of drag, based essentially on the boundary-layer concept; special aerofoil theories, including the theory of the propellor; high-speed gas flows became the subject matter of gas-dynamics, with its special field of the effects of pressure waves. More recent developments include flow acoustics, a special theory of flow phenomena in physiology, and the distinct field of turbulence research.

The development of special disciplines related to fluid mechanics research – of which the above is merely an incomplete and unsystematic listing – was both accompanied and actively promoted by a comprehensive social institutionalization of the field. A number of scientific bodies dealing with fluid mechanics had existed prior to the epoch of powered-flight, such as the Royal Aeronautical Society, founded in 1866. Two years after its own foundation in 1906, the German *Motorluftschiff-Studiengesellschaft* established the first specific research institute in the field, the *Göttingen Modellversuchsanstalt*. 1908 also saw the establishment of the *Luftschraubenstation Lindenberg* for the study of airscrew performance. Both organizations, and in particular the Lindenberg, received extensive finance from industry through the *Jubiläumsstiftung der deutschen Industrie*. 1912 saw the foundation of the *Wissenschaftliche Gesellschaft für Luftfahrt* (Scientific Association for Aviation), and the *Deutsche Versuchsanstalt für Luftfahrt* (German Aviation Experimental Establishment) in Berlin-Adlershof. The establishment of these organisations frustrated Klein and Prandtl's plans at the Göttingen Society *für angewandte Mechanik* (Association for Applied Engineering) for the foundation of a *Kaiser-Wilhelm-Institut für Strömungsforschung* (Kaiser Wilhelm Institute for Fluid Mechanics Research). However, military interest in aviation during the First World War brought Prandtl's plans to fruition at least as far as aeronautical engineering was concerned in the shape of the *Aerodynamische*

Versuchsanstalt (Experimental Establishment for Aerodynamics), Göttingen. In 1925 the *Kaiser-Wilhelm-Institut für Strömungsforschung* was eventually established, with the *Aerodynamische Versuchsanstalt* as an affiliate up to 1937. In 1936 these institutions were added to by the *Deutsche Forschungsanstalt für Luftfahrt* (German Aviation Research Establishment), located in Brunswick-Volkenrode.[18]

These organizations represent those research institutions located outside the scope of the university system. The institutionalization of research in fluid mechanics and aerodynamics also involved the establishment of its own technical journals, such as the *Zeitschrift für Luftschiffahrt und Flugtechnik* (1910), the creation of professorial chairs, and the formation of smaller university-based institutes, such as von Kármán's at the Technical University of Aachen. Developments in other countries followed much the same pattern, occasionally with a noticeable lag behind developments in Germany. In quantitative terms research into flow phenomena grew at an exponential rate during this period. Schlichting suggests a period of 10 years for the doubling of the number of publications dealing with the boundary-layer problem, and also found that the power of wind-tunnels increased ten-fold during the same period (see Figure 1).

The exponential growth in the wider literature in fluid mechanics can be seen in the expansion in the number of publications emanating from the *Max-Planck-Institut für Strömungsforschung* and the *Aerodynamische Versuchsanstalt* (see Figure 2). Since according to Price (1963), the law of exponential growth is valid across a broad range, no particular conclusions can be drawn from this fact alone. Nevertheless, the establishment of such a rate of growth for a given science is strong evidence for the scientific "arrival" of that field. In this case the rate of growth is determined by the existing stock of knowledge, since in mathematical terms, the law of growth implies that growth is proportional to what is already in existence.[19]

The development of research into flow phenomena in the twentieth century can be summarized as follows: the "take off" of the field took place via the broad public, industrial and military interest in aviation. This process was based on certain scientific achievements which enabled hydrodynamics, previously unproductive in terms of application, to be applied to practical problems.

In societal terms the process was characterized by an extensive degree of institutionalization and by exponential growth, and cognitively by the emergence of a large number of specialties and special disciplines within fluid mechanics research, each of which represented a response to specific external demands.

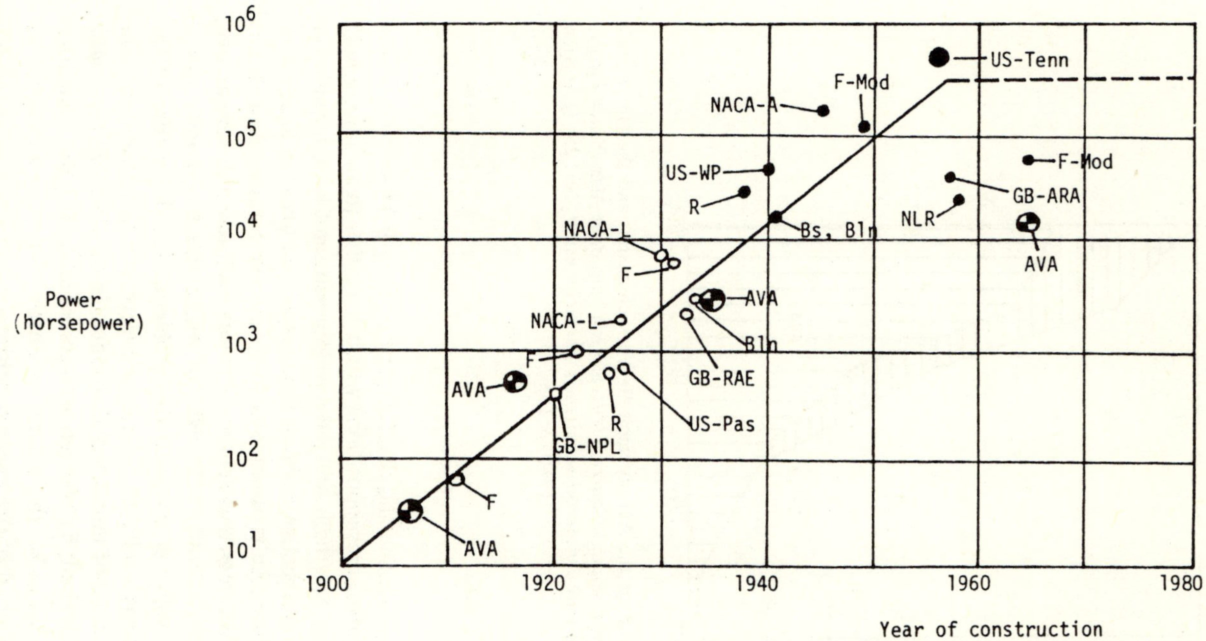

Fig. 1. Growth in wind-tunnel performance for major facilities. *Source*: Schlichting (1975), p. 167.

Key AVA = Aerodynamische Versuchsanstalt Göttingen
Bs = Brunswick,
F = France, GB = Great Britain, Mod = Modane, NACA = National Advisory Committee for Aeronautics (USA), R = USSR,
US–Tenn = Arnold Engineering Development Centre, Tullahoma/Tenn.

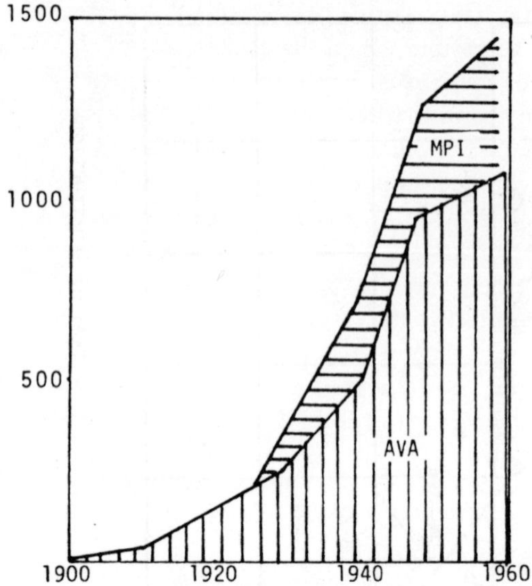

Fig. 2. Total publications of the Aerodynamische Versuchsanstalt and the Max-Planck-Institut für Strömungsforschung. *Source: Mitteilungen aus dem Max-Planck-Institut für Strömungsforschung und der Aerodynamischen Versuchsanstalt*, No. 25, Göttingen, 1960, p. 135.

2.3. *Analysis*

This section seeks to examine the development of each field with the aim of establishing the cognitive presuppositions which led either to an autonomous development or to an externally-steered development, and to determine the degree to which each respective research-field responded to external demands and was able to produce results of practical relevance. We consider four factors in this context: (1) the degree of consensus within each research community, i.e., the type and the amount of knowledge which this community has at its disposal for the solution of particular problems, (2) the nature of the theoretical development itself, i.e., – taken together with (1) above – the relation between the state of research and advances in the field, (3) the origins of the problems addressed by the research-community, (4) the evaluation of the results – specifically, the extent to which results are evaluated by criteria internal or external to the research field.

In order to allow a comparison to be made as each case-study is developed, we summarize the results of our consideration of these questions below. The only form of consensus prevailing in nineteenth century fermentation research was agreement as to how certain experiments were to be conducted. The overall scene was riddled with controversies; in fact, disputes between the different theoretical approaches served to hold the field together, and at the same time provide a mechanism for generating scientific problems. The results of research were almost exclusively evaluated within the field in accordance with their value as weapons in the disputes between theoretical approaches. In contrast to this, scientists working in the field of fluid mechanics operated within a general consensus as to the underlying principles in the field; there were no controversies, and research consisted in the development of fundamental theories into special theories which allowed the fundamental theories to be applied to the specific problems posed by practical application. Fluid mechanics was supplied with its problems by the "technical front", its results were corroborated directly within technical progress, and the "technical front" accordingly shifted forwards.

2.3.1. *Fermentation research*. The nineteenth century investigator into fermentation had no secure theory on which to base their work, although any work carried out in the field could be classified within the overall constellation defined by the three theoretical approaches – the catalytic, the sympathetic, and the physiological. Fermentation researchers were not, however, divided into distinct schools. Each theoretical approach was too incomplete and open to fundamental challenge from one of the rival positions to offer a sufficiently large framework within which the individual researcher could define themselves exclusively. The contact theory, for example, never specified whether the catalytic force was chemical, physical or biological in nature. And the physiological theory had to constantly take into account the possibility that processes such as nutrition and respiration could be explained in purely chemical terms; at the very least, the proponents of the physiological theory had to be prepared to subject their claims about physiological processes to chemical controls. Liebig's position was probably the most dogmatic – a fact connected to the *methodological* inspiration behind the theory, rather than its substantial hypotheses. Consequently, although any researcher was in principle open to the arguments of other approaches, the exigencies of carrying out actual scientific work necessitated taking up a particular position. The harshness which characterised some of the exchanges between different approaches may well be explained by the element of arbitrariness in this decision. The development of fermentation research was deeply influenced by such controversies; to cite one major instance, the friendship between Liebig and Berzelius did not survive the argument as to whether it was "scientific" to assume the existence of a particular catalytic force. Liebig and Pasteur argued over whether fermentation was a life-process (Pasteur) or a process indicating the death of organisms, in which organic substances were destroyed through purely chemical-mechanical processes (Liebig); Pasteur engaged in a long-running dispute with Frémy and Boutron over the issue of spontaneous generation, specifically whether living yeast cells could develop as the result of fermentation; Pasteur and Traube came into conflict over whether fermentation could be regarded as a process of respiration, and Pasteur also disagreed with Berthollet over a number of Claude Bernard's posthumously discovered papers, in which he had asserted that fermentation was possible without yeast cells. All these controversies, both the major and the less central, provided researchers with problems on which the field as a whole could set to work. Figure 3 illustrates the main elements and interconnections in and between these theories and conflicts in terms of the major personalities involved. Single lines indicate positive dependency, and double lines with arrow heads negative dependency – that is, the determination of the content of a position through criticism and refutation. Vertical lines indicate the progress charted by each of the three main positions. This line marks the principal site upon which facts were accumulated which supported each position, on which theoretical explanation was made and strengthened, and occasionally on which theoretical positions were modified.

Liebig's initial 1839 contribution was the clear result of the controversy between the existing positions of the contact theory and the physiological theory. These theories did not initially have any direct bearing on each other. Liebig set them in a constellation through his classification of fermentation as a manifestation of decomposition, along with decay and putrefaction. He was therefore committed to treating the physiological theory as being fundamentally absurd: his critique had three main elements, (i) empirical, (ii) a critique of the physiological theory's interpretation of fermentation and (iii) a logical critique. In empirical terms, he noted that yeast does not reproduce in a pure sugar-solution; this contradicted what might have been expected had fermentation been a process of the nutrition of individual organisms. Moreover, the physiological theory confused cause and effect: microscopic organisms appear because they feed on the *products* of fermentation – not because they are its cause. And thirdly, fermentation and decay only occur

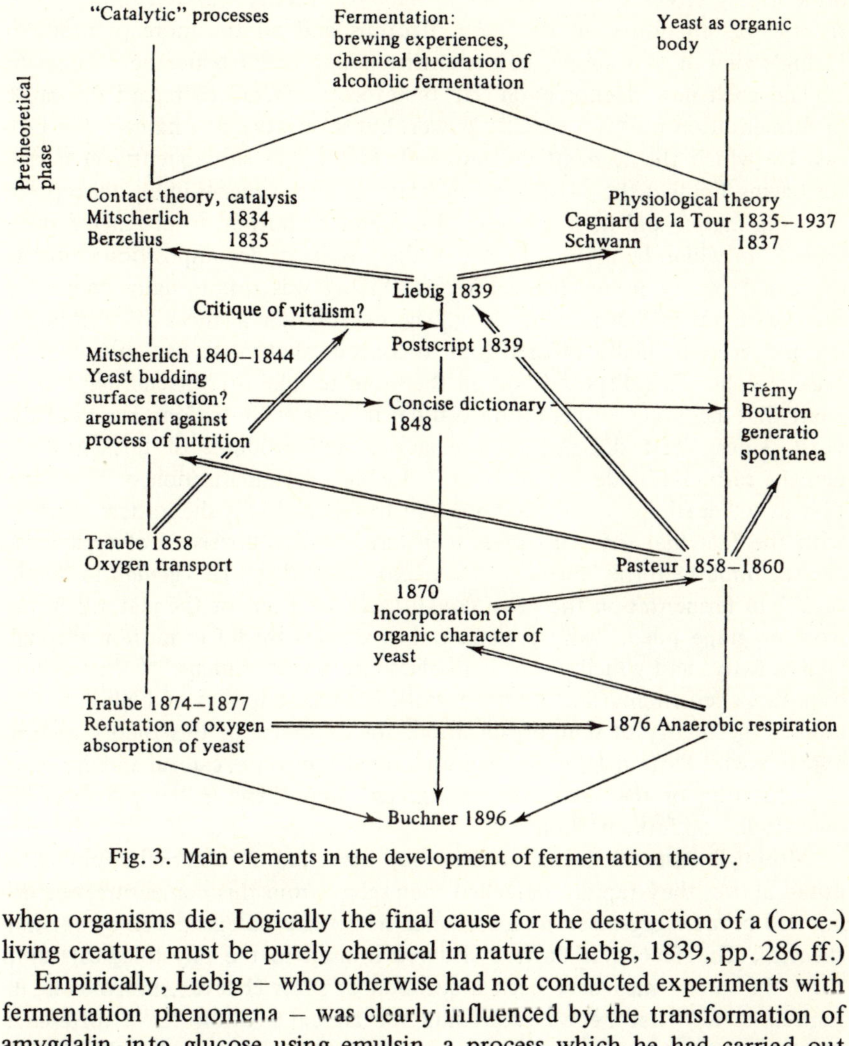

Fig. 3. Main elements in the development of fermentation theory.

when organisms die. Logically the final cause for the destruction of a (once-) living creature must be purely chemical in nature (Liebig, 1839, pp. 286 ff.)

Empirically, Liebig – who otherwise had not conducted experiments with fermentation phenomena – was clearly influenced by the transformation of amygdalin into glucose using emulsin, a process which he had carried out jointly with Wöhler (see Conant, 1952, p. 33). Together with Payen and Persoz's discovery of diastase in 1833, and Schwann's discovery of pepsin in 1836, this represented one of the first discoveries of a soluble ferment, that is, an identifiably non-living isolatable substance giving rise, nevertheless, to processes analagous to that of fermentation. This experimental background drew Liebig closer to the contact or catalytic theory: his rejection of this theory was therefore all the more rigorous and all the more pointed. In Liebig's view it was simply bad scientific policy to introduce new forces to explain each new phenomenon. He therefore preferred to regard the cause of fermentation not as a specific power, but as a process – namely the process by which the ferment decomposed. Liebig was subsequently criticised for having implied the existence of certain vitalist elements in his conception of living matter and its breakdown. He therefore hurried to produce a postscript[20] in which he made it clear that the process of decomposition could be explained purely in mechanical terms: Liebig was not initially convinced that yeast was in fact a living thing. The microscopic analysis of yeast budding and growth finally obliged him to concede that yeast was composed of organic cells. This disposed him all the more to take up Mitscherlich's argument that alcoholic fermentation could not involve a nutrition process of the yeast, since yeast did not require any of the products of fermentation, namely carbon-dioxide and alcohol.[21] Liebig's systematisation of fermentation as a "death process" required him to reconcile his theoretical position with the fact that the yeast present in fermentation consisted of organisms. He accomplished this through the *ad hoc* hypothesis of "*generatio aequivoca*": in fermentation the yeast *constituted* itself out of the material in the wort or grape juice, "whose elementary particles are set in motion through vital activity, and which order themselves into a yeast fungus."[22] This *ad-hoc* hypothesis led Frémy and Boutron, both French followers of Liebig, to formulate the theory of semi-organization: the existence of this theory in turn explains why Pasteur later devoted such attention to the careful and systematic refutation of the "*generatio spontanea*" hypothesis in the course of his own studies on fermentation.

Although Pasteur's initial studies may have had an external stimulus, as noted above, they rapidly detached themselves from this conjectural beginning and became organised in accordance with the problems posed by existing theories. The fact that Pasteur situated himself in the physiological camp from the outset may have been a question of sheer chance; nevertheless, he himself stated that he was searching for an explanation for a difference between the same chemical substance (laevo-rotatory and dextro-rotatory sugar) which he was unable to explain in chemical terms. His 1861 paper[23] shows very clearly that *what* he was studying was determined by the opposing positions on fermentation. For example, Liebig's argument against the view that fermentation was a growth-process rested on the observation that no yeast growth was discernible in sugar solution. Pasteur demonstrated that such growth was in fact possible if the solution was enriched through minerals, which the yeast needed in addition to carbohydrate (such a mineral/sugar-solution was produced by adding "yeast ash" to the sugar). Pasteur showed that the cellulose in the yeast multiplied in such a solution. At the same time, he used this finding to attack Mitscherlich's position that the yeast did not require any of the products of sugar-fermentation. Finally, he made a very detailed investigation of the migration of nitrogen during the process of alcoholic-fermentation. This study was clearly prompted by the fact that the release of nitrogen was currently regarded as a sign of organic dissolution (Liebig's "death process").

Moritz Traube's 1858 paper,[24] and his book on the same subject, took issue with all three theories. His principal objection to the physiological theory was that fermentation also occurred in the absence of living matter. Against the catalytic theory he contended that it was not in fact a theory in essence, but more a gathering together of a class of processes with certain analagous properties, for which a large number of different causes could be imagined. In fact, there was no certainty as to what the "catalytic force" actually consisted in, and one of the main directions of further development in this theory was the search for the individual causes underlying the catalytic effect. Mitscherlich, for example, considered that the surface concentrations on the catalyst represented such a cause, and he therefore tried to demonstrate that fermentation only took place on the surface of yeast cells.[25] In fact, Traube made a substantial advance in the theory of catalytic processes through his hypothesis that the ferment had to overcome a number of resistances in succession, such as elasticity, cohesion and molecular resistance.[26] He later identified one of the specific functions of the ferment as the transportation of oxygen, and distinguished oxydation and reduction ferments, in which the former carry freely available oxygen and the latter oxygen obtained through reduction. Traube used more accurate empirical investigation to refute Pasteur's view that yeast fermentation was a process of respiration: the yeast did not obtain its oxygen from the sugar.[27] His most severe criticisms were reserved for Liebig: his main argument was that Liebig's sympathetic theory implied that easily oxydizing substances ought to oxydize others. Traube used a number of specific experiments to show that this was not in fact the case.

The main arguments in Liebig's 1870 study, expressly intended as a polemic against Pasteur, were methodological in nature (as was Liebig's entire position in essence). Liebig considered it incorrect to attempt to explain chemical processes by physiological processes: "the proper method is ascertain the chemical and physical laws in both (*i.e., physiological processes of fermentation*)."[28] Liebig's work represented a step forward in its sophisticated adaptation to the changed empirical circumstances principally brought about through Pasteur's work, in which his studies of the retention of nitrogen compounds in the growth and ageing of the yeast were of particular importance. As far as Liebig was concerned, fermentation remained a process of decomposition: however, since he had to reconcile this basic position with the recognition that yeast actually grew during fermentation he shifted the process of decomposition to within the yeast cells themselves. Nitrogenous compounds in the cells broke down, followed eventually by the sugar which was bonded to them. The resultant soluble nitrogen compounds were reabsorbed by the cell with the sugar in its environment during growth and rebuilt.[29] This postulated cycle allowed him to retain his thesis of the parallel decomposition of proteins and sugars, without having to contradict the thesis that fermentation was a life-process of the yeast.

Pasteur's immediate reaction to Liebig's 1870 study was to demand that the disputed experiment be carried out before a commission of the Royal Academy (the experiment involved yeast growth in a mineral/sugar solution).[30] Pasteur's 1876 work offered a more precise specification of what the physiological process was considered to be – the consequences of which were the breaking-down of sugar into carbon-dioxide and alcohol; that is, a process of anaerobic respiration. This became the *definition* of fermentation as far as Pasteur was concerned, and allowed him to make an *ad-hoc* adjustment to Liebig's 1870 argument that there were a number of processes of decomposition which could take place without living cells. He pointed out that the ferment could consist of materials extracted from the yeast, or from other micro-organisms. Those processes of fermentation which took place without living cells he termed *non*-genuine fermentation.

Clearly, this latter position put Pasteur on the retreat. All that was needed to refute him was to isolate a substance from the yeast which was able to induce alcoholic fermentation without any living cells being present. Liebig had already intimated as much in 1870.

> It might be that the physiological process has no other relation to the process of fermentation than that through it material is *produced* within living cells, which through its own efficacy brings about the decomposition of the sugar and other organic atoms in the same way that emulsin works on salicyl and amygdalin; in such an instance the physiological process is necessary to produce the substance, but has no further connection with fermentation.[31]

The isolation of this substance, zymase, was accomplished by Buchner in

1896. His achievement unified the three positions. Liebig's methodological position had prevailed: physiological processes *were* reducible to chemical processes. The ferments were to be regarded as catalysts within the process of fermentation; they had a necessary function within the yeast cells, and were synthesized by these cells.

The problems of fermentation research in the nineteenth century were, as we have seen, generated by theoretical constellations. Similarly, what was considered to be scientifically acceptable, and what was in fact accepted was determined by the stage reached by the theoretical controversy. It is difficult to identify any clear extra-scientific influences related to practical concerns within this development: in particular, scientific findings were not evaluated by practical yardsticks. Although Pasteur's studies were clearly of major practical importance, especially in medicine and also in wine and beer-making – where his suggestions ran parallel to those he made for the prevention and treatment of infectious diseases – and although it is correct that Pasteur's investigations into the cause of disease were an element within his broader programme for explaining the connection between the life-processes of organisms and chemical processes – a process which grew out of his initial experiences with the souring of alcohol and milk – his proposals for the purification of beer during fermentation involved neither the application of his findings on alcoholic fermentation nor did they create new problems for fermentation research: it was simply a case of clarifying how micro-organisms other than yeast could be prevented from interfering with fermentation and producing damaging by-products by being kept out of the desired fermentation.[32] Despite their far-reaching practical consequences, Pasteur's studies were essentially autonomous both as regards the determination of problems and the evaluation of results.

2.3.2. *Fluid mechanics*. Nineteenth century fermentation research exhibited a basically positive relationship between the absence of theoretical consensus and the role of controversy as the main mechanism through which problems were generated. In contrast, fluid mechanics began from a general consensus as to the theoretical basis of the field: this feature also determined the nature of theoretical development with fluid mechanics. Before examining how problems were generated within the field, we begin by examining the nature of this relationship more closely.

Theoretical development within fluid mechanics represents an instance of finalization. As outlined elsewhere[33] one of essential preconditions for this is the possession of a fundamental and closed theory in a scientific field which can function as the basis for further theoretical development via specialization into problems related to practical application. Fluid mechanics met this criterion. From the mid-nineteenth century onwards a set of equations existed – so-called classical hydrodynamics – which were generally held to be valid for all fluids[34], or at least all those normally studied, and which encompassed all the related flow phenomena. According to the intrinsic criteria of science – the search for fundamental laws, explanation, consistency, etc. – the development of fluid mechanics had reached its objectives a hundred years ago. Internalists such as Truesdall[35] are therefore able to treat all subsequent development, marked by its relation to problems of practical application, as "the decline of classical mechanics". The central problem which impelled development onwards was the fact that the general theory – whose validity no one doubted – could not be applied, amongst others, to practical situations, and in particular to fluids with a high Reynolds' number – that is with a low but not completely discountable viscosity.[36] In the years immediately following the turn of the century these problems were resolved through such theoretical concepts as the boundary-layer, which did not alter or amend the original equations, but rather permitted these equations to be applied. We therefore term such concepts as the boundary-layer "principles of application" (*Anwendungsgrundlagen*). Such concepts are the means by which the general theory can be specialized into application-oriented theories, such as aerofoil theory, or the theory of drag.

The two principal questions which concern us here are, firstly, in what sense is it possible to refer to classical hydrodynamics as "closed" – or in other words, what accounted for the consensus as to its validity which allowed scientists after the mid-nineteenth century onwards to devote themselves to problems related to practical application? And secondly, what prevented the application of the general theory of these problems, and what was the theoretical status of the "principles of application" noted above?

Tackling the first of these questions involves a consideration of what classical hydrodynamics represents in theoretical terms. Figure 4 sets out the elements of the discipline. Classical hydrodynamics consists of four or five equations, and is a daughter discipline of general mechanics (an aspect not considered further here). The theory requires supplementing by thermodynamics where heat exchange in viscous flows is involved. The first equation, the continuity equation, expresses the conservation of fluid mass.

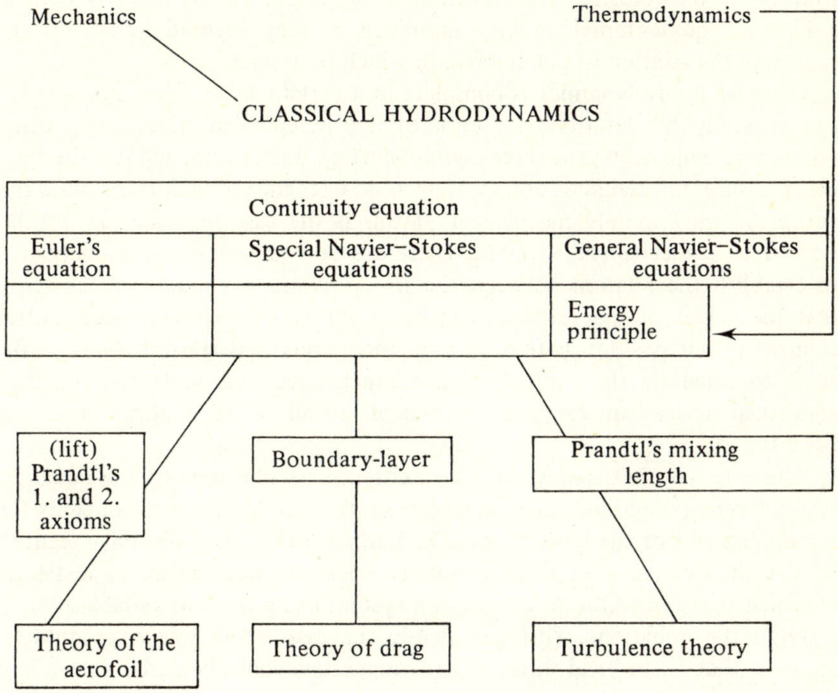

Fig. 4. Elements of fluid mechanics.

The three subsequent equations are equations for the three components of linear momentum. Euler's equations apply to frictionless, that is, ideal fluids, and the Navier–Stokes equations to fluids with internal resistance. Where viscosity is not a function of temperature it is possible to work with the special Navier–Stokes equations; otherwise the more general ones apply. If an exchange of heat takes place within the flow an additional equation is required, the law of the conservation of energy, in such a form that energy no longer has to be considered as innermechanical energy; a dissipation of energy takes place, and the equation contains a term for internal energy.

In what sense can this theory be considered "closed"? In the original formulation of the finalisation thesis we derived this term from Heisenberg. In Heisenberg's view a theory can be regarded as closed when it can no longer be improved by small changes. Improvement can only be achieved by a change in the entire theoretical structure, embracing a completely new stratum of phenomena. This position implies that closed theories possess a kind of quasi-eternal validity inasmuch as they succeed in adequately capturing the stratum of phenomena to which they refer.

Classical hydrodynamics is complete in a certain sense. Flow phenomena can generally be described in terms of five independent variables – three purely mechanical, i.e., the three components of momentum, and two thermal, i.e., pressure and temperature. As long as no exchange of heat takes place the entire system is purely mechanical. If this is the case pressure can also be treated as a mechanical variable. There are enough equations, namely five to establish the relation between the five variables –"enough" in the sense that the values of the variables can be established.[37] However, one central element in our position is that, in fact, these equations are not always sufficient to calculate the mutual relationship between the variables, and that additional devices are required – concepts to allow the application of the basic theory.

The internal consistency and the existence of solutions to the equation system were established some time before the development of concepts for the application of the basic theory. In Kantian terms the "objective reality" of the theory had been demonstrated. Scientists were aware of instances in which it was possible to represent a system using the five variables linked through the equations. This was clearly the case where viscosity could be ignored, that is, in ideal fluids. Euler's equations could be easily solved and there were no obstacles in the way of making applications. Prior to Prandtl's work there were even a number of applications for the Navier–Stokes equations, in particular for fluids with a low Reynolds number, i.e., high viscosity, or low-speed flow. In addition to the fact that classical hydrodynamics provided enough equations to solve for all the necessary variables, and that solutions did indeed exist for certain paradigmatic applications, the equation system also provided the appropriate and adequate *concept* for a given object-field. Whether this claim is valid is clearly central for Heisenberg's notion of closedness, which requires the capacity to delimit an object-field *independently* of the theory, and to offer grounds for believing that the theory will continue to apply within this object-field in as yet untested circumstances. If this requirement is met, then we must expect further theoretical development as a specialization into sub-domains of the general field covered by the general theory.

What is interesting, in fact, is that classical hydrodynamics has not only experienced theoretical advance through specialization, but currently displays the opposite development into greater generality by expanding into general rheology. This serves as a confirmation of the validity of classical hydrodynamics for a specific object-field, and allows a more precise definition of this field.[38] The development of hydrodynamics into greater generality took place because fluids such as baking dough or plastids were studied but which either for contingent or trivial reasons lay outside the sphere of applications originally conceived for classical hydrodynamics. This revealed that the formulation of classical hydrodynamics – more accurately, the Navier–Stokes equations – contained an assumption which empirically characterised a certain type of flowing medium: that is, the proposition that with a shear-flow, or "Couette flow", the shear stress is proportional to the velocity gradient (see Figure 5).

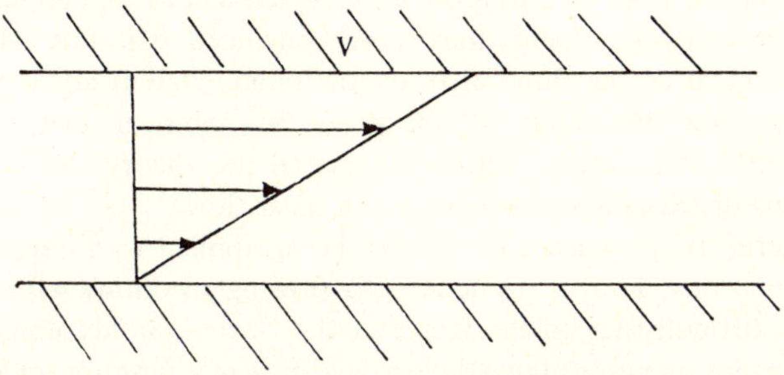

Fig. 5. Shear stress in a Newtonian fluid. Two parallel, flat plates between which a (Newtonian) fluid is flowing are moved with a relative velocity, ν. A linear velocity distribution is set up in the fluid: if D is the velocity gradient, p the shear stress and η the viscosity, then $D = p(1/\eta)$.

As is now recognised, this relation does not apply in many flowing media (see Reiner, 1969; Sherman, 1970). Those media in which shear-stress is proportional to the velocity gradient are known as Newtonian fluids.

This property is accommodated as follows in the Navier–Stokes equations: in order to account for the effects of viscosity a general stress tensor can be inserted into the equations for momentum. This does not have any empirical implications and is merely formal: however, the new, initially non-determined coefficients of the tensor mean that the equation system is underdetermined. By observing certain requirements of symmetry, the empirical relation which characterizes Newtonian fluids allows the reduction of the formal momentum equations to such a degree that they take on an empirical meaning. This takes place either by letting the coefficients of the stress tensor disappear, or by representing them as functions of viscosity and the velocity components (more precisely, the second derivative of these functions).[39]

The results of this operation are the Navier–Stokes equations. These relations clearly apply to a specific type of fluid, the Newtonian fluid. The aim of a general rheology would be to find equations analagous to the Navier–Stokes equations for the constitutive relations of other fluids.[40] They would all be interlinked through the more general mechanical approach.

Classical hydrodynamics is therefore closed inasmuch as it provides an adequate concept for a specific field of objects. The empirical relation which is constitutive for this field, that is which defines the objects belonging within in it, itself determines the form of the theory. One additional question is whether this judgement could have been made before the discovery of other fluids. Scientists were well aware that the Navier–Stokes equations had been formulated on the basis of a simplification which had certain empirical implications.[41] On the other hand, they were convinced that this relation held good for the most of the fluids they encountered, such as air, water and oil, or could convince themselves of this since the empirical relation could be tested for without having to have recourse to the theory (for example, by measuring the distribution of velocities in a shear flow).

What, therefore, prevented the theory being applied to a large number of practical situations, and in particular, to flowing mediums with a low viscosity? The difficulty lay in the fact that the Navier–Stokes equations were non-linear; more precisely, the viscosity effect was a function of the second derivative of the velocity components. This is, of course, a purely technical reply. Did this mathematical difficulty correspond to anything on the physical side – that is, are there any factors in the behaviour of fluids which might account for the problem? Does viscosity, therefore, lead to a type of contingent behaviour in fluid which although not contradicting the general theory cannot be simply deduced from it? Consider the following example from the theory of drag: flow resistance for very low viscosity was normally calculated by initially assuming that viscosity in fact disappeared, and that the fluid could be assumed to be an ideal fluid. However, Euler had already shown that a body, a sphere for example, would not encounter any resistance at all in an ideal fluid. In contrast, empirical study revealed that fluids with a very low viscosity produced a diminishing but finite resistance. This was the Euler–D'Alembert paradox. Mathematically this means that the second approximation in the calculation of resistance is not steadily connected to the first (viscosity = zero), physically that within this transition a special effect must reveal itself.

The difficulty in applying the general theory to the problem of low viscosity can be approached from two directions: either equations can be employed which are non-linear for certain terms because of the physical existence of a discontinuity effect; or alternatively, the mathematical equations cannot be solved because as viscosity approaches zero a discontinuity occurs. The problem of applying the general equations can also be correspondingly approached in two ways: either a concept has to be found for the particular effect observed, or a mathematical theory has to be developed which can offer solutions through approximations with discontinuities, that is, a theory of singular disturbances. The first path was followed by Prandtl who developed his concept of the boundary-layer on the basis of empirical investigation and mathematical reflection. By working out the concept of the boundary-layer for the boundary-layer theory, Prandtl was able to capture the physical effects which were responsible for the mathematical discontinuity. The second path has been adopted more recently in attempts to mathematically construct a general theory of singular disturbances. Viewed from the standpoint of such a theory Prandtl's boundary-layer concept would appear to represent an intuitive grasp of the first stage of approximation. Whichever method is chosen, it is evident that additional theoretical concepts are required before a general theory can be applied to specific practical concerns.

Classical hydrodynamics represents a closed theory inasmuch as it offers an adequate conceptualization of a delineable field of objects, namely the Newtonian fluids. The application of the theory to certain practical problems nevetheless required additional concepts not contained in the general theory. Three examples were noted in Figure 4; the boundary-layer, Prandtl's First and Second Axioms, and Prandtl's concept of the mixing-length. These concepts allow the transition to be made from the general theory to application-oriented special theories. Two examples illustrate this principle; the transition to the theory of the aerofoil, and the transition to the theory of resistance. Turbulence theory is excluded here since Prandtl's version has simply remained one amongst many, and because the statistical character of such a theory means that it is essentially analogous to the other examples, but also throws up additional problems.

The fact that bodies encounter drag in fluids was "in principle" covered by the Navier–Stokes equations; however, this had little relevance for practical problems involving drag since the equations could only be solved for fluids with a high Reynolds number, and which were not usually fluids of any practical significance.

The approach embodied in Euler's equations for fluids in which viscosity was presumed to disappear led to the Euler–D'Alembert paradox cited above.

Helmholtz had opened up one possible way ahead through his concept of surfaces of discontinuity (1868), in which the sharp edges of bodies in a flow produce surfaces on which the increase in the speed of flow was treated as being finite. According to Bernoulli's equation this implied a difference in pressure on the forward and reverse side of the body. Kirchhoff and Lord Rayleigh provided preliminary solutions for the resistance experienced by a body in an ideal fluid.[42] However, it was also evident that Helmholtz's basic approach was theoretically deficient and failed to match what could be observed empirically.

In 1904 Prandtl followed Helmholtz by trying to tackle the problem of resistance on the basis of the fundamental equations of hydrodynamics through the introduction of a new concept, the boundary-layer, which was intended to apply to fluids with a very low internal friction. This concept allowed the mathematically simpler Euler equations to be used, but treated them as a limiting case for the equations for a medium with very low viscosity, so that the effects of drag, which were causally related to viscosity, could be incorporated into the theoretical apparatus.

This was possible if the flow around a body in a fluid of low viscosity was treated as if it were a flow in an ideal fluid, with the effects of viscosity only being considered for a very thin layer on the edge of the boundary-layer. This gave the resistance on the surface of the body, or friction drag. What was more important, however, was the possibility of "boundary-layer separation" caused by the difference between the free flow and the delayed flow in the proximity of the surface. In a free flow this produces vortices which are responsible for pressure or "slipstream" drag. This concept allowed the phenomenon of fluid resistance to be fully explored mathematically.

The concept of the boundary-layer enabled the equations of hydrodynamics to be generally applied to problems of resistance: this does not mean, however, that the theory of drag represents an application of general hydrodynamics to the problem of fluid resistance. Rather, the theory of drag is an extension of the theoretical core of general hydrodynamics through the theoretical innovation of the boundary-layer concept.[43] It should also be added that the boundary-layer does really "exist" – that is, it can be observed.

The second example of a special theory considered here is that of the aerofoil. Figure 4 illustrates two concepts which served as the transition from general hydrodynamics to the theory of the aerofoil: the first, the concept of lift, and hence of circulation, and the second, Prandtl's axioms. Lift is not a concept found in general hydrodynamics, although it is easily accommodated within it. There are strong rounds for viewing lift as a concept determined through a technical interest, namely the desire to fly. As far as general hydrodynamics is concerned there is no "up" or "down". On the other hand bird- or insect-flight are clearly natural phenomena which could have become part of the subject matter of hydrodynamics without this technical interest. In practice, it was the first attempts at flight, and in particular Lilienthal's, which introduced the notion of lift. Earlier attempts to understand bird-flight (Leonardo, Borelli, Marey)[44] were based on aero-elasticity and action/reaction, rather than concepts specific to flowing media. The concept of lift, a force which in a flowing medium has its origin solely in the shape of the object around which the flow passes – a force acting vertically to the direction of flow – can be traced back to Lilienthal. The link between this concept and the mathematical notion of circulation (which was already known in vector analysis) was established by Kutta and Joukowski, and led to the first theory of the aerofoil – albeit a two-dimensional one. However, this was not yet a theory of real aerofoils, since it was based on the assumption of an infinitely long wing-span. Against a background of qualitative conceptualization by Lanchester, Prandtl developed the first theory for real aerofoils: like his earlier theory of the boundary-layer, Prandtl's aerofoil theory rendered classical hydrodynamics (Euler's equations) applicable through their extension by a special concept, so that the frictionless fluid postulated by classical hydrodynamics could be viewed as limiting instance of an infinitely small friction. This was formulated in two axioms and two theorems.

Axiom I. On the lines of confluences (lines along which two flows, previously divided by the body unite) vortex sheets can arise.
Axiom II. On sharp and projecting edges of bodies, infinite velocities do not appear, or at least appear with the maximum limitation.
Theorem I. Projecting edges of the bodies are, for a flow meeting them transversely, always lines of confluence (and therefore an origin of vortex sheets).
Theorem II. The vortex strength in each new section of the gradually forming vortex sheet takes that value which is necessary to maintain the velocity finite at the edge, or to limit, as far as possible, infinite values.[45]

In brief these theorems mean the following: according to the general theory, infinite velocities had to arise at sharp edges at which the flow rejoins. Prandtl's axioms state that in nature such infinite velocities are prevented by the formation of vortices, and additionally offer a method for determining the strength of these vortices.

The significance of this for aerofoil theory is as follows: during take-off a starting-vortex forms at the trailing edge of the wing. In order to conserve the momentum of the whole system (of air flow), a vortex of the same strength but with opposite rotation must form elsewhere. This is the Kutta–Joukowski vortex around the aerofoil. Moreover, since according to Helmholtz's theorem, a free vortex can never end in a medium but must either continue into infinity or terminate against a boundary, the bound-vortices on the aerofoil must continue as free vortices at the wing-tips. This means that, together with the starting-vortex, a vortex ring is created which during the course of flight is pulled out into a horseshoe shaped system of vortices through the stripping off of the starting-vortex. The vortices which form at the ends of wings are responsible for the drag associated with lift, so-called induced drag.

Prandtl's concept allowed the application of the general theory to a specific problem, that of the aerofoil. As with the boundary-layer, this theory also embraced an empirically observable phenomenon – the vortex system around an aircraft's wings. This phenomenon cannot, however, be derived from the general theory without additional conceptual tools. Although it is contained within the general theory, the theory on its own is insufficient to allow the conclusion that nature assists in the prevention of infinite velocities of flow through the formation of vortices.

Theoretical development in fluid mechanics therefore took place through a process of differentiation. Although Figure 4 illustrates this as a branching pattern, this requires some qualification. Clearly, the individual special theories are also related laterally: for example, the problem of drag played a major role in the theory of the aerofoil. In addition, sometimes the advance of one special theory made use of the conceptual apparatus of another: this can be seen in the study of the functioning of an aerofoil at sub- and supersonic speeds, where it becomes difficult to subsume the problem in question into one particular theory. The complications which this produces for the theory of science can be minimised if they are considered from the standpoint of the "non-statement view" of theories, where the issue is not that of the subsumption of a problem to a theory *qua* assertion, but rather the application of a conceptual instrument to a problem. In the above example of supersonic flight, the instrument of gas dynamics is applied, and specialized into an aerofoil theory for this purpose.

We now turn to the mechanisms for the generation of problems within fluid mechanics, and the processes for the evaluation of results. Once again, we concentrate on the development of aerodynamics.[46] The situation at the beginning of the century has already been described by Schlichting above: on the one hand a universally recognised but essentially useless theory, and on the other a stock of traditional practical problems which had been placed at the door of empirical hydraulics, most of which were concerned with resistance (such as problems in hydraulic engineering, shipbuilding, but also areas such as ballistics).

The theoretical breakthroughs which paved the way for modern fluid mechanics were prompted by practical problems. Kutta wanted to achieve a theoretical understanding of Lilienthal's empirical trials, and developed a two-dimensional aerofoil theory. Prandtl was drawn to the study of flows in the proximity of boundaries because of a problem encountered as an engineer.

I had placed a cone-shaped pipe in a large pneumatic system in he engineering works in Nuremberg with the aim of regaining pressure: instead of an increase in pressure a separation of the flow took place. I kept asking myself why the flow had separated from the edge, instead of flowing along it, until the "boundary-layer theory" supplied the solution three years later.[47]

The subsequent progress of fluid mechanics research was determined for the most part by the techniques of aircraft construction and control: in turn, the development of the aircraft was, from the very beginning, an object towards which scientific efforts were actively directed. Lilienthal and the Wright Brothers proceeded scientifically inasmuch as they carried out systematic measurements and experiments: the Wright Brothers even used a wind-tunnel. Once the bases of aircraft construction were understood, flow theory was employed to advance its design. This was formulated in programmatic terms in 1910 by Finsterwalder.

The significance of the issues of aerodynamics has never been doubted as far as aviation is concerned: however, these questions increase in significance in proportion to the extent to which economic exigencies make themselves felt – when it is no longer a matter of achieving speed and endurance records.[48]

Although at this initial stage records were the product of "ethical forces", to which Kohler later attributed the first successes in aviation,[49] it is noticeable that Finsterwalder stresses the economic aspect as being the area requiring the *scientific* treatement of the problems of flight. In addition to increasing speeds, altitude and pay-load, research was also directed at reducing drag in relation to lift, and consequently cutting down the amount of power needed to propel the aircraft. This imperative is clearly visible in the three fundamental tasks which Prandtl saw confronting wing-theory.

(1) Given the distribution of the lift over the span, together with G (*air-density*) and V,

to determine the wing form reproducing this lift distribution as well as the drag of the same.
(2) Given the wing form, to determine the lift distribution and the drag.
(3) Given the total lift and the wing-span, as well as G and V, to determine that distribution of the lift on the span, for which the drag becomes minimum.[50]

The necessity to use multiplanes because of the limited power and wing-span available spurred on the investigation of the mutual effect of the wings. This in turn stimulated the modern theory of the propellor, which defines a propellor as an assemblage of a number of aerofoils.[51]

The results of this research were translated relatively swiftly into engineering techniques for the construction of aeroplanes and airships. The shape of airships, for example, soon came to be influenced by investigations into those shapes likely to present the least resistance, and in particular by the realisation that the shape of the back of the body in a flow had a large determining effect on the overall drag. Wing-theory enables profiles to be constructed which were both technically effective as well as being economic. The specification of the power required to drive an aircraft forwards was facilitated by the recognition of the role of induced drag, that is, that drag systematically associated with lift. The insights gained in propellor theory led to the construction of the variable-pitch propellor which enabled aircraft to fly over a very large range of speeds. The problem of stalling was amenable to investigation via the boundary-layer theory, which revealed the cause of stalling to be the separation of the boundary-layer from the upper surface of the wing with a consequent increase in pressure and loss of lift. By suitably designing the profile of the wing, and developing methods for boundary-layer control, it was possible to maintain lift at high angles of attack.

This example of the practical utility of fluid mechanics provides a characteristic instance of the manner in which problems were generated and selected within the discipline. In principle, the possibility and effectiveness of boundary-layer control had already been recognized in Prandtl's first work on the theory of the boundary-layer (1904). However, it became a "hot subject" for research once the relation between stalling and the boundary-layer became apparent.[52] Boundary-layer stability, laminar and turbulent boundary-layers, boundary-layer separation and boundary-layer control became a "research front". This conformed to the general pattern observable in the development of problems in fluid mechanics research: it is the technical front – itself the product of preceding scientific development – which determines the scientific front. For example, as flying speeds gradually approached the speed of sound, it became evident that the onset of the problems which began to manifest themselves in this range of speeds could be postponed by adopting swept-back wings. This, however, demanded a new wing-theory. Once the transonic, and later, the supersonic region was reached, it was necessary to develop a theory for wings based on gas dynamics. A new type of drag also emerged, "wave drag", which required investigation, along with the interaction between the boundary-layer and the compression wave. This interaction produced a new type of stalling phenomenon, "pressure wave stalling". Breaking the sound-barrier also created the problem of flight-noise, producing in turn a new discipline, flow acoustics.

These examples indicate the relationship between the evaluation of scientific knowledge within technical contexts, and the generation of problems out of scientific findings in the field of aviation. The same principles can also be seen in other sectors. For example, the increase in the power of turbines and ships' propellors made the phenomenon of cavitation into a scientific problem. Flight stability also became a subject for fluid mechanics as the size of projectiles increased.

Fluid mechanics research represents a type of scientific development whose theoretical history is strongly influenced by external, that is technical, problems. This development was made possible because of the prior existence of a fundamental theory and access to concepts which, in the more complex instances such as the boundary-layer theory, could be termed "principles of application". These concepts allowed the fundamental theory to be developed into special theories, which then have a direct relevance for practical questions.

4. CONCLUSION

The conclusion to this study consists of three parts: (i) a summary of the results of the case-studies as regards the question of the cognitive conditions for autonomization and finalization, (ii) a systematization of some of the methodological aspects of the procedure adopted in the study, (iii) a consideration of a number of further questions raised by the study.

(i) Our basic thesis was that science contains the cognitive preconditions for both autonomization and finalization. The case-studies were able to demonstrate what this means concretely. Fermentation research represents an instance where there would appear to be an implicity close practical relevance. Nevertheless, development in this field was determined by the *internal* constellations of problems and theoretical claims. Scientists looked on almost spellbound at the discussion over fermentation theory. It was this discussion which determined whether a problem was considered worthwhile pursuing; and whether or not a particular study was regarded as making a significant contribution was measured in terms of its value within this theoretical controversy. The fact that practical problems could be solved was more of a side-effect, and was certainly not the force behind progress in the field. In fact, the very openness of the situation excluded a direct approach to practical issues. Such a direct approach would also have been distinctly unpromising as long as no one really knew which theory was "right". The basis for the autonomous development of fermentation research was the existence of opposed theoretical positions, which defined the scope of the work of the individual scientist and provided the criteria of relevance against which solutions were judged.

In contrast, fluid mechanics possessed a clear and thoroughly non-controversial foundation. Scientific breakthroughs were considered to be those accomplishments which allowed this foundation to be practically applied. These conditions represented the cognitive conditions for finalization, in which external goals – in the case of fluid mechanics from the sphere of technology – determined the further advance of the theory. Scientists were constantly on the heels of technical development and the "research front" constantly shifted to those areas where technical progress made the application of theory into an interesting proposition.

(ii) It is, of course, impossible to prove the validity of a theory using two case-studies. And since the effort involved in producing such studies precludes any significant increase in their number, the present study should not be seen as aspiring to such a standard of confirmation. In the context of theory-formation in the history of the sciences such case-studies should be seen rather as "proof of the objective reality", to use Kantian terms, of the theories. Theories, that is complex conceptual structures, are shown to capture empirical reality. On the one hand this provides an exemplary illustration of how an aspect of reality is "opened up"; that is, we present a concrete segment of the history of science. On the other hand, the theory is given additional refinement through the pressure to correspond more closely to reality, and may only become fully transparent in discrete parts. In this respect the studies cited here have served to clarify the question of endogenous problem-generation, how the concept of "closed theories" may be approached with greater precision, what type of theoretical developments take place on the basis of closed theories, and how fundamental theories are related to daughter theories.

(iii) By their nature case-studies go beyond what was initially theoretically anticipated. To conclude we therefore draw attention to some issues which unexpectedly arose during the course of the study.

A conventional distinction exists between the fundamentals of a science and its application. The finalization thesis had always maintained that special theoretical developments led onto applications. The case-study of fluid mechanics in fact revealed that a "middle-range" layer of theories exist which allow the application of basic principles. We termed these theories, "principles of application". As the example of the theory of the boundary-layer showed, the development of such theories can be a complex and wide-ranging undertaking. We term such a process "basic research", even though it does not strive to discover "what holds the world together" but obtains its meaning solely from the context of its possible applications. However, the development of the boundary-layer theory is not itself the application of the fundamental theory but rather the presupposition for such applications. The problem of the application of fundamental theories still remains largely a closed book as far as the history and theory of science is concerned, despite the evident practical relevance of such a subject. This is the sphere in which science policy operates, and in which most scientists are active. The neglect of achievements in this area – leading to the omission of such significant scientists as Prandtl or von Kármán from lexica of "great scientists" – is utterly misplaced. It is also hazardous from the point of view of science policy simply to throw everything which is not esoteric research into the vat marked "applied research". A better understanding of the nature of the problems involved in applying fundamental theories from the perspective of science-theory, together with an idea of what strategies are available for tackling these problems could make a contribution towards changing this practice.

There is therefore a good case to be made for looking at other research areas with problems of application using the same approach employed here for fluid mechanics: fusion research and solid-state physics represent possible candidates, as areas in which magnetohydrodynamics and quantum physics respectively are examined for possible practical applications.

We stated that in fluid mechanics research, the scientific front was for the most part determined by the technical front. This claim naturally conceals numerous problems. Even if it were possible to formulate *one* clear notion of a research front for the autonomous phase during the period of theoretical controversy, it would not be possible to obtain such a concept for the finalization phase simply by direct reference to the technical front. Since the scientific end-product – despite steadily decreasing translation times – never enters into the technical context of application without any mediation, its value cannot be determined simply by its "fit" to the technical front. This much at least is clear: during the finalization phase, the scientific front cannot be defined simply by reference to scientific criteria; what constitutes a technical front, and how this front interacts with the scientific front must become a subject for research.

NOTES

[1] It should be noted that this study deals only with the natural sciences, although much of what is said also relates to other types of science.

[2] See Böhme, van den Daele, and Krohn (1976).

[3] Böhme (1975).

[4] Pasteur (1922), p. V.

[5] Pasteur (1857), pp. 3–13.

[6] The canal-building project had a very stimulating effect on French scientists: one Commission which was involved in considering such questions included d'Alembert, the Abbé Bossuet and Condorcet. The experimental results gathered by the Commission were published by D'Alembert under the title *Nouvelles expériences sur la resistance fluides* in 1770: see Némenyi (1962/66, p. 76) and Flachsbart (1932, p. 28).

[7] That is, Liebig's "elementary analysis": see Volhard (1909), I, pp. 213–239.

[8] E. Mitscherlich, *Gesammelte Schriften*, 1896, pp. 334–400: J. J. Berzelius, 'Einige Ideen über eine bei der Bildung organischer Verbindungen in der lebenden Nature wirksame, aber bisher nicht bermerkte Kraft', in Wöhler (ed.) *Jahresbericht über die Fortschritte der physischen Wissenschaften*, Tübingen, 1836, pp. 237–245.

[9] See previous note.

[10] Liebig (1839*a*), pp. 250–288.

[11] Liebig (1839*b*), p. 367.

[12] Fruton (1972), p. 73.

[13] Buchner (1897), pp. 117–124.

[14] Fruton (1972), pp. 73 ff.

[15] Other branches of fluid mechanics have retained a more traditionally academic course of development, such as stability theory in England under H. M. Taylor.

[16] For historical development see Giacomelli and Pistolesi (1934), von Kármán (1956) and Rouse and Ince (1957).

[17] Schlichting (1965), p. 1. See too Truckenbrodt (1968): It should be noted that it was principally phenomena involving the effects of viscosity which could not be handled using the theory: Lamb's 1906 textbook shows that there were a number of practical problems which could be dealt with theoretically. However, the 100 pages (out of a total of *c*. 800) devoted to internal friction do not contain any references to such problems.

[18] For institutional developments in Germany see Betz (1954), Kohler (1935), Betz (1957), Tollmien and Vogelpohl (1961), Wendel (1975), pp. 193–201.

[19] The growth of literature in the field of fermentation research in the nineteenth in fact contradicts de Solla Price's general law.
[20] Liebig (1839*b*), pp. 363–368.
[21] Mitscherlich (1843), cf. Mitscherlich (1844), pp. 374–376.
[22] Liebig (1848), p. 232.
[23] Pasteur (1861).
[24] Traube (1858), p. 331.
[25] Mitscherlich (1841).
[26] Traube (1858).
[27] Traube (1874) and (1877).
[28] Liebig (1870), p. 47.
[29] Liebig (1870), pp. 30–32.
[30] Pasteur (1870); Liebig did not respond to this demand.
[31] Liebig (1870), p. 6.
[32] Pasteur's suggestions for purifying beer were in fact a failure in practice: 'This method (breeding pure strains of yeast and suppressing the development of bacteria. *Author's note*) was still completely insufficient and, as Hansen later demonstrated, actually incorrect in one respect, since it favoured the growth of "wild yeasts" . . . ' (Weinfurter, 1960).
[33] See Böhme, van den Daele, and Krohn (1976).
[34] See below.
[35] Truesdell (1976), pp. 127 ff.
[36] The Reynolds number is defined as

$$R = D \cdot \frac{V}{n} \cdot L$$

where D is the density of the medium, V is the speed of flow, L the length of the body around which the the flow takes place and n the viscosity of the medium. On its own a high R does not imply a low viscosity, but rather gives an index of dynamic similarity for various combinations of speed, size and viscosity.
[37] However, the equation system is not complete in the sense that one can demonstrate that there are solutions that are real and unambiguous.
[38] This development also characterizes other completed theories. The theories which supersede them confirm their validity for a a limited domain (generally in relation to their originally intended sphere of application), as in the relationship between the theory of relativity and classical mechanics.
[39] Oswatitsch (1959), pp. 47–52.
[40] See Truesdell (1976), p. 129.
[41] Lamb writes in his 1906 textbook, for example: "It will be noted that the hypothesis made above that the stresses p_{xx}, p_{xy} . . . are *linear* functions of the rates of strain *a, b, c* . . . is of a purely tentative character, and that although there is considerable a priori probability that it will represent the facts accurately in the case of infinitely small motions, we have so far no assurance that it will hold generally. It has however been pointed out by Osborne Reynolds that the equations based on this hypothesis have been put to a very severe test in the experiments of Poiseuille and others. . . . Considering the very wide range of values of the rates of distortion over which these experiments extend,

we can hardly hesitate to accept the equations in question as a complete statement of the laws of viscosity." (The fact that Lamb uses the term 'rates' rather than 'velocity gradient' is a result of a differential conception of the equation.) It is interesting that Lamb does not consider a deviation in the case of other materials but rather for other ranges of speed: this reveals genuine historical blindness on his part.

[42] Giacomelli and Pistolesi (1934), p. 332 ff.; Kármán (1956), pp. 28–30.

[43] See Sneed (1971) and Stegmüller (1973). According to the concept of "theory" developed by these authors, the relation of a fundamental theory and a special theory, which is our particular concern here, can be expressed as the transition from a core of a theory to its extension. In our instance here, the core of the theory would be characterised by the predicate "*X* is a classical hydrodynamics", the respective extensions would be achieved by addition of the boundary-layer concept, the axioms, and the concept of "Prandtl's mixing path".

[44] Giacomelli and Pistolesi (1934), pp. 336 ff.

[45] Prandtl (1918), cited in Prandtl (1963), p. 328.

[46] See Note 15.

[47] Prandtl (1948), p. 90.

[48] Finsterwalder (1910).

[49] Kohler (1935), p. 179.

[50] Prandtl (1918), translated in Giacomelli and Pistolesi (1934), p. 389.

[51] The nineteenth century theory of the propeller was a theory based on the idea of the "airscrew"; see Kármán (1956), pp. 186–192.

[52] Betz (1949), p. 257.

BIBLIOGRAPHY

General Literature

Böhme, G.: 1975, 'Die Ausdifferenzierung wissenschaftlicher Diskurse', in Stehr, N. and König, R., *Wissenschaftssoziologie*, Sonderheft 18 of the KZfSS, Cologne, pp. 251–253.

Böhme, G., van den Daele, W., and Krohn, W.: 1976: 'Finalization in Science', in *Soc. Sci. Information* **15**, 307–330.

Price, D. J. de Solla: 1963, *Little Science, Big Science*, Columbia University Press, New York, London.

Sneed, J. D.: 1971, *The Logical Structure of Mathematical Physics*, Reidel, Dordrecht.

Stegmüller, W.: 1973, *Probleme und Resultate der Wissenschaftstheorie und Analytische Philosophie, II*, Springer, Berlin/Heidelberg/New York.

Wendel, G.: 1975, *Die Kaiser-Wilhelm-Gesellschaft 1911–1914*, Akademie-Verlag, Berlin.

Literature Relevant to Fermentation Research

Berzelius, J. J.: 1836, 'Einige Ideen über eine bei der Bildung organischer Verbindungen in der lebenden Natur wirksamen, aber bisher nicht bemerkten Kraft', in *Jahresbericht über die Fortschritte der physischen Wissenschaften*, Tübingen, 237–245.

Buchner, E.: 1897, 'Alkoholische Gährung ohne Hefezellen', in *Bericht der deutschen chemischen Gesellschaft*, Vol. 30, 117–124.

Conant, J. B.: 1952, *Pasteur's Study on Fermentation*, Harvard University Press, Cambridge/Mass.

Fruton, J. S.: 1972, 'From ferments to enzymes', in Fruton, J. S., *Molecules and Life*, J. Wiley, London, pp. 22–86.

Liebig, J.: 1839*a*, 'Über die Erscheinungen der Gährung, Fäulnis und Verwesung und ihre Ursachen', *Annalen der Pharmacie* **30**, 250–288.

Liebig, J.: 1839*b*, 'Über Gährung, Verwesung and Fäulnis. Nachtrag zur Abhandlung', *Annalen der Pharmacie* **30**, 363–368.

Liebig, J.: 1848, 'Gährung', in Liebig, J., Poggendorff, J. C., and Wöhler, F. (eds.) *Handwörterbuch der reinen und angewandten Chemie*, Vieweg, Brunswick, pp. 217–233.

Liebig, J.: 1870, *Über Gährung, Quelle der Muskelkraft und Ernährung*, Winter, Leipzig/Heidelberg.

Mitscherlich, E.: 1834*a*, 'Über Ätherbildung' in *Gesamte Abhandlungen* (1896), Berlin, pp. 394–400.

Mitscherlich, E,: 1834*b*, *Lehrbuch der Chemie*, Vol. I.

Mitscherlich, E.: 1841, 'Chemische Zersetzung und Verbindung vermittelst Kontaktsubstanzen und Zusätze', in *Gesamte Abhandlungen* (1896), Berlin, pp. 504–522.

Mitscherlich, E.: 1896, 'Über die Gährung', in *Gesamte Abhandlungen* (1896), Berlin, pp. 537–546.

Paolini, C.: 1968, *Justus von Liebig. Eine Bibliographie sämtlicher Veröffentlichungen*, Winter, Heidelberg.

Pasteur, L.: 1857, 'Mémoire sur la fermentation appellée lactique', in *Oeuvres complètes*, Vol. 2, pp. 3–13.

Pasteur, L.: 1860, *Die Alkohol-Gaehrung*, Lampert & Co., Augsburg (1871).

Pasteur, L.: 1870, 'Note sur une mémoire de M. Liebig, relatif aux fermentation', in *Oeuvres complètes*, Vol. 2, pp. 361–366.

Pasteur, L.: 1876, 'Études sur la bière, ses maladies, causes qui provoquent, procédé pour la rendre inaltérable, avec une théorie nouvelle de la fermentation', in *Oeuvres complètes*, Vol. 5, Chap. VI, pp. 187–218.

Pasteur, L.: 1922, *Oeuvres complètes de Louis Pasteur réunies par L. P. Pasteur Vallèry-Radot*, Vols. I–VII, Masson, Paris.

Traube, M.: 1858, 'Zur Theorie der Gährungs- und Verwesungs-Erscheinungen, wie der Fermentwirkungen überhaupt', in *Poggendorfs Annalen der Physik und Chemie*, Vol. 103, 331 ff.

Traube, M.: 1874, 'Über das Verhalten der Alkoholhefe', in *Bericht der deutschen chemischen Gesellschaft*, Vol. VII, pp. 872–887.

Traube, M.: 1877, 'Über das Verhalten der Alkoholhefe in sauerstoffgasfreien Medien', in *Bericht der deutschen chemischen Gesellschaft*, Vol. X, pp. 510–511.

Volhard, J.: 1909, *Justus von Liebig*, 2 Volumes, J. A. Barth, Leipzig.

Weinfurter, F.: 1960, 'Geschichtliches über die Kenntnisse der Hefen', in *Die Hefen in der Wissenschaft*, Vol. I, H. Carl, Nuremberg, pp. 1–19.

Literature Relevant to Fluid Mechanics

Bellmann, R.: 1967, *Methoden der Störungsrechnung in Mathematik, Physik und Technik*, Oldenburg, Munich/Vienna.

Betz, A.: 1949, 'Ziele, Wege und konstruktive Auswertung der Strömungsforschung', in *Zeitschrift der VDI*, Vol. 91, pp. 253–258.

Betz, A.: 1954, '50 Jahre Aerodynamik', *Zeitschrift für Flugwissenschaft* **2**, 105–113.

Betz, A.: 1957, 'Aus der Geschichte der Aerodynamischen Versuchsanstalt Göttingen' in *Jahresbericht der MPG*, Göttingen, pp. 40–59.

Finsterwalder, S.: 1910, 'Die Aerodynamik als Grundlage der Luftschiffahrt', *Zeitschrift für Flugtechnik* **I**, 6–10 and 30–31.

Flachsbart, O.: 1932, 'Geschichte der experimentellen Hydro- und Aeromechanik insbesondere der Widerstandsforschung', in *Handbuch der Experimentalphysik*, Akademische Verlagsgesellschaft, Vol. IV, pp. 1–61.

Giacomelli, R. and Pistolesi, E.: 1939, 'Historical Sketch of Aerodynamics', in Durand, W. F. (ed.), *Aerodynamic Theory*, Springer, Berlin, pp. 305–394.

Helmholtz, H.: 1868, 'Über discontinuierliche Flüssigkeitsbewegungen', *Monatsbericht der Königlichen Akademie der Wissenschaften zu Berlin*, 215–228.

Kármán, T.: 1956, *Aerodynamik. Ausgewählte Themen im Licht der historischen Entwicklung*, Interavia, Geneva, 1956.

Kohler, N.: 1935, 'Die Luftfahrtforschung in den letzten 25 Jahren und ihr heutigen Stand', *Zeitschrift der VDI* 79, 1179–1191.

Kutta, M.: 1902, 'Auftriebskräfte in strömenden Flüssigkeiten', in *Illustrierte Aeronautische Mitteilungen*, 133ff.

Lamb, H.: 1906, *Hydrodynamics*, The University Press, Cambridge.

Mitteilungen aus der Max-Planck-Institut für Strömungsforschung und der Aerodynamischen Versuchsanstalt Nr. 25, Bibliography of publications 1907–1957, Göttingen, 1960.

Némenyi, P. F., 'The main concepts and ideas of fluid mechanics in their historical development', *Archive for the History of the Exact Sciences* **2** (1962–66), 52–86.

Oswatitsch, K.: 1959, 'Physikalische Grundlagen der Strömungslehre', in Flügge (ed.), *Handbuch der Physik*, Vol. VIII, pp. 1–124, Springer, Berlin/Göttingen/Heidelberg.

Prandtl, L.: 'Über Flüssigkeitsbewegung bei sehr kleiner Reibung', in Prandtl, L., *1961, Gesammelte Abhandlungen*, Vol. II, Springer, Berlin/Göttingen/Heidelberg, pp. 575–584.

Prandtl, L.: 1918, 'Tragflügeltheorie 1. Mitteilung', in *Gesammelte Abhandlungen*, Springer, Berlin/Göttingen/Heidelberg, pp. 322–345.

Prandtl, L.: 1948, 'Mein Weg zu Hydrodynamischen Theorien', *Physikalische Blätter* **4**, 89–92.

Prandtl, L.: 1961, *Gesammelte Abhandlungen zur angewandten Mechanik, Hydro- und Aerodynamik*, 3 Vols., Springer, Berlin/Göttingen/Heidelberg.

Rouse, H. and Ince, S.: 1957, *History of Hydraulics*, Dover, New York.

Reiner, N.: 1969, *Rheologie in elementarer Darstellung*, Hanser, Munich, 1969.

Schlichting, H.: 1965, *Grenzschicht-Theorie*, G. Braun, Karlsruhe.

Schlichting, H.: 1975, 'Ludwig Prandtl und die Aerodynamische Versuchsanstalt (AVA)', *Zeitschrift für Flugwissenschaften* **23**, 162–167.

Sherman, P.: 1970, *Industrial Rheology*, London.

Tollmien, W. and Vogelpohl, G.: 1961, 'Max-Planck-Institute für Strömungsforschung in Göttingen', in *Jahresbericht der MPG* **II**, 726–745.

Truckenbrodt, E.: 1968, *Strömungsmechanik*, Springer, Berlin/Heidelberg/New York.
Truesdell, C.: 1976, 'History of Classical Mechanics', Parts I and II, *Die Naturwissenschaften* **63**, 53–62, and 119–130.
Zierep, J.: 1967, *Vorlesungen über theoretische Gasdynamik*, G. Braun, Karlsruhe.

RAINER HOHLFELD

CANCER RESEARCH. A STUDY OF PRAXIS-RELATED THEORETICAL DEVELOPMENTS IN CHEMISTRY, THE BIOSCIENCES AND MEDICINE

1. INTRODUCTION

This study examines how and under what conditions scientific disciplines establish practice-oriented research in the boundary zone between biology and medicine, and in particular considers a number of attempts at the direction of theory-construction and research towards the problem of cancer.[1]

The cancer-problem is a highly topical concern, both socially and politically. Alongside cardiacal and circulatory disease, cancer is the major killer disease in the industrialized countries. Previous successes in combatting disease have produced higher life-expectancies; since cancer is primarily a disease of old-age, the fight against it is now central to the maintenance of health for those of middle-age and upwards. As a consequence, the fight against cancer has been treated as a priority in both health policy and science policy.

With the enactment of the National Cancer Act 1971, the US Administration under Nixon began a "crusade against cancer", with a proposed annual research budget of $1.4 billion to be administered in a coordinated programme by the National Cancer Institute.[2] In West Germany efforts to achieve a national organization and coordination of cancer research first got underway after 1976.[3]

Irrespective of the specific reason – its importance for health policy, the particular challenge to science posed by the search for an explanation of tumour growth, or the extensive scope of the phenomena involved – the problem of cancer has engaged, and continues to engage, the attention of a large number of scientists. An examination of the case of cancer, and cancer-research, therefore offers the advantage of observing the relation between theory and practice over a wide spectrum of disciplines.

We begin by suggesting one possible interpretation of the cancer-problem, with the aim of providing a guide-line through the subsequent discussion. The problem of explanation of malignant tumours, or cancers, should be understood as the search for the conditions, factors and mechanisms which induce cells to grow when they should not grow.[4] Focus on these factors therefore takes in both a characterization of carcinogens in the environment,

G. Böhme et al., Finalization in Science, ed. by W. Schäfer, 93–126.

and their metabolism within the body. An understanding of the mechanisms which produce cancer, that is the onset of uncontrolled growth following the transformation of a normal-cell into tumour-cell, requires a theoretical elucidation of the regulation of cell-functioning, cell-growth, and cell-differentiation. These areas constitute the central problems of cell biology.

However, it is not sufficient to define the problem of cancer simply as the disruption of the regulation of cell-growth. The emergence of a tumour-nucleus does not necessarily imply its growth into a tumour: there are indications that the growth of tumours is to some extent subject to control by the immune-system. In addition, the overall disposition of the organism has a bearing on how effectively the body's internal control systems, such as the immune-system, function. As far as the issue of combating cancer is concerned, this interpretation implies the possibility of intervention at any stage during the emergence and spread of the disease.

The course pursued by scientification in the area of the cancer problem can be analysed from two different angles. In the first, cancer itself is taken as the point of reference, and attention is focused on that research which deals directly with cancer: in the second, the point of reference is development in physics, chemistry and biology, with attention focused on how these developments relate to the problem of cancer. If the first approach is taken, cancer-research will appear as a research-field whose programme is structured by the phenomenon of cancer and the problems encountered in its associated treatments. Cancer-research consequently appears as an aggregate of research activities, whose centre is represented by experimental cancer-research, moving forward via induction and the exploration of tumours in animal and human systems and the elaboration of various concepts to explain the phenomenon of cancer. The fundamental natural-scientific disciplines make their contribution to progress in experimental cancer-research by the transfer of their discoveries, explanations and methods into the realm of practice.

In the second approach, the starting-point is taken as the development of chemistry, biochemistry, genetics and cell biology – all of which proceed in accordance with research-programmes conceived of and existing independently of the problem of cancer. The relevant issue is then when and how these disciplines deal with the cancer problem, and consequently at what point they acquire the capacity to construct a bridge across to experimental cancer-research. Should this happen, individual facets of the cancer-problem will become associated with research fronts in chemistry, biology or biochemistry, simultaneously advancing the development of theory in these disciplines and creating a bond between cancer-research and disciplinary developments.

This study takes the second of these approaches,[5] and consequently centres on the possible relations between theoretical developments in the natural sciences and the central questions of experimental cancer research. The theoretical aim of such a study is the specification of the relation between those theoretical and practical criteria and norms not purely oriented to the acquisition of knowledge which are intrinsic to science, and those which are extrinsic – that is, social and practical. Research programmes may be termed *internal* if they are directed towards the development of fundamental theories for generalized biological objects. In contrast, a specific and practical concern such as cancer will constitute an *external* research objective. Although research within a discipline may be related to such an external programme, the issue here is whether this relationship to a practical problem is dependent on the exigencies of internal disciplinary development, or whether the problem area itself generates new criteria for the further progress of the discipline.

This study therefore concentrates on that aspect of the scientification of the cancer-problem related to theoretical development in the fundamental, natural-science disciplines. Nevertheless, leading scientists expect any theoretical solution to lead to a substantial breakthrough in the prevention or treatment of cancer by a medical "high technology".[6] This expectation is based on the perceived limitations of existing conventional strategies: steel, beam and drug. The limitations suffered by these approaches are primarily a function of too late a diagnosis, the side-effects of therapeutic measures, and difficulties in identifying and treating secondary tumours (metastasis). A causal explanation of cancer would offer the prospect of a truly effective early-diagnosis, a biological elimination of malignant cells, anticipation of associated consequences and a reversion of primary damage.[7] The consequence of such an approach would be that effective and reliable control of the disease in all its manifestations will only be possible once the relevant molecular mechanisms are understood: the precondition for this is the molecular biology of cell differentation and cell development.[8] However, such a perspective does not automatically imply that science policy should automatically favour that type of cancer-research pursued through the disciplinary development of theory, the location of research on the newest fronts of a particular discipline, or in fact, that there is any necessity to settle on a theory at all. The fight against cancer has enjoyed a number of major successes which have not been built on a theory, and could not have been built on a theory.[9]

Any analysis of the conditions under which theoretical developments may become oriented to the cancer problem requires some initial assumptions as to those factors which determine theoretical development in the relevant disciplines. Section 2 below presents and offers a definition of the context within which subsequent material can be interpreted. Section 3, which depicts the relationship between scientific disciplines and cancer, begins by considering the development of special theories on the basis of mature fundamental theories as one form in which research can acquire a specific orientation. The specific approaches dealt with are those based on histology, chemistry, genetics and biochemistry. This ordering also corresponds to the key phases in the history of the theoretical understanding of cancer. A closely related topic is that of the issue of those types of problem-orientation which are feasible for disciplines still in the paradigmatic phase: molecular biology is taken as an instance of the relationship between the most advanced research fronts in contemporary science and external problems. Finally, some consideration is given to that form of cancer-research which in contrast to fundamental theoretical development represents a direct "bed-side" approach. So far, it is this type of research which is still responsible for the little progress made in cancer treatment over recent decades.

2. THE DEVELOPMENT OF CHEMICAL AND BIOLOGICAL THEORIES: DEFINITIONS AND ASSUMPTIONS[10]

Chemistry, biochemistry, genetics, cell-biology, developmental biology, anatomy and histology are the disciplines involved in tackling the problem of cancer.[11] One method for analysing the extent to which theoretical development can be oriented towards practical problems is to start with those classes of phenomena within which theoretical explanations and prognoses for solving the practical problem are required. Two such classes can be discerned for cancer: biological *processes* and biological *structures*.

A breakdown of different processes yields the following disciplines within the bio-sciences: biochemistry, or chemical physiology (investigation of metabolic processes within organisms); genetics (investigation of heredity); developmental biology (investigation of differentiation and morphogenesis). This breakdown is ordered in terms of the degree of complexity: for example, the phenomena of heredity are more complex that metabolic processes which can be genetically controlled. In turn, the phenomena of differentiation and development of organisms which are determined by environmental factors and not simply by the stock of genes are more complex than genetic phenomena. If, in addition, the levels of organisation of biological systems are also considered, than the following hierarchy emerges: bacterial cells – higher cells (eucaryotes) – cellular systems – organisms – systems of organisms (populations)[12] (see Figure 1).

The levels of organization represent the degree of complexity of stable, ordered systems. Theories relating to the structures and laws of classes of phenomena – as presented here – will only be valid if they can be empirically shown to be applicable to the various levels of complexity of biological systems. Appropriate research programmes for the dynamics of disciplines are those which are determined by the methods, and in particular by the theories with which classes of phenomena are described and explained. Such theories establish structural units for the analysis of a class of phenomena and define basic laws for the structure and functions of these units.

The development of a theory passes through characteristic phases[13] for which it is possible to specify the conditions for transitions from a theoretical development established through an "internal" research programme to a development directed by "external" problems. The initial phase of the development of a discipline is customarily characterised through a preponderance of empirical strategies (the explorative phase). This is followed by a phase of theoretical dynamics in which the basic explanatory concepts of the relevant class of phenomena are worked out (paradigmatic phase). During this phase the research programme of the discipline is determined by the theoretical approaches and their implications. If the theoretical bases become stabilised and the theory becomes "mature" as a consequence, the theory can be further developed for special systems in this class of phenomena through a process of theoretical specialisation (post-paradigmatic phase). The development of such special theories can occur in a direction suggested by "external" goals (finalization): as such, it constitutes a bridge between disciplinary basic research and practice.[14]

In addition to characterizing disciplines and paradigms in terms of processes, structures and dimensions, and describing the dynamic of theories through the three-phase model, the determination of scientific progress in the bio-sciences also requires that *inter*theoretical and *inter*disciplinary development be taken into account. This development can be approximately described in terms of a move towards the theoretical unification of physics, chemistry and biology, and can be interpreted as a paradigm-shift for the biosciences in the sense used by Kuhn.[15] This intertheoretical development overlaps the various organisational levels of the classes of biological phenomena, and is directed at the theorisation of complex biological phenomena using molecular approaches derived from chemical theory. Biochemistry and molecular genetics represent instances of such a theoretical programme.

degree of complexity — Structure / Function	Bacterial cell (procaryotes)	Higher cell (eucaryotes)	Cellular systems	Organism	Population
Metabolism	Bacterial biochemistry	Cellular biochemistry	Biochemistry of cellular systems, e.g., blood cells	Biochemical integration, e.g., humoral regulation of glucose-metabolism	Biochemical ecology
Heredity	Genetics of bacteria	Cytogenetics	Developmental genetics	Integration of different genetic systems	Population genetics
Development	Morphogenesis of the bacterial cell e.g. sporulation	Cellular differentiation	Morphogenesis, embryology	Ontogenesis	Phylogenesis

Fig. 1. Diagram of biological subject areas relevant for cancer. The scheme is ordered by levels of growing complexity of biological structure (from left to right) and of biological function (from top to bottom)

Molecular genetics is currently the *clearest* representative of the unity of chemistry and biology. It initially explained the phenomena of heredity at the level of bacterial cells.[16] From there, development led to the molecular explanation of heredity in higher cells: the molecular genetics of eucaryotes, and molecular concepts of the genetic regulation of cell-differentiation and cell-development.

It remains an open question as to whether a new paradigm will have to be developed for the phenomena of morphogenesis and the integration of the central nervous system. Whether this is so or not, the biological phenomena will be the last to be embraced by a development of "microtheories for increasingly complex phenomena in ordered systems."[17]

3. CANCER APPROACHES BASED ON MATURE THEORIES

3.1. *Classification of Tumours in Pathological Anatomy*

The discovery of the cell as the basic building-block of tissues and organs was made possible by advances in the optical microscope in the nineteenth century.[18] An exhaustive description of the structure of cells was possible, within the constraints set by the limits of resolution for optical microscopes of that vintage. The theory of microscopic structure – paralleled in medicine by anatomy and in biology by morphology – describes the construction of organisms in terms of the diverse forms assumed by the basic structural unit "cell". This process of description was completed in the nineteenth century. All cell-types, their composition and all the microscopically distinguishable cell-structures of significance for classification are now known.[19]

Microscopic anatomy provides the basis on which structures which deviate from normal cell make-up can be identified, therefore allowing the recognition and description of pathological phenomena as a special field of the phenomena of cell-structures. Malignant growth represents one sub-field of such pathological phenomena: Johannes Müller was the first to have employed the cellular approach in the description and classification of such growth, and in 1836 he constructed a taxonomy of tumours which played a central role in the diagnosis and treatment of cancers in the nineteenth century.[20] The continuing empirical validity of the contribution made by the cellular taxonomy of tumours can be attributed to the precise definition and classification of tumours which it has produced.[21]

The taxonomy of tumours developed by pathological anatomy plays a crucial practical role in combatting cancer as an indispensable aspect of diagnosis. Cell-tissue examination allows an identification of the type of tumour, and its stage of development, together with a possible prognosis. In instances where cell-examination enables an early diagnosis, this approach can yield high rates of success in the remission of cancers.[22]

3.2. *Chemical Theories of Carcinogenesis*

Cellular pathological anatomy allows the identification of the stage at which a normal cell becomes a cancer cell. However, as a descriptive science it cannot provide an explanation of the mechanism by which this transformation takes place. In contrast to such descriptions, a *theory* brings with it the expectation of an identification of those factors which prompt and condition the processes to which it applies, and consequently some type of schema of how these factors operate.[23]

The first theory of cancer to be proposed on the basis of a mature discipline was the classical theory of the generation of cancer by chemicals (chemical carcinogenesis). The establishment of a relationship between chemistry and the problem of cancer was made possible by a discovery in experimental cancer research made in 1915 in which the Japanese scientists Ichikawa and Yamagiwa succeeded in inducing skin-tumours with tar.[24] This experimental system for generating cancer prompted intensive and wide-ranging research,[25] and marked the beginning of a new era in cancer research – that of chemical cancer research.[26]

The progress and refinements in the analytical techniques of organic chemistry provided the foundation upon which it was theoretically possible to identify and isolate carcinogenic factors. Nevertheless, the cancer-inducing component of tar, the cyclic hydrocarbon 3,4-benzpyrene was not in fact identified until 1932.[27] This class of substances subsequently served as the model for a chemical approach to carcinogenesis.[28]

The theory built on the foundations which had been laid down by the theories of organic chemistry. At the time at which benzpyrene was isolated, the mechanisms by which aromatic ring compounds reacted were already known. Laws therefore existed for the phenomena of organic chemistry which, in principle, had embraced the relevant structural unit – the carbon compound – and its interactions. The claim made by the special chemical theory of carcinogenesis is that the active, carcinogenic chemical principle is a particular chemical structure.[29] This structure was considered to be common to the polycyclic hydrocarbons. The thesis that a relation existed between a defined chemical structure and the capacity to induce tumours had to be revised however; a very small modification of the structure, which had no effect on its chemical reactivity, reduced its carcinogenic properties. It also became apparent that a variety of chemical structures could induce carcinomas. The active principle could not therefore be reduced to a homogeneous chemical structure.[30]

The further development of the chemical theory of carcinogenesis traced the capacity of certain substances to induce tumours back not simply to their structure, but to the chemical reactivity of an established double-bond in polycyclic hydrocarbons.[31] Once again, however, it transpired that chemical reactivity and carcinogenic activity did not exist in a one-to-one correlation. The concept of reactivity therefore failed to supply any generalizable theory of chemical carcinogenesis.

The significance of the special chemical theories lay in the fact that they revealed the limited scope for the chemical approach in the understanding of carcinogenesis. Self-evidently, any theory which treats the cell as a "black box" and focuses solely on the relation between the cell and its environment cannot adequately tackle the mechanisms which produce cancer. The inadequacy of the chemical theory was a pointer to the fact that the central causes of the cancer-process had to be looked for not outside the cell, but within it, and therefore had to be conceived in biological terms.[32] However, the idea of a highly reactive group of molecules reacting with the components of a cell continued to play a guiding role in chemical cancer research, and reappeared in the form of biochemical theories of cancer.[33]

Although the attempt to use organic chemistry as the basis for a special theory of the specific phenomenon of carcinogenesis via chemical substances may not have succeeded, this type of research did have a considerable relevance for medical practice. The feverish search for the active principle produced as "fall out" the understanding of a large variety of carcinogens and suspected carcinogenic substances,[34] knowledge of which was applied in cancer-prevention, especially with occupational cancers.[35]

3.3. *The Somatic Mutation Theory of Tumour Induction*

The shortcomings of the chemical theory led to the realisation that the mechanism of carcinogenesis could not be understood without some knowledge of how the cell functions. The decisive step into the cell had to be taken, and the "black box" opened. One step in this direction was made by the geneticists through the chromosome theory of heredity, in which the phenomena of heredity were conceptualized in terms of "invisible" units of genetic information and recombination, the genes, localized within the chromosomes. What was significant about the chromosome theory of heredity was that it could be elaborated through genetic experimentation with the fruitfly (drosophila).[36]

This theory provided the basis for the first genetic approach to carcinogenesis. Boveri had observed anomalies in chromosomes which were linked with the growth of tumours as early as 1912. These observations led him to the hypothesis that mutations in chromosomes were responsible for carcinogenesis.[37] In advancing this position he established the first link between malignant growth and the carriers of the cell's genetic information, the chromosomes.

The discovery that ionizing radiation could not only induce carcinomas but also mutations[38] added weight to the genetic hypothesis. The idea that the factor which prompted the cancer also affected the cell's genetic apparatus additionally provided the opportunity to fill two explanatory gaps left open by previous research. Tumours did not exhibit remission once the tumour-inducing factor was removed: in fact, they continued to grow. This had to mean that the information relevant to the cell's transformed state was *stabilized* and *transmitted* from cell to cell as the tumour grew through cell proliferation. A theory of heredity on the basis of somatic cells (body cells) could explain this phenomenon: mutations in the genetic apparatus of the body cells, which are passed on to following generations of cells, change the cell's genetic information and lead to a loss of control over cell growth.[39]

The mutation theory of carcinogenesis can be regarded as a special development of the chromosome theory of heredity, experimentally verifiable for germ-cells, and extendible in one step to somatic cells.[40] The somatic mutation theory of carcinogenesis consists of a number of propositions about the specific phenomenon of somatic cell mutation; this, in turn, comprises those mutation events which lead to a change in the growth-behaviour of the somatic cells. When originally formulated the mutation hypothesis could not be tested as techniques were not available for identifying and localizing somatic mutations. Using microscopy it was nevertheless possible to associate a change in a chromosome (Chromosome 22 – the "Philadelphia" chromosome) with a cell-transformation for one type of blood cancer, chronic myelogeneous leukaemia.[41] However, this was not in itself sufficient to explain cell-transformation, since the question as to whether the change in the chromosome was the *cause* or *consequence* of the malignant growth remained unresolved.[42]

The fact that the unresolved problems of the mutation hypothesis remained inaccessible to empirical testing prevented them from serving as a basis for practical application. Nevertheless, the theory was important in bringing about a shift in the approach to cancer towards the realization that the factor which prompted malignant growth had to affect the cell's genetic apparatus.

3.4. *The Biochemical Approach to Cancer*

The object-field of biochemistry embraces metabolic processes within organisms, but unlike physiology – which regards such transformations as balancing processes and compares the input and output of an organic system's various metabolisms – biochemistry is concerned with the molecular mechanisms through which those molecules characteristic of living things are transformed via the agency of biocatalysts, the enzymes. The objective is the specification of the metabolic pathways of processes of synthesis and breakdown which lead either to chemically bound energy or the building of cell structures.[43]

Based on enzyme theory as a paradigm[44] and having explained certain key metabolic chains within the cell, biochemistry has reached a stage of theoretical maturity. The theory has enabled a research programme to be conceived in principle for any unknown metabolic process, with the purpose of producing a biochemical special theory for the process under investigation.

The first biochemical approach to cancer was proposed by Warburg in 1925[45] in the context of his studies to trace the individual biochemical steps in the respiratory chain. During his studies of cell-respiration using various types of cells, Warburg discovered that tumour cells converted sugar to lactic acid (fermentation) in circumstances where normal cells would have broken the sugar down into carbon-dioxide and water (respiration). It is a characteristic feature of cancer cells that fermentation also takes place in a oxygen-rich environment. Warburg termed this phenomenon "aerobic glycolysis".[46]

This empirical discovery served as the foundation for a theory of cancer: a large number of non-proximate causes of cancer (carcinogenic influences) ultimately converge into one common cause – the irreversible damaging of respiration. In the subsequent process of selection the only survivors are those damaged cells which can compensate for the loss of energy from respiration by deriving energy from fermentation. Energy from fermentation represents an archaic form of energy production typical of undifferentiated cells both in phylogenesis and embryology. The substitution of fermentation for respiration is therefore associated with a reversal of cell differentiation. Moreover, the undifferentiated cells are uncontrolled in growth behaviour and consequently form cancer cells.

Warburg succeeded in specifying the differences between the respiratory chain of normal cells and that of tumour cells, and in developing a specific biochemical model of the energy metabolism of tumour cells. His cancer theory made further claims however: Warburg claimed to have found an explanation for the first step in the emergence of a tumour, although he was not in fact able to supply any criteria to establish whether aerobic glycolysis was the cause or consequence of malignant degeneration. The *sequence* of steps in cell-transformation could not be exposed to empirical proof in Warburg's time.

In the subsequent discussion and testing of Warburg's hypothesis it became clear that the contention that aerobic glycolysis was the cause of carcinomas could not be sustained. Increasing numbers of tumours were found which either exhibited this deviation in the enzyme patterns of their respiratory chain at a very late stage, in a different form, or not at all.[47] The one aspect of Warburg's theory which was verified and continued to hold good was the discovery that the cells of some types of tumours exhibited a deviant energy metabolism at later stages of development.

Further work in biochemical tumour research yielded additional determinations of the biochemical differences between tumour cells and normal cells. Greenstein, for example, advanced the hypothesis that tumour cells converged towards a more or less unitary enzyme pattern.[48] Potter's catabolic deletion hypothesis claimed that the development of tumour cells was characterized by the disappearance of catabolic enzymes.[49] Attempts similar to Warburg's to assemble such separate characteristics into a theory of carcinogenesis were not, however, successful. Any generalization derivable from one tumour was refuted by newly discovered, or induced, tumours. As a consequence only the secondary changes in the tumour cells, and not the initial conditions for carcinogenesis, were actually captured by these special biochemical models of tumour metabolism.

The step towards the exploitation of the microtheoretical potential of tumour biochemistry for therapeutic practice was made possible by the further accumulation of knowledge about the metabolic chains within the cell. One of the principal strategies consisted in the development of very efficient anti-cancer drugs based on the difference in the nucleic-acid-metabolism. The nucleic-acid-metabolism is the site at which the building blocks for the replication of the genetic material are prepared. Tumour cells make much greater use of metabolite supply paths which are only of secondary importance in non-transformed cells. The biochemical explanation of this metabolism allows the strategic construction of "metabolism blockers" (antimetabolites). The antimetabolite approach consists in the construction of structural analogues to the building blocks of metabolism which block the essential chains of synthesis by being irreversible bonded to the enzymes. The approach was developed following the discovery of sulphonamides in combatting bacterial infectious diseases. Sulphonamides are structural analogues of folic acid, one of the building blocks of a bacterial vitamin. By bonding irreversibly to an enzyme they block the synthesis: this constitutes their function as antimetabolites.[50]

In the specific case of the nucleic-acid-metabolism, the use of this approach involves the construction of chemical compounds which bond irreversibly onto the enzymes of the tumour's particular type of nucleic-acid-metabolism, consequently blocking the multiplication of the cell's chromosomes and stopping cell-growth.[51]

One biochemical theory has also enabled a preventive strategy to be developed. Carcinogens are converted by enzymes in the cell. The end-product of this special metabolism of carcinogens in the cell, the ultimate carcinogen, finally reacts with a particular class of cell-molecules and induces the transformation of the cell. According to the theory, the ultimate carcinogen possesses a reactive electrophilic centre which can attack the electron-rich centres of the nucleic acids and proteins.[52] A further step of specialization also allows the antimetabolite approach to be applied to this metabolism by aiming to block the cell-enzymes from guiding the synthesis of the reactive carcinogens.

3.5. *Synopsis*

Special theories of carcinogenesis, together with some theoretical strategies for combating cancer, have been formulated on the basis of mature theories. The hypothesis of a particular, cancer-including structure or a highly reactive group was based on theories of organic chemistry; the hypothesis of carcinogenesis through somatic mutation presupposed the chromosome theory of heredity; the various theories of tumour metabolism, and the antimetabolite approach, represent special ramifications of biochemical theories.

The characterization of these approaches as "*special* theories" and "theoretical developments" could provide fuel for controversy. What is intended here, for example in the biochemical instances, is essentially the experimental explanation of a particular metabolic pathway by reference to the existing methods and principles of biochemistry. However, this process of explanation is a research programme which aims at the determination of a series of reactions within cell-metabolism and its description in the formulae of chemical structures. This metabolic pathway represents a theoretical model which cannot be derived from known theories. The development of this model requires more than the insertion of empirical values into the function of a known model. Although possibly at a lower level of generality, it is comparable with the special theories of physics which deal with the empirical determination of particular forces and the specification of conditions for the application of the principles of the general theory.[53]

The order in which these theories have appeared reflects twentieth century development in the natural sciences, in which chemistry, biochemistry and genetics have attained a new level of theoretical explanation. This development can be considered as a path to an increasingly differentiated theoretical understanding of increasingly complex phenomena. One development step consists in a shift in the dimensions of the structural units in which phenomena are conceptualized. In the course of this development macrotheories of complex phenomena have been replaced by microtheories: the classical laws of chemical reaction have been replaced by electron theories of chemistry; classical Mendelian genetics, which operated at the level of the organism as a whole, has been replaced by the chromosome theory of inheritance which operates with sub-cellular units (chromosomes); and chemical physiology, which conceptualizes metabolic processes in organisms as input–output balances has been replaced by biochemistry, which explains the molecular mechanisms of metabolism.

This development of research in medicine, chemistry and the bio-sciences corresponds with that of the theoretical approaches to the cancer-problem. The path from the descriptive approach of histology and the causal model of chemical carcinogenesis to the genetic and biochemical approaches carries the problem of cancer into the cell itself, and increasingly expresses it in microtheoretical concepts. The history of these cancer theories is simultaneously the history of the cutting down of their *general* claims to their *particular* contributions to the understanding of cancer.[54] Histology showed that the phenomena of cancer had to be considered at the level of the cell. Chemistry elucidated the influence of factors in the cell's environment on carcinogenesis. Biochemistry explained some special characteristics of the metabolism of tumourous cells in advanced stages. Genetics made it possible to understand the irreversibility and transmission of the transformed state. Mutation theory approaches the status of a causal explanation most closely, but is as yet untestable as such. The initial step of cell-transformation remains an unsolved problem. The negative results of previous theories of cancer indicate that the explanation of cell-transformation is to be looked for in the processes of cell-growth controlled by its genetic apparatus. In terms of its general causes, cancer is formulated as a problem for molecular theories of the regulation of cell functions. However, cell genetics has not developed any mature molecular theory for its field; it is still at the stage of acquiring a paradigm. The possible relation between theoretical development at this stage to the cancer-problem is discussed in the following section.

4. CURRENT CANCER RESEARCH IN CELL AND DEVELOPMENTAL BIOLOGY

4.1. *The Molecular Biology of Higher Cells*

Development in molecular cell biology[53] has not yet produced a mature theory. No theoretical solutions have yet been found for its fundamental problems: the questions of cellular growth, control and differentiation, the development of cells in cellular systems and the emergence of forms and figures during "morphogenesis", to take the most intriguing cases. The molecular explanations of these processes is one of the fundamental tasks of present-day biology. In this connection, the development of theory for the organizational level of individual cells differs from that of cell-systems, since the former can draw on the paradigm of molecular genetics, through which a basic theoretical understanding of molecular structures and mechanisms of genetic information has been obtained for the individual cell at the organisational level of bacteria (procaryotic cells). The molecular paradigm of genetic information was proposed by Watson and Crick in 1953, and implemented and verified in the subsequent ten years for the genetics of bacteria cells.

The paradigm contains the explanation of the structure and function of genetic information (DNA-double helix, genetic code), the mechanism for the identical reduplication (replication), and the transcription and translation of genetic information into the characteristics of the cell (the "central dogma": DNA makes RNA, RNA makes protein) together with the regulation of this gene expression explained by the "operon model".[56]

The theory of molecular genetics is expected to be valid for all processes of heredity, although as yet it has only been successfully applied to bacteria cells. Its application to higher cells (eucaryotes) would represent a step beyond the scope of the previous theory into fields of greater complexity. The issue is one of extending the range of applications of molecular genetics to this field.

The molecular genetics of bacteria serves as a model here for the outlining of problems and research strategies. The central problems are already posed:[57] the division of the cell (compartmentalization) into nucleus and cytoplasm has a major implication for the concretization of the central dogma. The processes of the transcription of DNA into RNA and the synthesis of proteins on the RNA (translation) which run simultaneously in bacteria cells are separated structurally and temporally in eucaryotic cells. The first step takes place in the nucleus, the second in the cytoplasm. As a consequence, complex problems of transportation and regulation have to be resolved which are not encountered with bacteria cells. The chromosome is not the sole carrier of genetic information in the cell; certain sub-cellular structures, such as mitochondria and chloroplasts (responsible for energy production and photosynthesis) possess "mini-chromosomes" with altered compounds of the genetic code. The interaction between the various genetic systems in the cell is still not understood.

The functional state of the cell is established through the regulation of the pattern of gene expression. The question is whether the regulation of protein synthesis in eucaryotic cells takes place at the same level as that in bacteria cells.

No definite theoretical model exists for dealing with the next higher organizational level – the molecular biology of cell systems – comparable to the molecular biology of the single higher cell.

The concern of the molecular biology of cell systems is the regulation of cell functions within associations of cells and their integration and differentiation in systems – in short, the "social behaviour" of cells. Since the development of a cell from a totipotent state, as represented by the fertilized egge, the zygote, to a functional differentiation inevitably requires the cooperations of cells,[58] differentiation, development and morphogenesis in organisms take place at the level of cell *systems*. The question of differentiation and development is consequently at the core of the biology of cell systems; developmental biology defines itself as that field which is centred on this problematic.[59]

The development of an organism leads from a single cell, the fertilized egg, to a complex and coordinated pattern with different sub-systems of cells, each responsible for specific functions. In the course of their development cells take particular paths of differentiation. Such paths also exist in the mature organism: cells which are used up in the life-process itself, such as blood and skin (epithelia) have to be replaced. The path of differentiation of a cell from a pluripotent initial cell to a differentiated cell with certain functions[60] is a model for the process of the progressive commitment of cell functions and is described via the concept of the stem-cell. The key problem of the molecular biology of developmental processes is consequently how the cascade of progressive commitment or gene expression is achieved and maintained as a result of the interaction of factors emanating from "nature" (the genetic programme) and from "nurture" (the environment or "epigenetic" factors).

Neither molecular cell biology nor developmental biology have access to mature theories which could function as the basis for special theories of cancer. Nevertheless, "premature" hypotheses are formulated within the context of this theoretical development which conceptualize carcinogenesis as a problem of cell genetics and cell differentiation. These hypotheses are central to the current understanding of the cancer-problem.

The molecular new formulation of the mutation theory of cancer links chemical carcinogenesis with cell genetics. Carcinogens are metabolized in the cells into ultimate carcinogens which initiate mutation, and can therefore be considered as promutagens.[61] Mutation leads to a change in the gene products which control cell-growth. The loss of growth control in turn produces cell-transformation.[62]

The theory of the induction of tumours by viruses (viral oncogenesis) also postulates that the viral gene products interfere with the regulation of cell growth.[63] Molecular developmental biology interprets cancer cells as the result of "errors of commitment".[64]

4.2. *Experimental Linkages between Basic Research in Cell and Developmental Biology and Cancer Research*

The empirical verification of special cancer theories in cell genetics and developmental biology cannot take place prior to the resolution of the fundamental questions of the theory of cell-development and cell-differentiation. Nevertheless, it is possible for research in cell-biology to have a reference to cancer which does not presuppose the existence of a mature theory; for example, basic disciplinary research and cancer research may *coincide* through the conscious selection of tumour cells or systems of cell-transformation as the experimental tools for cell-biology research. Additionally, the results of basic research may be *transferred* to experimental systems which are relevant to cancer (ultimately, to human tumour cells). Both instances present different relations between basic research in cell-biology and experimental cancer-research, and are considered in the examples discussed below.

4.2.1. *Tumour virology*. Tumour virology has been one of the dominant approaches in cancer research since the early 1950s.[65] The precedent was set by Peyton Rous who successfully used viruses to induce tumours in 1910.[66] Rous discovered that sarcoma tumours in chickens could be induced by particles smaller than bacteria. However, microbiology was not then methodologically or theoretically sufficiently advanced to explain his discovery. Tumour virology only became a theoretical challenge (in the early 1950s) when methods became available for breeding viruses in tissue culture and making virus particles visible by electron microscopy.[67]

The strategy pursued by tumour virology consists in observing the interaction between the virus and the cell at the genetic level, with the aim of identifying the gene product or cellular event(s) which initiate the process of cell-transformation.[68] Within their research programme, molecular biologists and virologists attempt to produce an experimental elucidation of this process, on the assumption that in viral carcinogenesis a definable segment or the whole virus genome is transferred into the cell. Since viruses have a limited number of genes, it is hoped that a full specification of their functions in the cell will be possible.[69] Out of these one, or more, genes or gene products should be responsible for cell-transformation.

The mechanisms by which viruses and bacteria cells interact are explicable in terms of molecular genetics in which the extensive use of virus systems (phages) has played a major role. Tumour virology therefore makes extensive use of the tehcniques and concepts of phage-genetics.[70] Nevertheless, tumour virology cannot be regarded as a specialisation of the mature theories of molecular genetics. In contrast to the bacterial chromosome into which the phage-DNA is integrated, the chromosomes of eucaryotic cells have a much more complex structure. Questions of how the virus-genome is integrated into the chromosome and how it interferes with the regulatory processes within the cell in general – in fact, most of the problems of virus-cell interaction remain unresolved.

Tumour virology represents a *special* subject area within cell-biology, dealing with a particular class of biological carcinogens, tumour viruses, which in animals have been shown to be capable of inducing a certain number of tumours. This induction takes place through mechanisms which can be interpreted in terms of molecular genetics.

Tumour virology *coincides* with certain key research fronts of cell biology in the study of the unresolved questions of the interaction between virus and cell in the regulation of cell-growth and differentiation. Viral carcinogenesis (oncogenesis) implies that viral genes or gene products produce a defect in the maintenance of a particular state of differentiation of the cell which embraces growth. The molecular explanation of the mechanism by which tumours are induced at the same time represents one aspect of how growth is regulated.[71]

The investigation of the *disturbed* function of growth regulation is one strategy for understanding its *normal* functioning. The selection of the virus induction of tumours as an experimental system in cell-biology offers a number of advantages as far as such a research strategy is concerned. The phenomena of cell-transformation, or the shift away from the cell's differentiated state, is at the heart of one of the central phenomena of cell-biology. At the same time, the genetic material involved is limited and can, in principle, be decoded using the methods of molecular genetics. As far as cancer research is concerned, the use of this experimental system which allows both special and fundamental questions to be pursued simultaneously, has the advantage that the solution of the problems of cell-biology will directly allow the development of a special theory of viral tumour induction.[72]

4.2.2. *Cell biology using systems of tumour cells*. Not only viral induced tumour cells have been of great significance for basic research in cell-biology; certain tumour cell-lines have a particular advantage as far as research is concerned – they can be rapidly, easily and permanently grown in cultures.

The technique of growing cells in culture in a synthetic nutrient medium outside the organism ("in vitro") represented a major advance for the biochemical analysis of cell functions. It enabled a chemical definition of the conditions of cell-growth to be made, allowed relatively "clean" systems to be maintained and permitted the separation of sufficient quantities of cells for analysis. From its inception, cell and tissue-culture research has therefore been involved with tumour cells.[73]

Tumour cells have been used to examine problems of cell-function, cell-growth and cell-differentiation.[74] The phases of the cell-cycle and the regulation of chromosome division have also been amenable to analysis using systems of tumour cells. Certain tricks developed in tissue-culture technology have enabled normal cells and tumour cells to be fused.[75] Such experiments investigate the interaction between the cell-nucleus and cytoplasm and the dominance in the regulation of the gene activities of each nucleus. The understanding of these interactions is a prerequisite for reestablishing growth control in transformed cells.

Comparing tumourous and normal cells reveals differences in the structure of the cell-surface which are important in the recognition of signals and which are absent in tumorous cells.[76] Both cases therefore represent areas of coincidence between research strategies in cell-biology and those in tumour cell research.

4.2.3. *The system of blood-forming cells*. The system of blood-forming cells (hematopoietic system) has also proved to be a favourable experimental system for investigating the mode of development of stem-cells into differentiated cells and the molecular mechanisms through which an equilibrium of resting stem-cells, growing, differentiating and dying cells is established. One advantage possessed by this system from the standpoint of research is its limited complexity. There are fifteen to twenty different types of cells, and four paths of differentiation:[77] differentiation into the phagocytes (granulocytes and monocytes; myeloblastic series); the path to the blood platelets (thrombocytes), which play a central role in blood-clotting (thromboblastic series); the path to the red blood cells (erythrocytes; erythroblastic series); the path to the cells of the lymphatic system (lymphocytes) which are responsible for the immune defence (lymphoblastic series).

Because this cell-system is "fluid", it is not burdened with processes of morphogenesis which creates complications in solid tissues.[78]

This system is of major importance in cancer research. Among those causal mechanisms which can induce uncontrolled growth in the blood-forming cells are errors of differentiation processes which may occur at some point along these developmental paths. Myeloblastic and lymphoblastic leukaemia are the product of the myeloblastic and lymphoblastic series,[79] erythroblastosis and its associated tumour anaemia are the product of the erythroblastic series.[80] In principle, it is therefore possible to describe the specific damage which produces uncontrolled growth by explaining how the system functions. On the other hand, the use of systems which exhibit pathological symptoms is an important research strategy in the search for an understanding of how the normal differentiation of blood cells takes place. The blood systems allow research to be undertaken with clinically relevant tumours. As far as basic research is concerned, the use of human blood cells does not have any negative implications. This example therefore represents an instance of the coincidence of basic research in cell-biology and *clinical* cancer research.[81]

4.2.4. *A coincidence of tumour research and research into early developmental phases: the teratoma.*[82] The final case of the coincidence of cancer research and basic research in cell-biology which we propose to deal with here is that of research carried out on a tumour found in the testes and ovaries of certain mammals: the teratoma system. This tumour is produced through a disruption of cell-differentiation at a very early stage of development. It can best be understood as an embryonic development which has run completely of control and which begins from a cell of the germ line without fertilization (that is, parthogenetically). The cells of this tumour can be cultured and replanted in certain tissues where they give up their carcinomic properties and begin to differentiate.

This system is therefore ideally suited for the study of epigenetically regulated processes in differentiation. Beginning with a cell-culture of teratoma cells, it would be possible, for example, to analyse the pathways of nerve, skin and connective tissue cells in an "in vitro" system. As far as tumour research is concerned, this model provides a striking demonstration of the conceptualization of carcinogenesis as "errors of commitment".

4.3. *Transfer Research*

If the choice of an experimental system relevant to cancer ceases to be compatible with the efficient exploration of fundamental questions this system will cease to have a place in basic research: its disadvantages will be defined in terms of experimental parameters. For example, systems of cells whose transformation leads to epidemiologically significant carcinomas derived from epithelia are often referred to as "dirty systems". They do not facilitate the idealised representation of isolated effects since many of their structures are superimposed, and the results difficult to reproduce. Developmental biology therefore works with simple, reproducible, so-called "clean" experimental systems, such as the established and permanent lines of tumour cells.[83] Such systems allow more rapid progress than the use of more complex carcinoma cells. As a consequence, a hierarchical relationship can arise between basic research in cell biology and cancer research. Basic research pursues its own goals, and cancer research is obliged to develop any of the possible applicable results which cell biology produces. Oriented research becomes "transfer research" – it transfers the results from basic research to clinically relevant systems.[84]

For example, the culturing of epidemiologically relevant tumours (such as lung cancers) or the "in vitro" breeding of tumour cells from individual patients[85] both represent an important basis on which specific therapeutic programmes can be elaborated. Such "in vitro" systems are necessary for measuring the growth parameters of tumour cells, which indicate the required dosage, duration, concentration and scheduling of chemotherapy.[86] Such highly necessary and important special research in cell-biology does not however coincide with basic research, which examines the cell-cycle and growth of cells (cell-kinetics) by using "pure" systems such as highly artificial permanent tumour cell lines or yeast cells.[87]

Cancer research makes use of the results and methods of this research, such as the phases of the cell-cycle as described by cell-kinetics, and modifies them for studying and breeding cells from human tumours.[88] Such transfer processes can be seen everywhere where the results of basic research must be adapted to the specific conditions of clinically relevant systems. Results or sub-theories which can be obtained in simple systems must be made empirically applicable to human beings, with the appropriate modifications, because a man or a woman is not a tissue culture.

This type of research corresponds most closely to what is often easily devalued as "applied" research. Research is applied to practical goals (applied, that is, in the sense of "related to practice") but is at the same time isolated from the disciplinary research fronts, and consequently from the dynamics of theoretical development. It does not produce theories itself, but uses previous theories, and "applies" them. Nevertheless, this research work must be carried out if the cancer patient is to be helped.

4.4. *Synopsis*

The above sections dealt with a number of approaches in cancer-research within cell- and developmental-biology. The molecular biology of the higher cells is currently in a phase of theoretical development which, according to our model, can be termed the paradigmatic phase. Theoretical development is based on a research programme which proceeds from the paradigm of molecular genetics: this paradigm encompasses the mature theory of the molecular mechanisms of the central regulating functions in the bacterial cell, the mechanism for the translation of the genetic programme into cell characteristics (instruction) and the identical reduplication (replication) of this programme. With its molecular terminology and defined structural units (DNA-double helix, central dogma), this paradigm supplies a conceptual framework which is considered to hold good for all levels of complexity of biological organisation and the fields of heredity and development, and which allows previously separated fields of phenomena to be dealt with within a common theoretical approach.

The paradigm defines a theoretical programme which extends over all the bio-sciences, defining the state of theoretical development within the disciplines, and serving as the vehicle through which the trend in the modern natural sciences, in which increasingly complex phenomena are explained in micro-theoretical terms, and where possible, subsumed under a common paradigm, makes its advance.

The results and dynamics of cell- and development-biology are of key importance for cancer research. Hitherto non-understood but central aspects of the cancer problem which have been tracked down to the regulation of cell functions can now be conceptualized at a molecular level, allowing previously unconnected approaches to carcinogenesis to be linked together.

The loss of control over cell-growth is interpreted as a disruption in the regulation of the activity of the cellular genome, in which the mechanisms of chemical and biological carcinogenesis (viral oncogenesis) become conjoined. In addition, possible defects in the instruction mechanism which are not necessarily attributable to the cell genome, but are related to the spread and transmission of chemical signals in the cell's environment (epigenetic mechanisms) can also lead to a change in cell functions.

The interpretation of the cancer cell as a cell whose differentiation has been blocked or misdirected demonstrates that malignant growth represents a kind of derailment of normal growth. The problem therefore has less to do with a quite specific pathogenesis than which the stabilization and control of very complex equilibria in cell-growth and differentiation. The pathology of cancer is fully reflected in physiology, the etiological process can be seen to be as complex as life itself.

Since cell and developmental biology are still in the paradigmatic phase, a goal-oriented translation of hypotheses relevant to cancer into a programme of empirical research is either not possible or would be identical to tackling the unsolved fundamental problem of cell biology. The state of development and dynamics of cell biology do not permit the separating out of strategies which are both specific to cancer and yet not fundamental to the field as a whole. Current basic research in cell and developmental biology is equally relevant for cancer as well as other biological phenomena. A specific orientation to the cancer problem of this type of basic research is in principle not possible. Nevertheless, there are some opportunities for the practical orientation of basic research. Fundamental problems can be tackled by using experimental systems (tumour systems) which simultaneously represent particular aspects of the cancer problem (coincidence). However, as soon as tumour systems lose their methodological significance as far as basic research is concerned, cancer research and basic research will diverge. Research into cancer systems will become transfer research, in which the results of the disciplinary research fronts are recast and adapted for externally relevant systems.

5. PURE EMPIRICAL CANCER RESEARCH: HUMAN BIOLOGY AND MEDICINE

It has recently become apparent that a conclusive understanding of the cancer phenomenon and practical steps to combat it require the solution of theoretical problems which are not touched on by cell and developmental biology. These are the problems of the integration of sub-systems in an overall system such as the human body and its associated mechanisms for regulation and coordination. Many indications point to the fact that chronic illnesses are diseases of the system; that is, their genesis and spread through the organism hinge on the functioning of the control systems, the overall disposition of the organism, and finally, the material and social environment.[89] One basis on which an understanding of the systemic aspects of disease in humans can be constructed is that offered by human biology. Research in human biology is closely meshed with medicine, and is therefore extensively a directed field.

The molecular approach to biological systems currently offers the first steps towards a theory of the differentiation and morphogenesis of cell systems and organisms. Fields of higher complexity, such as the coordination and integration of biological sub-systems into a functional, stable and reproducible system, or still further, the interplay between organism and environment, are still outside the scope of the molecular approaches. The molecular biology of these phenomena is still in an explorative phase. There is no clear theoretical research programme structuring the dynamics of the phenomena embraced by human biology. As a consequence, human biology now appears to be almost exclusively a domain of medicine – that is of research oriented towards practical objectives.[90] This convergence of human biology and medicine is also the area for the realisation of a form of problem-orientation typical of the explorative phase, which we have termed the "functionalization" of research.[91] The empirical strategies which predominate in this phase are pre-set by purposes defined by technical, for example, medical goals. In terms of problems of implementation, this strategy requires the least amount of internal scientific structuring.

Functionalist strategies at the level of human biology play a major role in cancer research: the classical therapeutic strategies of surgery, radiology and chemotherapy ("steel, beam and drug") have come up against limitations.[92] But no established theoretical findings exist for early-diagnosis, the specific inactivation of secondary cancers, reversion of cell-transformation and activation of the specific immune defence against tumour cells – all areas in which breakthroughs are awaited. Advances in diagnosis and therapy are therefore thrown back on a "step-by-step strategy", the application of empirical trial-and-error methods. These methods are used in attempts to identify through parameter variation those factors which influence the organism as a whole under certain circumstances, and which produce desired effects.[93] In this way, long-term multicentred clinical trials are used to develop new therapeutic programmes. Combinations of forms of treatment – surgery, radiotherapy and chemotherapy – applied in varying doses, concentrations and combinations at different times are tested in patient-groupings to determine the optimal programme of therapy. This strategy has produced some notable successes, such as in the treatment of infantile leukaemia.[94]

A quasi-natural trial-and-error procedure also constitues the basis of epidemiology. In this instance, however, it is not the influencing factors which are known but the effects – the malignant growth of cells. The possible agents inducing or promoting cancer can be identified by a comparative statistical approach, in which population samples distinguishable by occupational or living conditions are selected and compared. This requires the isolation of a bundle of environmental factors which can be clearly correlated with a high cancer risk group.[95]

The field of human biology has an overwhelming medical orientation, due at least in part to the large gap between the theoretical research fronts of the bio-sciences and the science of human biology. Apart from the ethical constraints imposed on experiments with people, this system is a highly complex experimental system, and inappropriate for precise analysis. Biologists have therefore left this field to medical researchers. The human being is just beginning to figure within the molecular biology approach, leading to a type of research termed "biomedical research".[96] Research strategies in medicine are not determined, however, by the criteria which apply to the academic strategies of biological disciplines but are governed by practical goals. The exigencies of medicine do not permit a wait for theoretical insights, but compel the pursuit of any possible likelihood of "pre-theoretical" scientification. As far as the human systemic diseases – including chronic illnesses and cancer – are concerned, no theoretical solution is "around the corner", either at the level of cell-systems or the organism as a whole. The construction and implementation of functionalized research strategies complementary to the explanation-oriented research of molecular biology therefore constitutes the current task of cancer-research.

NOTES

[1] Some of the material used originated in journal literature, some from overviews of the state, development and history of the cancer-problem and cancer research. *Cf.* in particular, Bauer (1963), Berenblum (1957), UICC (1973), Greenstein (1947), Maugh and Marx (1975), Oberling (1959), Pilgrim (1974), Potter (1964), Shimkin (1973), Süss *et al.* (1973), Tooze (1973), Triolo (1965), Triolo and Shimkin (1969), Watson (1976), Chapter 18 and 20. For additional bibliographical information on the history of cancer research, see Shimkin (1974). The remainder of the material, and especially the bases for the systematic interpretation, originated with interviews with researchers carried out by the author in the following fields: molecular genetics, cell-biology, virology, immunology, chemical carcinogenesis, tumour cell-biology, epidemiology, clinical cancer research.

[2] US Department of Health, Education and Welfare (1974*a*), p. 11.

[3] *Cf.* the Federal Government's reply to a parliamentary question on the state of cancer research in Federal Germany (Deutscher Bundestag, 1976); the emphasis on cancer research in the *Rahmenprogramm Forschung und Technologie im Dienste der Gesundheit* (1976), p. 21; and the constitution of the *Nationale Stiftung für Krebsforschung* in 1976 (Stifterverband, 1976, pp. 22–24).

[4] This definition stresses the theoretical rather than practical aspect. As I show in the course of the paper, the interpretation of the problem changed with the expansion in knowledge. Cf. Watson (1970), p. 589.

[5] For an approach to the pattern of cancer research from the perspective of political direction see Hohlfeld (1979).

[6] *Cf.* 'Finalization Revisited' in this volume.

[7] *Cf.* Fliedner (1976), pp. 3–4. This assessment only applies to those processes which commence after cell-transformation, and not those which lead up to it. Strategies for cancer-prevention would not be affected. This system is therefore allocated a pace-maker role in cancer research: its investigation is understood as decidedly "basic cancer research" (Bundesministerium für Forschung und Technologie, 1975, p. 61).

[8] *Cf.* the US National Cancer Programme, which considered it essential to plug "one of the largest gaps in modern biology" through achieving an understanding of the precancerous stage via the explanation of the mechanisms of cell-differentiation and the factors responsible for the life-long stability of this differentiated state: US Department of Health, Education and Welfare (1974*b*), III-3.

9 For example, successes in classical research in carcinogenesis (see below pp. 100–101), and clinical cancer research into leukaemia (see below p. 117).

10 An acquaintance with the empirical material will probably assist in following this very abstract section: see Sections 3–5 below.

11 The concept of a "discipline" is of course vague. It can be defined in sociological terms through the process of education and training, or through the communication structrues established between scientists; alternatively, recourse can be made to the object-field of research, or the theories and methods which organise research as the key characteristic (*cf.* Kuhn, 1962; Hagstrom, 1965).

12 *Cf.* Polanyi (1968), Gatlin (1972), pp. 107 ff.

13 See the more detailed description of the phase-model of science in the Introduction to this volume.

14 *Cf.* 'Finalization Revisited' in this volume.

15 Kuhn (1962).

16 See Stent (1969) for an excellent presentation of the theoretical development of molecular genetics, which can be accommodated without any difficulty to the phase-model of scientific development.

17 *Cf.* Stent (1969), pp. 66 ff; Crick (1970*b*).

18 See Oberling (1959), p. 159.

19 A full description is not yet an explanatory theory in the sense required to meet the prerequisites of being "theroretically mature". Nevertheless, despite its purely descriptive character, the completeness and the relevance for cancer research and treatment of microscopic anatomy merits its inclusion at this point.

20 Müller (1836).

21 Oberling (1959), p. 160.

22 For example, the early diagnosis of cervical cancer can be achieved through a smear and examination of cells: surgery leads to high rates of success in obtaining a cure (see UICC, 1973, pp. 71–72 and pp. 222 ff.).

23 I.e., the *etiology* of the process.

24 Yamagiwa and Ichikawa (1918).

25 Henschen (1968), p. 450.

26 Epidemiological research had already provided some indications that chemical substances could induce cancer. The phenomenon of occupational cancers, in particular cancer of the testes in English chimney-sweeps, cancer of the bladder in dye-workers, and skin-cancer in lignite-tar workers suggested this possibility (see Oberling, 1959, pp. 65–66).

27 Kennaway (1955).

28 Berenblum (1967), p. 46.

29 *ibid.*

30 *ibid.*, p. 48.

31 *Cf.* Süss *et al.* (1973), pp. 18 ff.

32 Berenblum: "It should be realized, however, that the mechanism of cancer induction is not a mere problem of chemical interactions. The way a normal cell becomes transferred into a cancer cell is essentially a biological problem (though naturally dependent on chemical changes), and any attemt to explain carcinogenic action in chemical terms only may be a self-defeating process of rationalization" (1967), p. 59.

33 See p. 105 below.

34 See in particular Bauer (1963), pp. 330 ff.
35 Bauer (1963), pp. 30 ff.
36 *Cf.* Sturtevant (1965), Chapter 7 and 8.
37 Boveri (1914).
38 The discovery that radiation could lead to breaks in the chromosomes and to chromosome-rearrangements was made quite early on. Muller obtained mutated drosophila sperm by means of radiation in 1927, with the lethal inactivation of the sperm. It was known that X-rays could induce cancer of the skin soon after the commencement of radiation therapy (see Sturtevant, 1965, pp. 70 ff.). It could therefore be supposed that carcinogenesis could take place via the induction of mutants. In chemical carcinogenesis, the trigger for the linking of the discipline with problems of cancer was a development in experimental cancer research. In genetic the opposite was the case: a discovery made within the discipline stimulated the association with cancer, which was already well-known to be producible through radiation.
39 As long as classical genetics prevailed such a theory would inevitably hang somewhat in the air, and was in fact formulated independently in a number of different contexts. Bauer, for example, notes it first formulation as being in 1923 (1963, pp. 521 ff.). Potter ascribes it to an American Cancer Committee in 1938 (1964, p. 1088), and Burnet formulates it in a modified way without indicating any previous references (1957).
40 Since germ cells and somatic cells do not differ in principle as regards their structure and make-up, it should not be assumed that the extension of the chromosome theories to this class of cell signifies the transition to another level of complexity.
41 Casperson *et al.* (1970); Berenblum (1967), p. 112.
42 The crossing and recombination of chromosomes of somatic cells had to await the more recent techniques of cell-biology, such as cell-cultural and cell-fusion: this has allowed the localization and identification of mutants (*cf.* Tooze, 1973, pp. 143 ff.).
43 *Cf.* for example, Lehninger (1977), pp. 4–9. The biochemical paradigm is restricted in comparison to molecular genetics (see Section 4 below) in that it deals with all metabolic processes in the organism without considering the information aspect. The term molecular biology is often used synonymously with molecular genetics, although the phenomena which it embraces cannot always be clearly assigned to the biochemical or the molecular genetic paradigm.
44 *Cf.* Kohler (1975), pp. 288 ff.
45 Warburg (1925).
46 See the detailed discussion in Bauer (1963), pp. 161 ff.
47 Potter (1964), p. 1087.
48 Greenstein (1947), pp. 197–202.
49 Potter (1964), p. 1087.
50 See Handler (1970), p. 636.
51 Heidelberger *et al.* (1958) developed the first effective cancer drug on this basis: the growth-inhibitor (cytostaticon) 5-fluorouracil.
52 Miller and Miller (1977).
53 See "Finalization Revisited" in the volume for a a discussion of the concept of the special theory in physics.
54 Potter notes for example in a historical study: "It is important to recognise how each of the older hypotheses has proved inadequate and how each retain elements that still command our respect", (1964), p. 1087.

[55] This term is intended to embrace those theoretical developments which, within the subject area of higher cells and cell systems, seek to explain the phenomena of heredity *and* development at these levels of organisation in molecular terms.

[56] The paradigmatization of molecular genetics is dramatically described by Stent (1969), a pioneer and eye-witness of this development (see Olby, 1974, for a deatailed science-history presentation of developments up to the double-helix). According to Stent the development of the paradigm can also be divided into phases:

(1) The *classical* period was represented by the chromosome theory of heredity – that is, macrogenetics.

(2) In the *romantic period* competing candidates for the role of paradigm can be distinguished – DNA and protein – as regards the material carriers of genetic information, the "genetic molecule". This phase was concluded with the hunt for the Nobel Prize (*cf.* Watson 1968). Watson and Crick won the race in 1953 with the structure of the DNA-double helix as the paradigm of molecular genetics.

(3) In the *dogmatic period* the central dogma was elaborated and the paradigm verified (unil *c.* 1963).

(4) In the *academic phase* special questions were answered.

In terms of our phase-model the romantic and dogmatic periods correspond to the paradigmatic phase, and the academic period to the post-paradigmatic phase of theoretical development.

[57] *Cf.* Watson (1976), Chapter 16.

[58] See, for example, Handler (1970), p. 150.

[59] Luria (1975), pp. 215 ff.

[60] See Section 4.2.3.

[61] *Cf.* Dulbecco (1976), p. 439.

[62] *Cf.* Hecker (1972), p. 61 who formulated an "information-loss hypothesis of tumourgenesis". However, he also includes other mutagenic mechanisms which lead to a "loss of cellular information", from which the loss of growth-control eventually results.

[63] *Cf.*, for example, Dulbecco (1976), p. 438, Maugh and Marx (1975), p. 12.

[64] Luria (1975), p. 248; Rubin (1976) seeks an epigenetic explanation principally to relativise the somatic mutation theory. Foulds (1969) counts as one of the pioneers of an epigenetic theory of cancer.

[65] One of the founders of molecular tumour virology even refers to a "turn of the tide" in cancer research (Gross, 1970, p. 930).

[66] Rous (1910).

[67] *Cf.* Tooze (1973), pp. 14 and 17.

[68] *Cf.* Weiss (1976).

[69] *Cf.* Handler (1970), p. 151 Baltimore (1976), p. 635. For cell-genetics the main handicap in analysing gene-functions is the sheer immensity of genetic material (10^6 genes). However, viral gene-products and genes can be identified using the classical methods of molecular genetics. Hopes exist for using techniques of gene-surgery to transplant genes from eucaryotes into bacteria cells, where those will multiply (a process termed "cloning") and allow an analysis of the synthesis of function of these gene-produces (*cf.* Dulbecco, 1976).

[70] Analogous to the interaction of bacteria and bacteriophages it was postulated at a relatively early stage that viruses could be integrated into the cell-chromosome in a process analogous to lysogeny in bacteria cells (integration of phage-DNA into the

bacterial chromosome, and that this interaction between virus and host could be associated with the phenomenon of cell-transformation (*cf.* Gross, 1970, pp. 916 ff.; Dulbecco, 1976; Temin, 1976). Tumour virology therefore became the first model for a transfer of the molecular genetic paradigm to the eucaryotic cells: the possibility of a link with molecular biology explains in no small measure the dominance of the theory of viral oncogenesis in cancer-research after 1953 (*cf.* Gross, 1970, p. 930).

71 Dulbecco (1976), p. 437, for example, one of the pioneers in tumour virology assessed his 1958 decision in the following terms: "I set myself the task of determining the molecular basis of tumour induction." Baltimore (1976) points out the complementary aspect of the simultaneous research into the molecular basis of growth regulation.

72 Tumour virologists always stress the relevance of their research for cell biology (see Note 71 above). Looked in narrow terms this is beyond question. What is under dispute is more the question of whether tumour virology also represents a relevant model for human cancers (*cf.* Tooze, 1973, pp. 59 ff.; Watson, 1976, p. 683; Temin, 1976, p. 1080). So far a probable virus etiology for tumours has only been established for the class of herpes viruses (*cf.* Hausen, 1975), and a particular class of RNA viruses ("C-types") has been associated with leukaemia (*cf.* Maugh, 1975). The relevance of viruses for majority of human tumours is disputed (*cf.* Maugh, 1975, p. 336). Clinical oncologists constantly complain about the "touch of arrogance and aesthetics" in research problems in molecular biology and virology, the resolution of which is of hardly any clinical relevance (*cf.* Ford 1974).

73 One cell system used for the development of a number of key cell-culture techniques was derived from a carcinoma of the cervix – the Hela system (taken from the initials of the patient's name; *cf.* Tooze (1973), pp. 78 ff; Pollack (1973), p. 125.

74 *Cf.* Pollack (1973), pp. 123 ff. and 181 ff.

75 *Cf.* Stanbridge (1976).

76 See Imperial Cancer Research Fund (1977), p. 8.

77 From Watson (1976), pp. 571 ff. and Bucher (1970), pp. 266 ff.

78 For example, it has been possible to provide at least a partial answer to the central question of the differentiating factors which perform the role of biochemical "decision-makers" where developmental paths cross using the hematopoietic system. It has been possible to identify the proteins which initiate the step of differentiation to macrophages, thrombocytes and erythrocytes (see Watson, 1976, pp. 571 ff.).

79 *Cf.* Frick (1976), p. 462.

80 *Cf.* Sträuli (1976), p. 1080.

81 This sytem is therefore assigned a pacemaking function in cancer-research; its study is considered as "decidedly basic research" (Bundesministerium für Forschung und Technologie, 1975, p. 61).

82 Sherman and Solter (1975).

83 *Cf.* Pollack (1973) Chapter 1 and 2.

84 The prestige-gap between cell-biology and cancer research is ironically explained by cancer-researchers in terms of the "purer the system, the higher the IQ."

85 *Cf.*, for example, the cloning of Hodgkin-cells (Hodgkinson's disease is a tumour of the lymphoblastic series). See Boecker *et al.* (1975).

86 Gallmeier (1976), p. 4 pleads on behalf of a cytological-experimental basis of polychemotherapy.

87 *Cf.* Tooze (1973), p. 114.
88 *Cf.* Oettgen (1975), p. 204.
89 See UICC (1973), pp. 3–26; Epstein (1977).
90 In the classical bio-sciences human biology is virtually only important in comparative studies and for the history of the human species.
91 Van den Daele and Krohn (1975).
92 *Cf.* Kärcher (1974), p. 4; Deutscher Bundestag (1976).
93 The side-effects are also of course identified.
94 *Cf.* Maugh and Marx (1975), pp. 167 ff.
95 Frentzel-Beyme (1976), p. 13.
96 In this sense, the term is used by one of the leading protagonists of the philosophy of combining molecular biology and medicine to achieve a high technology in fighting disease: C. L. Thomas (1972).

BIBLIOGRAPHY

Baltimore, D.: 1976, 'Viruses, polymerases, and cancer', in *Science* **192**, 632–636.

Bauer, K. H.: 1963, *Das Krebsproblem*, Springer, Berlin/Heidelberg.

Berenblum, I.: 1967, *Cancer Research Today*, Pergamon, Oxford.

Boecker, W., Hossfeld, D., Gallmeier, W. and Schmidt, C.: 1975, 'Clonal growth of Hodgkin cells', *Nature* **258**, 235–236.

Boveri, T.: 1914, *Zur Frage der Entstehung maligner Tumoren*, Jena.

Bucher, O.: 1970, *Cytologie, Histologie und mikroskopische Anatomie des Menschen*, Hans Huber, Berne/Stuttgart/Vienna, Vol. 5, 7th Edition.

Bundesministerium für Forschung und Technologie: 1976, *Rahmenprogramm Forschung und Technologie im Dienste der Gesundheit*, Bonn.

Bundesministerium für Forschung und Technologie, Bundesministerium für Jugend, Familie und Gesundheit: 1975, *'Forschung und Technologie im Dienste der Gesundheit'. Überlegungen der Hochschulberatergruppe (HSG) zu einem 10-Jahres-Programm*, Bonn.

Burnet, M. F.: 1957, 'Cancer – a biological approach', *British Medical Journal* **1**, I. 'Processes of control', p. 779: II. 'Significance of the somatic mutation', p. 782: III. 'Viruses associated with neoplastic conditions', p. 841: IV. 'Practical applications', p. 844.

Casperson, T., Gahrton, G., and Lindsten, J.: 1970, 'Identification of the Philadelphia chromosome as a number 22 by quinacrine mustard fluorescence analysis', *Experimental Cell Research* **63**, 238 ff.

Crick, F.: 1970*a*, 'Central dogma of molecular biology', *Nature* **227**, 561–563.

Crick, F.: 1970*b*, 'Molecular biology in the year 2000', *Nature* **228**, 613–615.

van den Daele, W. and Krohn, W.: 1975, 'Theorie und Strategie – zur Steuerbarkeit wissenschaftlicher Entwicklung', in Weingart, P. (ed.): 1975, *Wissenschaftsforschung*, Campus, New York/Frankfurt, pp. 213–254.

Deutscher Bundestag, 7. Wahlperiode: 1976, *Antwort der Bundesregierung auf die Grosse Anfrage betr. Krebsforschung vom 9.2.1976*, Sachgebiet 212, Drucksache 7/4711.

Dulbecco, R.: 1976, 'From the molecular biology of oncogenic DNA viruses to cancer', *Science* **192**, 437–440.

Epstein, S.: 1977, 'Cancer and the environment', *Bulletin of the Atomic Scientists*, March 1977, 23–31.

Fliedner, T.: 1976, *Diagnostik und Therapie maligner Tumoren aus pathophysiologischer Sicht*, Zentrum für klinische Grundlagenforschung der Universität Ulm.

Ford, E.: 1974, 'Does cancer research have realistic aims?', *Nature* **249**, 299–300.

Foulds, L.: 1969, *Neoplastic Development*, Vol. I, Academic Press, London/New York/San Francisco.

Frentzel-Beyme, R.: 1976, 'Epidemiologisches Vorgehen in der Arbeitsmedizin bei der Beurteilung des Berufsrisikos für Krebs', in *Sozialpathologie – Epidemiologie in der Forschung*, Wissenschaftliche Jahrestagung der Deutschen Gesellschaft für Sozialmedizin e. V., Gentner, Stuttgart, pp. 111–124.

Frick, P.: 1976, 'Blut', in Siegenthaler, W. (ed.): 1976, *Klinische Pathophysiologie*, Thieme, Stuttgart, pp. 444–480.

Gallmeier, W. M.: 1976, 'Polychemotherapie', in *Kongressbericht*, Deutscher Krebskongress Hamburg February 1976, Deutsche Krebsgesellschaft e. V.

Gatlin, L. L.: 1972, *Information Theory and the Living System*, Columbia University Press, New York/London.

Greenstein, J. P.: 1947, *Biochemistry of Cancer*, Academic Press, New York.

Gross, L.: 1970, *Oncogenic Viruses*, Pergamon, Oxford.

Hagstrom, W. O.: 1965, *The Scientific Community*, Basic Books, New York/London.

Handler, P. (ed.): 1970, *Biology and the Future of Man*, Oxford University Press, New York/London.

Hausen, H.: 1975, 'Oncogenic herpes viruses', *Biochimica et Biophysica Acta* **417**, 25–53.

Hecker, E.: 1972, 'Krebsforschung – heute und morgen', *Ruperto Carola* **52**, 51–63.

Heidelberger, C., Griesbach, L., Cruz, O., Schnitzer, R. J., and Grunberg, E.: 1958, 'Fluorinated pyrimidines. VI. Effects of 5-fluorouridine and 5-fluoro-2'-deoxyuridine on transplanted tumors', *Proc. Soc. Exp. Biol. Med.* **917**, 470–474.

Henschen, R.: 1968, 'Yamagiwa's tar cancer and its historical significance', *Gann* **59**, 447–451.

Hohlfeld, R.: 1979, 'Strategien gegen den Krebs – Die Planung der Krebsforschung', in van den Daele, W. *et al.*: 1979, *Geplante Forschung*, Suhrkamp, Frankfurt, pp. 181–238.

Imperial Cancer Research Fund: 1973, *Scientific Report to the Council by the Director of Research for the Year 1973*, London.

Kärcher, K. H.: 1974, 'Arbeitskreis für Geschwulstbehandlung', in Ott, G. *et al.* (ed.): 1974, *Standardisierte Krebsbehandlung*, Springer, Berlin/Heidelberg.

Kennaway, E.: 1955, 'The identification of a carcinogenic compound in coal-tar', *British Medical Journal* **2**, 749–752.

Kohler, R. E.: 1975, 'The history of biochemistry: a survey', *Journal of the History of Biology* **8**, 275–318.

Kuhn, T. S.: 1962, *The Structure of Scientific Revolutions*, University of Chicago Press, Chicago/London.

Lehninger, A. L.: 1977, *Biochemie*, Verlag Chemie, Weinheim/New York.

Luria, S. E.: 1975, *36 Lectures in Biology*, The MIT Press, Cambridge, Mass.

Maugh, T. H.: 1975, 'Leukemia: a second human tumor virus', *Science* **187**, 335–339.

Maugh, T. H. and Marx, J. L.: 1975, *Seeds of Destruction: The Science Report on Cancer Research*, Plenum Press, London/New York.

Miller, J. A. and Miller, E. C.: 1977, 'The concept of reactive electrophilic metabolites in chemical carcinogenesis', in Jollow, E. D. (ed.): 1977, *Biological Reactive Intermediates*, Plenum Press, New York.

Müller, J.: 1836, 'Über den feineren Bau der Krankhaften Geschwülste', *Berichte über die Bekanntmachung geeigneten Verhandl. der Königl. Preuss. Akad. der Wissenschaft Berlin* **I**, 107–113.

Oberling, C.: 1959, *Krebs, das Rätsel seiner Entstehung*, Rowohlt, Hamburg.

Oettgen, H. F.: 1975, 'Beiträge der Immunologie zur Kenntis der Krebskrankheiten', *Internist* **16**, 199–205.

Olby, R.: 1974, *The Path to the Double Helix*, Macmillan, London.

Pilgrim, I.: 1974, *The Topic of Cancer*, Thomas Y. Crowell, New York.

Polanyi, M.: 1968, 'Life's irreducible structure', *Science* **160** 1308–1312.

Pollack, R.: 1973, *Readings in Mammalian Cell Culture*, Cold Spring Harbor Laboratory.

Potter, R.: 1964, 'Biochemical perspectives in cancer research', *Cancer Research* **24**, 1085–1098.

Rous, P.: 1910, 'A transmissible avian neoplasm: sarcoma of the common fowl', *Journal exp. Medicine* **12**, 696 ff.

Rubin, H.: 1976, 'Carcinogenicity tests', *Science* **191**, 241.

Sherman, M. J. and Solter, D. (eds.): 1975, *Teratomas and Differentiation*, Academic Press, New York/San Francisco/London.

Shimkin, M. B.: 1973, *Science and Cancer*, US Dept. of Health Education and Welfare, DHEW Publication No. (NIH) 7–568.

Shimkin, M. B.: 1974, 'History of cancer research: A starter reading list and guide', *Cancer Research* **34**, 1519–1520.

Stanbridge, E.: 1976, 'Suppression of malignancy in human cells', *Nature* **260**, 17–20.

Stent, G. S.: 1969, *The Coming of the Golden Age*, The Natural History Press, Garden City, N.Y.

Sträuli, P.: 1976, 'Neoplasien', in Siegenthaler, W. (ed.) *Klinische Pathophysiologie*, Thieme, Stuttgart pp. 1074–1086.

Sturtevant, A. H.: 1966, *A History of Genetics*, Harper and Row, New York/Tokyo.

Süss, R., Kinzel, V., and Scribner J. D.: 1973, *Cancer. Experiments and concepts*, Springer, Berlin/Heidelberg.

Temin, H. M.: 1976, 'The DNA provirus hypothesis', *Science* **132** 1075–1080.

Tooze, J. (ed.): 1973, *The Molecular Biology of Tumor Viruses*, Cold Spring Harbor Laboratory.

Triolo, V. A.: 1965, 'Nineteenth century foundations of cancer research advances in tumour pathology, nomenclature, and theories of ontogenesis', *Cancer Research* **25**, 75–105.

Triolo, V. A. and Shimkin, M. B.: 1969, 'The American Cancer Society and cancer research, origins and organization', *Cancer Research* **29: 9** 1615–1641.

UICC (Union internationale contre le cancer), Committee on Professional Education (ed.): 1973, *Clinical Oncology. A Manual for Students and Doctors*, Springer Berlin/Heidelberg.

US Dept. of Health, Education and Welfare: 1974*a*, *Operational Plan. National Cancer Program*, DHEW Publication No. (NIH) 75–777.

US Dept. of Health, Education and Welfare: 1974*b*, *Digest of Scientific Recommendations for the National Cancer Program Plan*, DHEW Publication No. (NIH) 74–570.

Warburg, O.: 1925, 'Über den Stoffwechsel der Carzinomzelle', *Klinische Wochenschrift* **12**, 534–536.

Warburg, O.: 1947, 'Ideen zur Fermentchemie der Tumoren', *Abh. dtsch. Akad. Wiss., math. naturwiss. Kl. Nr. 3*, Berlin.

Watson, J. D.: 1968, *The Double Helix*, Weidenfeld and Nicholson, London.

Watson J. D.: 1970, *Molecular Biology of the Gene*, W. A. Benjamin, London/Amsterdam, 2nd edition.

Watson, J. D.: 1976, *Molecular Biology of the Gene*, 3rd edition.

Weiss, R.: 1976, 'Molecular analysis of the oncogene', *Nature* **260**, 93.

Yamagiwa, K. and Ichikawa, K.: 1918, *Cancer Research* **3**, 7 ff.

PART II

THEORETICAL CONSIDERATIONS

SUMMARY OF CONTRIBUTIONS

The conceptual and programmatic studies of *Finalization Revisited* and *Normative Finalization* together with the essay on *The Scientification of Technology* elaborate our theoretical framework of critical science studies.

Finalization Revisited is a reflection on, and elaboration of the theoretical approaches underlying the empirical studies. The first part offers a theoretical analysis of the conditions for finalization, principally through a more precise specification of the concept of "mature theories", and a clarification of the concept of the "application" of a general theory. The second part deals with the implications of the concept of finalization for science policy and discusses the technical and possible normatively critical potential of finalization as compared with other forms of scientific direction. Finally, it raises the question of the future social structure of a theoretical science oriented towards external goals.

The Scientification of Technology seeks to portray the historical process of the mutual interpenetration of science and technology. Up until the 19th century science and technology followed relatively separate paths, although the unity of both forms of knowledge-interests had already been formulated in the Renaissance. Despite this rhetorical unity, it was not possible to unify the technical demands on knowledge and the theoretical interests in the explanation of nature into one system of knowledge. It was only in the 19th century – following the Industrial Revolution – that technical sciences emerged which aspired to levels of explanation matching those of the natural sciences, and natural sciences, which attained a level of technical concreteness. In the 20th century this relation has become so close, that the distinction between science and technology can only be made analytically.

Normative Finalization reconstructs and evaluates the basic hypotheses of the theory of finalization and develops the programmatic notion of a social natural science. Proceeding from the original premise of a possible and necessary linkage of cognitive structures and social needs in ecology, it opposes the threat of a retraction of key theoretical positions on the one hand and of a political suppression of normative reasoning in science and science studies on the other. It argues that a mediation of true knowledge

G. Böhme et al., Finalization in Science, ed. by W. Schäfer, 129–130.

and rational interests is both feasible and imperative if modern science is to gain what early modern science failed to achieve: conscious and enlightened normativity.

GERNOT BÖHME, WOLFGANG VAN DEN DAELE AND RAINER HOHLFELD

FINALIZATION REVISITED

I. *Problems in the Philosophy of Science*

1. THEORETICAL MATURITY AND CLOSED THEORIES

The finalization thesis postulates that in the course of its development a science (discipline, or field of knowledge) reaches a state of maturity as a result of which it can be said to have completed its work. The fundamental problems of a discipline are resolved when a general theory for its subject matter is formulated. Based on such a theory the discipline can proceed to tackle a large number of new problems. However, no internal criteria of scientific progress are provided which might enable a rational choice to be made between these problems.

How can such a mature state be identified? In our original formulation we explained scientific maturity using Heisenberg's concept of closed theories.[1] Much of this essay is therefore concerned with this concept. Initially, however, we concentrate on the phenomenon which prompted Heisenberg's idea, and which also underpins our own contention that by solving its fundamental problems a discipline can arrive at a stage of completion.

1.1. *The Phenomenon of the Continuing Validity of Theories*

(a) Scientists repeatedly express the conviction that all the essential work in a certain field has been done. The physicist Friedrich Hund, for example, claims that, "nowadays we possess a physics which, in principle, appears to be completed for a middle range of phenomena", namely those dealt with in optics, mechanics, thermodynamics, electricity and chemistry. "The age-old problems of motion, light and materials are solved, then. The present day is concerned with the exploration of sub-atomic phenomena."[2] The biologist Gunther Stent considers the problems of heredity to have been finally understood: molecular genetics has arrived at its "academic" period in which it exploits existing knowledge within specialized research and is unlikely to produce any more major surprises as far as the basic principles of the discipline are concerned. For Stent, the unsolved basic problems of biology consist of embryology, the emergence of life, and the molecular mechanisms of the higher nervous system.[3]

G. Böhme et al., Finalization in Science, ed. by W. Schäfer, 131–172.

(b) It is a historical fact that theories are confirmed as valid by successor theories which supersede them. Examples can be seen in the relation between Kepler's and Newton's theories of planetary motion, Newtonian mechanics and relativity, and thermodynamics and statistical mechanics. This takes place through the restriction of the domain of intended applications,[4] through explaining the validity of a theory from the standpoint of the more advanced one, or through a form of generalization such that the older theory assumes the status of a special case of the more general.

(c) Theories whose internal potential for development has been exhausted retain their paradigmatic significance in pragmatic contexts. In particular, they remain the basis for the development of technologies, where they are regarded as suitable instruments for dealing with particular classes of empirical questions. As far as solving such pragmatic problems is concerned, the improvements represented by their succcessors are irrelevant. The orbits of space vehicles are still calculated using Newtonian, not Einsteinian, mechanics; and in chemical engineering, phenomenological thermodynamics and not statistical mechanics are used.

The older theories achieve the status of "classical" theories. Truesdell[5] describes this process for mechanics.

> The word 'classical' has two senses in scientific writing; (1) acknowledged as being of the first rank or authority, and (2) known, elementary, and exhausted ("trivial" in the root meaning of that word). In the twentieth century mechanics based upon the principles and concepts used up to 1900 acquired the adjective "classical" in its second and perjorative sense, largely because of the rise of quantum mechanics and relativity. "Fundamental" in physics came to mean "concerning extremely high velocities, extremely small sizes, or both". Physicists gave less and less attention to classical mechanics because they thought nothing more could be learned from it and nothing new discovered about it, although of course they continued to use it in the design of the experimental apparatus with which they claimed to controvert it. At about the same time "applied" in mathematics came to refer not to the object studied but to the originality and logical standards of the student, again in a pejorative sense.
>
> Engineers still had to be taught classical mechanics, because in terms of it they could understand the machines with which they worked and could devise new machines for new purposes. Research in mechanics came to be slanted towards the needs of engineers and to be carried out largely by university teachers who regarded mathematics as a scullery-maid, not a goddess or even a mistress. Leading exponents of applied mechanics were Ludwig Prandtl (1875–1953) and Geoffrey Ingham Taylor (born 1886).

The phenomena noted in (a)–(c) represent a stability in the life of theories which points beyond Kuhn's conception of theoretical paradigms, and in contrast emphasizes the constancy of theories through revolutionary changes.

It would therefore be misplaced to explain scientists' awareness of *this* type of stability solely in sociological terms, as a product of socialization into established paradigms. Kuhn's description of scientific development does lend itself to such an interpretation. In many fields scientists are indeed socialized not to problematize the fundamentals of their discipline, but rather to use them as the precondition for the exercise of normal science.[6] It is not, however, correct to assume that,

> this training explains why the answers provided by a paradigm are able to fix themselves so firmly in scientific thinking, so firmly sometimes that some theories are regarded as being closed and of eternal validity.[7]

Such an account still leaves unexplained why belief in the validity of closed theories is able to persist through scientific revolutions which change the conceptual basis of the field.

Is it possible to provide a theoretical legitimation for the phenomenon of theoretical stability – stability able even to survive scientific revolutions? The concept of "closed theories" offers an approach to this question.

1.2. *The Concept of Closed Theories*

In his paper 'Der Begriff "abgeschlossene Theorie" in der modernen Naturwissenschaft'[8] Werner Heisenberg considers the experience of physicists during the transition from Newtonian (and by extension, from the disciplines of classical physics in general) to quantum theory. Physicists found it increasingly difficult to fit the disciplines of classical physics to certain empirical evidences without undertaking substantial changes to the disciplines' conceptual structures. However, the new theories of relativity and quantum theory which arose as a result did not lead to the abolition of the superseded theories, although "strictly speaking" these were no longer correct from the standpoint of the new theories: rather, the new theories either required the older theories as presuppositions in some aspects, or were able to explain why the older theories were valid for certain areas of experience. This phenomenon led Heisenberg to designate the theories of classical physics as "closed", and to assert their eternal validity.

> Closed theories are valid for all times; whenever experiences can be described with the concepts of these theories, no matter how far in the future, the laws of these theories will prove to be correct.[9]

Heisenberg then extended the concept to cover other theories: "in addition to Newtonian mechanics, Maxwell's theory with the special theory of relativity, then thermodynamics and statistical mechanics, and finally (non-relativist) quantum mechanics, atomic physics and chemistry." [10]

Aside from the trivial requirement that a theory should not be internally contradictory, Heisenberg's criteria for inclusion in this category are that the theory should adequately describe a specific field of experience, and that it should figure as one of the "preconditions for further research". This latter criterion refers to Bohr's contention that the classical concepts were indispensable in the construction of quantum theory: [11] it asserts both the phenonomenon of continuing validity and suggests a reason for it. Theories that are the precondition for further research, specifically in the case of the relation between classical and quantum theory supplying the semantics for later theories, may be superseded in the course of scientific advance, but not annulled.

The first criterion also suggests a reason for the claim of continuing validity. If one were able to specify what it means to say that a theory supplied an adequate concept for one, and precisely *this* one field of experience, it would be possible to demonstrate its continuing validity. Heisenberg himself does not implement this strategy but rather tends to emphasise the logical properties of a closed theory, and the historical arguments which lead to a firm belief that it is in fact closed. For example, he employs the concept of "compactness" to indicate the logical-conceptual consistency of such theories, and emphasizes the high degree to which they have been confirmed. In the final analysis, he appeals to the competent judgement of the experienced scientist.

A more systematic approach to the notion of closed theories is envisaged by von Weizsäcker.[12] He proposes to reformulate Heisenberg's criterion that closed theories properly represent a particular kind of experience in transcendental terms. Following Kant, closedness might be explicated as the *relative* universality and necessity of a theory. Universality and necessity would have to be demonstrated as holding in relation to certain presuppositions (for example, rules of access, types of phenomena) that provide a transcendental characterization of precisely that field of experience for which the closed theory is meant to represent the adequate concept. Von Weizsäcker has not, however, pursued this argument in subsequent works in detail, although he has sought to elaborate its most extreme consequence; namely, to establish the structure of physics itself from the conditions of the possibility of experience.[13]

The arguments given by Heisenberg and von Weizsäcker do not seem sufficient to justify the concept of closed theories and the claim for their continuing validity, nor do their criteria fit all their examples. Arguments which apply for the closed character of classical mechanics – the elaboration of its domain and its status as a presupposition for later theories – do not apply in the same way for quantum mechanics. The belief that the latter theory is closed as well can only be inferred from its high degree of confirmation, its compactness and the competent judgment of physicists. This seems to suggest that closedness can only be determined with hindsight, once the domain of validity and the status of a theory have been understood within the more comprehensive framework of a later theory. The strategy outlined by von Weizsäcker could be stronger here. However, the presuppositions of physics explicated by him so far only relate to the conceptualization of experience in general (decidable alternatives, principles of symmetry, temporal structure), and do not characterize, as he originally formulated it, "a particular kind of phenomena."[14]

1.3. *Systematic Objections*

Although the legitimation of the notion of closed theories cannot itself be regarded as a closed issue, the criticisms which have been directed against it do not actually attack its weak points. Instead, they merely deny the phenomenon which this concept was intended to capture.

The first argument is that offered by fallibilism. Popper concludes from the obvious impossibility of establishing the truth of general propositions from empirical data, that "we will have to get accustomed to the idea that we must not look upon science as a 'body of knowledge', but rather as a system of hypotheses; that is to say as a system of guesses and anticipations which in principle cannot be justified but with which we work as long as they stand up to tests, and of which we are never justified in saying that we know that they are 'true' or 'more or less certain' or even 'probable'."[15]

Viewed from this standpoint, adoption of the concept of closed theories is at best a naive epistemological optimism, and at worst, a dangerous form of dogmatism.[16] It contradicts the fallible character of scientific theories. Two points can be made in response to this objection.

(1) The objection itself is based on the dogmatism of fallibilism. The fact that the truth of theories cannot be determined empirically does not imply that it cannot be established at all. This would only be the case if strategies of justification other than empirical ones – and in particular transcendental

ones – were inconceivable.[17] At best Popper justifies the postulate of regarding theories as provisional as a maxim of research strategy; he does not demonstrate that these theories indeed *are* provisional. The question whether theories can be justified is open.

(2) On the other hand, we are not forced into a dogmatic belief in the possibility of a final justification of theories – that is, the sheer presumption that transcendental strategies will be successful. The concept of closed theories is intended to elucidate the phenomenon of the stability of theories through scientific revolutions. This phenomenon remains a reality which the philosophy of science has to take account of, even if we share Popper's scepticism as to the justifiability of scientific theories. In fact, most scientists would allow for the possibility of a future revision or even total supersession of the paradigmatic theories with which they operate. However, this possibility remains an abstract, philosophical proviso with few, if any, consequences for the actual practice of scientific work. In contrast to the methodological postulates of fallibilism, paradigmatic theories in science are not regarded as provisional hypotheses but as the unproblematic preconditions for further development. This assessment persists through scientific revolutions. In fact, superseded paradigms do not merely remain valid as handy instruments for constructing prognoses, which Popperians would also concede,[18] but serve as the conceptual basis for further theoretical development.[19] Science treats paradigmatic theories as closed theories. The philosophy of science has to acknowledge this fact.

Dogmatic fallibilism is an ideological defence of the belief of scientists in science as the endless frontier. Popper's systematic doubt as to the possibility of the final justification of theories is translated into the positive assertion of a "constantly increasing number of fundamental problems."[20] It is no accident that such a hydra-theory of scientific development has been formulated in an epoch in which the possibility has been glimpsed that the scientific enterprise may in fact be finite. The modern scientist is not socialized into the role of sage, as someone who knows something and is therefore useful to society, but rather as the producer of a constantly growing body of new knowledge.[21] His self-image and his social legitimation rest on the supposition that the process of the acquisition of knowledge will never be brought to a close. The positive experience of many scientists in this century that a large number of fields of possible knowledge have been completely explored – at least as far as their fundamental problems are concerned – generates a kind of *horror vacui*. Popper quite correctly notes that, "It is not his possession of knowledge of irrefutable truth that makes the man of science, but his resilient and recklessly critical *quest* for truth."[22] The possibility that this search could lead to its goal induces an ambivalence in scientists who subscribe to this view: they may prefer the absurdity of the scientific enterprise to the fact that it may be finite. "The development of science is characterized by the fact that every problem which is solved throws up new, fundamental problems."[23] "With every step we take forward, with every problem we solve, not only do we discover new and unsolved problems, but also find that those areas which we believed to represent firm ground are, in fact, treacherous and unstable."[24] Popper's theory of science and Camus' *Myth of Sisyphus* are the product of *one* epoch: doubts about the futility of human action become stylized into the heroism of the absurd.

The second argument raised against the concept of closed theories is based on Kuhn's thesis of scientific revolutions. Kuhn reconstructs the history of science as a sequence of traditions of normal science, which change as a result of scientific revolutions in which previously dominant paradigms are superseded by new variants. This historical model is then used to prove the general claim that every theory will be replaced by another.[25]

Kuhn seems to justify such a claim. In his model, the outlook for any normal science (if not the strategy) is ultimate breakdown of the dominant paradigm and replacement by a new one. Yet Kuhn's treatment of the status of superseded theories remains unsatisfactory. There are, indeed, historical examples of old theories being completely overthrown and subsequently condemned to obscurity – such as the displacement of phlogiston theory by the modern oxygen theory of combustion. However, such a total displacement of theories is not the norm – although Kuhn seems to suggest that the fate of the phlogiston theory can be generalized. In fact, far from being completely overthrown theories may, as we have seen, attain classical status. Communities of scientists continue to use these theories as paradigms, although they may cease to be located on the more spectacular theoretical fronts of scientific development. Impressed by the twentieth century revolutions in physics, Kuhn clearly concentrates his gaze on these fronts and concludes that relativity had displaced Newtonian physics, overlooking the fact that classical mechanics retains its paradigmatic significance, although – as the case-study of fluid mechanics illustrates[26] – in a much more technically oriented area of research. It is this continuing validity of theories that provides the starting-point for the notion of closed theories.[27]

1.4. *Closed Theories and the Dynamics of Theory*

The historical stability of theories captured by the notion of closed theories obliges us to consider a picture of scientific development different to those outlined by Popper and Kuhn. Popper feels that no theory is immune from the likelihood of revision, and Kuhn presents scientific development as resembling, like political history, a never-ending chain of upheavals in which accumulation of knowledge occurs only at the level of practical problem-solving. One must, however, acknowledge that in a certain sense it is, indeed, possible to speak of "true", that is, definitive knowledge. The issue is therefore one of producing a conception of the development of science in a way which properly integrates the persistence of theories.

Closed theories do not stop developing – they do not, as it were, die. Between Newton and Hamilton and Jacobi (a period extending from 1685 to the mid-nineteenth century), classical mechanics passed through a development in which its logical structure was extended and elaborated, and its mathematical form refined. Two other lines of development can be distinguished apart from these formal possibilities. The one transcends existing theories. It leads to more general theories within which previous theories can be embedded, the explanatory approaches of these theories (such as particular forces, concepts of elements, constants) can themselves be explained, and the domains, for which the theories provide reasonable approximations defined. The relation between the various theories of mechanics (Kepler, Newton, Einstein) is a case in point. The other line of development is towards the specialization of existing theories for complex applications. Scientific development in a growing number of fields no longer consists in proceeding from existing theories to more general ones but in working with them, applying them to real, complex phenomena.[28] Kuhn subsumes this dimension under the rubric of "normal science". Stegmüller, following Sneed, explicates this term as "disposition over a theory".[29]

The two dimensions of development cited here can be broadly accommodated to Weisskopf's distinction between "intensive" and "extensive" research.

> Intensive research goes for the fundamental laws, extensive research goes for the explanation of phenomena in terms of known fundamental laws . . . Solid-state physics, plasma physics, and perhaps also biology are extensive. High energy physics and a good part of nuclear physics are intensive. There is always much less intensive research going on than extensive. Once new fundamental laws are discovered, a large and ever-increasing activity begins in order to apply the discoveries to hitherto unexplained phenomena. Thus, there are two dimensions to basic research.[30]

The notion of "disposition over a theory" embraces both the application of quantum theory to solid-state properties, or of general hydrodynamics to real flows, as well as the application of physics and chemistry to the biological phenomena of heredity and sense-perception. However, the two instances embody different scientific dynamics which are not adequately elucidated in the concepts of extensive research or normal science.[31]

In the former case, the fundamental theory is applied to systems for which the theory has claimed to be valid. Real flowing media are a part of the domain of general hydrodynamics. The theory has been developed and established as a paradigm for precisely this type of phenomena. Disposition over a theory as a paradigm for an object-field means applying it to special phenomena or systems within this field. It is only in this instance that the claim holds that, "when a new experimental discovery is made, its theoretical explanation is usually found very quickly" and that the "fundamental principles are understood, and the task of research is to discover precisely how they apply to real objects or systems."[32] We term this dynamic *post-paradigmatic development.*

In the second case, the theory is applied to systems which do not fall within the domain for which the theory was developed as a paradigm. What is really at stake therefore is the attempt to establish a new paradigm for an object-field by subsuming this field under a theory of another object-field. The theory is used as a hypothesis, as a heuristic: its validity is not yet established in principle – in fact, the problem is precisely whether the domain of the theory can be extended to encompass the new object. The research programme of reductionist physics belongs within this dynamics. It strives to explain complex phenomena, for example in biology, with theories which have been validated as paradigms for physical and chemical systems.[33]

In what follows, we concentrate on the post-paradigmatic development of science. The specification of mature theories for the complex objects within their domains is the ultimate state scientific disciplines will achieve, and represents a significant "endless frontier" for scientific development.[34] Two points are of relevance:

(1) Post-paradigmatic scientific development is theoretical development. The application of general theories to the complex phenomena within the domain requires the development of special theories, and therefore represents basic research in that research area (see Section I.2 below).

(2) Research during the post-paradigmatic phase cannot fall back on criteria internal to science for guidance – for example, to determine which problem-areas or types of phenomena should be made the subject of theoretical

specialization. Once the fundamental theory of an object-field has been formulated, research can no longer rely on the aim of providing the simplest and most general explanation possible as the main guideline for further development. This guiding function can now be taken over by external goals (see Section II.1.3 below).

2. FUNDAMENTAL THEORIES, SPECIAL THEORIES, AND THE PROBLEM OF APPLYING THEORIES

2.1. *Theory-application as a Problem in the Philosophy of Science*

One aim of the finalization thesis was to describe a particular type of relationship between science and (external) goals – in which special theories are developed based on fundamental theories. Finalization is a "process through which external goals for science become the guide-lines of the development of the scientific theory itself."[35]

Attempts have been made to nullify this claim by invoking the traditional distinction between basic research and applied research, corresponding to the difference between the formation of a theory and its application. If a strict dichotomy is assumed between these two activities, then basic research will be independent of external goals by definition, and it would clearly be true, but trivial, to regard applied research as being linked with such goals. Theoretical development in the form postulated by the finalization thesis would therefore be either absurd or trivial. However, the specific claim made by the finalization thesis *is* that something does exist between pure basic research and straightforward application, that there is such a thing as oriented basic research, that there are specific fundamental principles for the application of a theory, and that the application of theories represents theoretical development.

The philosophy of science has not as yet provided a great deal of explanation as to what the application of theories actually consists of. As a consequence, it is also unclear how the difference between applied and basic research is to be conceptualized. Radnitzky even goes as far as to deny that this represents any kind of problem at all.[36] For Agassi application is simply using a theory as an algorithm; it is an "exercise in deduction".[37] Sneed uses a specialized concept of the application of a theory which derives its meaning from the broader framework of his position: applications of a theory are systems to which the laws of a theory apply. For Sneed, therefore, the problem of application relates to the demarcation of the domain for which a given theory is valid and not to the issue of how a general theory can be applied to specific phenomena – how it is possible to actually "work" with it.[38]

For Sneed, the question of the application of a theory is answerable by an existential proposition: a theory has to be shown to have a model. That is to say, a potential partial model (an observable state-of-affairs for which one can ask whether the theory applies) which can be extended by the addition of theoretical functions into a model of the theory (that is, for which one can show that the theory is valid). No full explication is made of what the existence of models means in this connection. Whereas the first formulation suggests that there must be a real system which is subsumable to the concept (which according to Sneed is the essence of the theory), the second formulation can also be interpreted as meaning that the existence of the model simply implies logical consistency, since in Sneed's view a partial, potential model only enters the language of theory through non-theoretical, i.e., measuring, functions.[39] Whichever is the case Sneed seems to hold that the application of a theory merely implies something like proof that a model for the theory does exist – in Kantian terms, proof of the objective reality of the concept of the theory. However, Sneed also clearly feels that application requires the extension of the structural core of the theory through special laws and constraints.[40] We now turn to this latter aspect of the problem of application.

By application of a theory we mean the question of how it is possible to actually operate with it. For example, how general hydrodynamics can be used to explain the phenomena emerging in the construction of a turbine, how the laws of chemistry can be used to explain the links between soil-characteristics and plant-nutrition so that an artificial fertiliser can be devised, or how the basic propositions of biochemistry can be used to understand the mechanisms of metabolic disease, and consequently develop appropriate medicines.

2.2. *What Does it Mean to Say "A Theory is Valid for an Object, but Cannot Be Applied to It"?*

We have to start from the curious phenomenon that it is possible to be convinced that a theory is valid for a particular system without being able to apply it to that system. Classical hydrodynamics, for example, existed for a long time before it could be applied to practical problems. Currently, although it is recognized that the behaviour of plasmas must conform with the laws of magnetohydrodynamics, this theory cannot yet be applied to the plasma phenomena which are relevant to the building of a fusion reactor. The general validity of quantum theory for problems of solid-state physics does not mean that it can be directly applied to surfaces or semiconductors. The attainment of a fundamental theory for an object-field does not, therefore, mean that this theory can be applied to individual objects from this field, or by extension, that the theory can be used to solve practical problems.[41]

How should this phenomenon, with which the practicioner will be quite familiar, be treated by the philosophy of science? Doesn't the validity of theory *T* for system *S* mean that *T* can be applied to *S*?

This problem crops up irrespective of whether one subscribes to the standard view of theory as a system of propositions ("statement view") or, following Sneed, treats theories as predicates, concepts ("non-statement view").[42] In the former case, the assertion that *T* is valid for *S* can mean that all propositions of *T* are true as propositions about *S*. However, it is impossible to see how that claim can be made without *T* being applicable to *S*, since the proof of validity must be obtained by introducing the boundary conditions for *S* into *T* and by deriving correct predictions for *S* after initial conditions have been inserted. That is to say, validity must be demonstrated through the application of *T*. In the second case, the validity of *T* for *S* is even defined as an application of *T* to *S*, through which *S* is shown to be a model of *T*. For this view, as well, it is therefore difficult to understand how a theory can be valid for a system but not applicable to it.

In the case of fluid mechanics, for example,[43] researchers were able to assure themselves that classical hydrodynamics was valid since the empirical systems which were of interest to them showed certain characteristic properties – namely, that the shear-stress for a homogeneous fluid flowing in parallel layers was proportional to the velocity gradient. Expressed more generally, the conviction of researchers that their media were indeed what was subsequently termed Newtonian fluids was based on three factors: firstly, they were convinced that their media could be tackled mechanically (crucial for the principle of linear momentum); secondly, they knew empirically that their systems were conservative (conservation of mass and energy), or could be isolated in such a way that neither mass nor energy was created or disappeared. Thirdly, they knew, empirically, that the above law applied for flow in parallel layers. Let us disregard the first of these factors for a moment since it simply displaces our problem to the question of the validity of mechanics in general for flowing media. The second factor seems to allow two interpretations: either the principle of conservation and incompressibility is treated as an empirical fact – in which case the conviction is similar to that in the third factor – or it is interpreted as the possibility of carrying out certain constructive operations with the media. If the latter is the case, one falls back to the empirical conviction that these operations can, in fact, be put into practice. In either instance, the conviction that the theory T (here classical hydrodynamics) holds for particular systems, S, rests on the knowledge that systems S have particular empirical properties, $e_1, e_2, \ldots, e_n$, or that certain empirical laws apply in these systems. And, as we have seen, it is precisely these empirical properties which, as basic assumptions about the system, allow the formulation of the theory, or to be more accurate, the elaboration of the very general approach of hydrodynamics into a theory capable of generating propositions applicable for special systems.

Sneed conceives of such a complementation of general theoretical frames as a logical addition of theory cores or as the extension of a core through empirical laws.[44] What is decisive is that empirical laws introduced into the theory as general assumptions are needed to develop the general conceptual frame into a practicable instrument.

However, this is not in itself sufficient to justify belief in the validity of T for S. S may, for example, possess additional empirical properties which might contradict T. What is crucial appears to be that the properties e_1 to e_n serve to define a *type* of object. The firm conviction that T holds for S is the product of this decision: these properties are the only empirical constraints on the theory T. On the other hand, the object S is solely defined through properties e_1 to e_n – additional empirical features which could also serve as a means of definition are treated as disturbances and boundary conditions.[45] The problem of application consists first in identifying those boundary conditions which determine a special object of type "T" (we are now entitled to use the name of the theory as a means of denoting the type of object). It is then a matter of finding a solution for the theory T given these particular boundary conditions. That T is valid for S, but that T cannot be applied to S, therefore means knowing on empirical grounds that S is of the object-type "T" without having a solution of T for the particular boundary conditions of S.

2.3. *Principles of Application*

The difficulty encountered in applying a theory comes from two sides: the boundary conditions and the complexity of the structure which the theory itself represents (in mathematical theories, the fact that the systems of equations are frequently unsolvable). The boundary conditions of systems selected on practical grounds are determined by what technical intentions are in mind. They may eventually allow a simplification of the mathematical structure of the theory so that solutions can be found. Frequently, however, they increase the complexity. This was the case in fluid mechanics because of the existence of boundaries to the flow, and later in aerofoil theory because of the finite length of the wing. A similar phenomenon is encountered in fusion research where technical and economic factors require particular plasma geometries.

Our case-studies show that the problem of solubility is approached from both the boundary conditions and the mathematical difficulties. Additional concepts are added to the theory, or: the theory is related to the application through a further theory that provides a concept for the behaviour of the object under certain types of boundary conditions or disturbances and allows the solution of the equations of the basic theory for these conditions. We have termed these concepts or theories "principles of application": examples are the concept of the boundary-layer in fluid-mechanics, or the band-concept in the quantum theory of solid-state matter. Viewed purely in logical terms, these principles for the application of basic theories are no more than extensions of the theory core. However, they should not be placed on the same level as the empirical laws which characterize a particular type of object. Nevertheless, there may only be ontological reasons not to carry the differentiation between types of objects and boundary conditions one step further and speak of a "Newtonian fluid flowing around a wing" as a new type of object. The actual effect of such principles is to allow the specification of the general equations of the theory or operationalization of their solution, so that they can be solved for specific boundary conditions.

How closely the situation in other fields corresponds to the analysis carried out for the theories of fluid mechanics will depend on the peculiarities of the field concerned, in particular, its degree of mathematization. Strict analogies will be found in the theories of plasma physics and solid-state physics. As far as the application of chemistry to agriculture is concerned, it could be conjectured that belief in the validity of chemistry was based more on methodological considerations rather than an empirical characterization of the object in question. Nevertheless, empirical knowledge of chemical regularities in physiology did show that physiological processes (such as plant-growth) have a "chemical side" for which it was meaningful to ask how the laws of chemistry could be applied.

It would, however, be incorrect to claim that the organic chemistry of quantitative transformations could be directly applied to agricultural processes. Such an application would have produced little more than a number of uninteresting statements about inputs and outputs, which would not have covered plant reproduction. However, "reproduction" is the very boundary condition which characterizes the systems dealt with in agricultural chemistry. It can only be included in chemical explanations after the cycle of minerals had been introduced: the concept of the cycle acts as a principle mediating the application of chemistry to agriculture.[46]

To summarize: fundamental theories cannot be applied to concrete problems without additional steps which entail the formulation of special theoretical developments. Principles for the application of theories allow fundamental theories to become differentiated into special theories. These special theories will frequently represent theories of a particular technology; they account for certain sets of boundary conditions specifiable in terms of external, technical goals.[47]

3. THEORETICAL MATURITY AS THE BASIS FOR GOAL-ORIENTATION

The finalization thesis identifies a particular type of scientific development – the orientation of a discipline to external goals through theory-formation. One of the necessary conditions for such an orientation was considered to be the theoretical maturity of the field, indicated by the formulation of basic general theories.[48] However, evidently not every case allows the demand that closed theories should exist. "Theoretical maturity" should be specified so as to imply that closed theories represent the optimal instance, but that other forms of theoretical stability are admissible.

3.1. *"Theoretical Maturity" in the Case Studies*

Since the finalization thesis has been formulated in reference to the concept of closed theories as developed for physics, its claims can best be exemplified in cases from the physical sciences. Fluid mechanics rests on a closed theory, classical hydrodynamics, which has been used for the development of technology-related special theories since the beginning of this century. Plasma-fusion research is able to build on the basis of magnetohydrodynamics and similar general theories. "In principle" these theories are valid for all plasma phenomena (just as "in principle" classical hydrodynamics embraced the behaviour of fluids); however, they cannot be immediately solved for all special cases. The boundary conditions and strategies which are determined by the project of building a fusion reactor lead to special theories.[49] Theoretical chemical engineering deals with processes implied in technical operations for which existing general theories in physics and physical chemistry can be elaborated.[50] Similar conditions might be expected to occur in the relationship between quantum theory and the individual branches of solid-state physics (e.g., the sciences of matter).

The pattern becomes less clear once examples from outside physics are considered. Agricultural chemistry could not be based on a closed theory in chemistry in the period around 1840. Structural chemistry, through which organic chemistry became a genuinely workable area, was in its infancy. And since there was not even consensus as to the theory of chemical bonding, inorganic chemistry likewise cannot be considered to have been closed at that time.

Both organic and inorganic chemistry contained unresolved fundamental problems, and both exhibited a long period of internal theoretical development. Nevertheless, it is possible to claim that a class of theoretically and empirically established laws did exist – those relating to quantitative reactions: conservation of mass, laws of simple and multiple combination and so on, which taken together could be regarded as constituting a theory of quantitative transformations. Although this theory sufficed for certain purposes, and retains its validity within these contexts, it could not be regarded as being closed. Specifically, it lacked any approach for considering the heat of reaction, and did not possess a law of mass action. The theory only sufficed to set out the empirical formulae for chemical reactions. As such it was clearly a central precondition for the application of chemistry to the problems of agriculture. Whether it was an adequate theoretical basis is another question.[51] In the case of cancer research only a few sub-fields exhibit paradigmatic theories comparable in terms of stability and compactness with the closed theories of physics. Such theories can be found in biochemistry, for example. By employing the principles provided by biochemistry for the molecular analysis of metabolic processes, it is possible to describe and explain the developmental stages of tumour cells and the transformation of carcinogens in the cell. Such explanations can become the starting point for the development of techniques for cancer therapy, such as appropriate drugs.

In contrast, no mature theories exist against which the central questions of carcinogenesis – the mechanisms by which normal cells become tumour cells, and the causes of errors of commitment – can be explored. Although molecular genetics represents a mature theory for the phenomenon of heredity and the translation of genetic programmes into cell-characteristics at the level of the *bacterial* cell (individual cells), cancer is not a problem afflicting individual cells but rather a disease of cell *groups*. The extension of the principles of bacterial genetics to higher cells is a programme of fundamental theory-formation which has not yet reached a conclusion – and which cannot therefore be oriented towards the cancer problem by developing special theories.[52]

3.2. *Criteria for Assessing the Maturity of Theories*

What the case-studies show is that it is not particularly expedient to use the stark definition of "closedness" if "finalization" is intended to embrace all those situations in which theoretical development can be seen to be goal-oriented based on a given level of theoretical achievement. We therefore propose to elaborate a concept of "mature" theory, or theoretical maturity, which takes its pattern from the concept of the closed theory, but is broader than the latter.

In this connection, it should be borne in mind that it is no easy task to establish the closedness of closed theories: two types of argument are used – the historic and the systematic. The historical arguments always have to be formulated as a reconstruction; one can show retrospectively that the advance of a theory took place through its differentiation into special theories; and one can show that the revolutionary changes in fact confirmed the validity of a superseded theory for a specific field of phenomena. As noted above, such confirmation may take place through restricting the range of the intended applications, or when the one theory provides the explanation or generalisation of the other. The systematic arguments for closedness must be based on a necessary relationship between theory and experience. One could elaborate them in a theory of constitution: a field of experience must be defined by the norms and rules under which it becomes experimentally accessible, and it then has to be proven that phenomena which can be provided under such constraints will exhibit only those features for which the conceptual apparatus of the proposed theory can be shown to be complete. Alternatively, one could show that the structure of the theory is determined by certain characteristic, but contingent, features of the object field.

What is the pragmatic significance of the concept of closed theories in assessing recent developments in science? Reconstructive arguments are sufficient to establish the objective validity of the concept (in a Kantian sense): the history of science does indeed contain instances of closed theories. This means that we can also expect to find closed theories in current science, although we may not yet be able to account for them reconstructively. The proof, however, that those theories which do not yet have their status explicated through a successor theory are indeed closed can only be obtained via systematic arguments. Since this systematic proof has not yet been provided for any single instance of a theory, one might question whether the concept of closed theories has any pragmatic, in contrast to historical, significance at all.

However, complete proof can never be demanded in a pragmatic context. Discourse with a pragmatic intent differs from theoretical discourse through the need to bring it to a conclusion so that action can proceed. The issue is not, therefore, that of complete proof but of sufficiently justifying one's assessment of a situation. We saw what such an argument might consist of in the example of fluid mechanics. Around the turn of the century, researchers in this field would have been justified in regarding classical hydrodynamics as being closed since the simplification on which the formulation of the Navier–Stokes equations was based proved empirically valid for all those fluids in which they were interested. Such arguments would be expected if a theory was considered to be closed, and hence it was thought legitimate to concentrate its further development on possible applications. One would, however, be aware that revolutionary transitions to other theories, or higher degrees of generalization may still occur.

In general three things can be said of a closed theory, even though it may not yet be possible to assess it in historical reconstruction: firstly, it will possess sufficient conceptual material to capture a particular field of phenomena; secondly, its validity will at least be proven for a number of instances; and thirdly, there are good reasons to expect that its validity extends to the whole category of phenomena in question. A concept of "theoretical maturity" must correspondingly contain, (i) a notion of the completeness of the concepts the theory provides, (ii) the existence of exemplary applications, (iii) grounds for believing that the theory in question is valid for a wider domain embracing instances of envisaged practical application. Of these, the requirement of the existence of exemplary application is the least problematic. It merely guarantees that the theory whose maturity is to be established possesses objective reality. In view of what has already been noted above, the belief in the general validity of the theory does not require a proof of general validity but must merely suffice in pragmatic terms. On the other hand, such a belief requires that the first condition of completeness be met – which is the most problematic of the three.

In view of these criteria, is it possible to regard the chemistry of Liebig's period as being "mature" enough for application to agriculture? In what sense is it possible to say that the elements of a theory of quantitative reactions which were at Liebig's disposal – the atomic hypothesis, fixed atomic weights, conservation of mass, law of simple and multiple weight proportions – were "complete"? They were evidently sufficiently complete to allow the setting out of the empirical formulae of reactions. Using this method, any chemical transformation could be analyzed in such a way so as to establish which elements were involved in the reactions, in what weight-ratios, and then, on the basis of the atomic weight and simple and multiple combining proportions to recalculate the transformed weights into equivalent weights. The conviction that this theory was also valid for processes in physiology and agriculture was to some extent a methodological belief on the part of Liebig: biological processes were reducible to chemical processes, and not vice versa. It was, however, also supported by metabolic experiments in which the burning of plants and analysis of their constituents on the one hand, and control of the environment on the other, proved the chemical character of biological processes, at least in principle.

However, the usefulness of a pragmatic characterization of theoretical maturity is quite limited. Although such a concept may suffice to establish whether it is meaningful to shift from general theories to application-related special theories, it cannot say whether it is possible or meaningful to proceed with the further development of the fundamental theories themselves.

II. *Science Policy Considerations*

The termination of the theoretical dynamic of the paradigmatic phase enables a scientific discipline to become oriented to external, social goals. The internal development of the science creates a space in which the social direction of science at the level of basic research can take place. What is the significance of this form of direction for science policy? What are its implications for the institutions of scientific research?

1. FINALIZATION AND DIRECTION

A glance at the reality of oriented research, such as in the industrial and state-military sphere, reveals that theoretical development under the guidance of external goals (finalization) is neither the only nor the predominant form of the social orientation of science. In order to direct the course of science it is not necessary to wait for theoretical maturity to be achieved: it is often impossible to wait that long. Special problems press for scientific solutions irrespective of the fronts of research and the potential for theory-building in the disciplines. Wherever the pursuit of goals, such as the provision of energy, combatting disease, or raising the productivity of labour might be hampered by the limitations of our technological capabilities, demand for science represents a possible way out.

The fact that the definition of the demand for science is not regulated by the fronts of science, but by social goals, can be seen in the objectives set for research within the US National Cancer Program. For instance, "Develop the means to cure cancer patients and to control the progress of cancer" (objective 6) is a goal of health policy formulated without any consideration for its amenability to a scientific solution.[53] The participation of scientists in the formulation of goals and associated research programmes generally tends to prevent social needs becoming translated into questions that are not scientifically workable. It may even influence the definition of needs to ensure that they conform to the research interests of scientists.[54]

External demand can be translated into scientific questions at different levels. One may address the problem to be solved directly. This always seems to be a possibility in principle: experiments can be undertaken, hypotheses proposed and tested. One may, however, also address general preconditions or fundamentals for a direct solution. One example of the former in the field of cancer research would be experimental therapy, such as radiology; and of the latter, investigations into the cell-kinetics of tumours, and studies of the biochemical characterization of cancer cells. The different levels of approach are distinguished in terms of their generality, the proximity to the theoretical dynamics of the discipline, and the specificity of their relationship to the external problem. This is reflected, for example, in the selection of experimental systems which can be worked with at different levels: human organs, animal systems, cell cultures.[55]

Which levels offer themselves depends on the structures of the science. Which levels are chosen is a matter for science-policy, and depends on evaluations of the speed and probability of reaching the goal by various strategies, and the need to find not only empirical solutions covering specific cases, but theoretical solutions which have a general validity. However, what is absolutely necessary is to link the external problem to existing science – that is, to scientificate it in methodological, if not also in theoretical terms. How this can be done hinges on the structures of the disciplines concerned.

In the following we discuss the options available for an external orientation of science at the various phases of a discipline's development. Each phase offers it own particular opportunities for orientation and higher stages of development can make use of the forms of orientation which emerged in previous phases. Developed disciplines therefore have a number of alternative possibilities of development. The significance of finalization as a theoretical strategy for science policy has to be considered in the light of these alternatives.

1.1. *Orientation of Science in the Explorative Phase*

By explorative phase we refer to the initial stage of the scientific development of a discipline, characterized by the dominance of such empirical strategies as the experimental screening of the object-field, establishment of effects, descriptive and classificatory procedures.[56]

The exploration of an object-field can be carried out subject to external goals. The "screening" of the research can be guided by pre-given aims, such as the search for possible causes for desired effects, or the classification of phenomena in accordance with their significance for external goals. This form of orientation may be termed "functionalization". Functionalization is usually understood to mean the complete and direct subordination to a given aim. In science, we define research as functionalized when it is obliged to pursue external goals using *empirical* means. In the actual practice of the direction of science, empirical strategies are frequently preferred as a means of avoiding the lengthy elaboration of explanatory theories with the internal scientific dynamic implied in such an approach.[57]

Functionalization is possible with virtually no scientific requirements. The only condition is that the phenomena in question are accessible to objective empirical procedures – that is, methods oriented to reproducible and controllable observation.[58] Historically, the scientification of an object-field which gives rise to a new discipline often begins with functionalized explorations.

One example of functionalized research in the explorative phase is research into the pathogenesis of the human system in the medicine of chronic diseases. No paradigmatic theories yet exist for the study of the "human being" as a total system with its complex mechanisms of regulation, and relationship to its environment. In clinical investigation of chronic disease, the exploration of this system is bound to the goal of optimising therapies; for example, by systematically varying the parameters "dosage", "duration" and "combination" in chemotherapy. The aim of the searching procedure, the desired effect, consists on one hand in the reduction of damaging side-effects, and on the other, increase in the therapeutic specificity of medicines. This type of clinical research can be seen as a functionalized strategy in the explorative phase of the biology of the human system.

The planning of "fall out" represents a further type of external orientation during the explorative phase. In this instance a research field which is considered relevant for external purposes is picked out for exploration. For example, a fresh water ecosystem can be equally well investigated in waters which are either relevant or irrelevant as far as environmental protection is concerned. However, choice of the former enables results (effects, data) to be produced which may be of direct use for external aims.

1.2. *Orientation of Science in the Paradigmatic Phase*

The dynamics of theoretical development during the paradigmatic phase of a discipline do not allow an orientation to external goals: theoretical development pursues an internal research programme, which is directed towards the elaboration and confirmation of the fundamental explanatory theories of the object-field. This programme generates the questions for research and indicates the relevant experimental systems according to internal criteria of science. The science cannot be subject to orientation at the level of the research fronts defined by such a programme. There is no way of allocating the unresolved problems of the theoretical programme to external problems, or of deciding the degree to which the possible strategies for paradigmatic research might contribute to theoretically solving the problems. Only when a paradigmatic theory has been achieved is it possible to define the theoretical deficits specific to an external problem.

The theoretical dynamic of the paradigmatic phase is basic research, relevant to external problems by virtue of the fact that it creates the potential for new technological developments but is not itself basic research oriented toward specific external problems.

Science policy can only promote, not direct, such research. This can be seen in the treatment of cell-biology in the US National Cancer Program. The understanding of the differentiation and stability of cells represents one of the fundamental fronts of modern biology, and is recognisably a precondition for a causal understanding of carcinogenesis. The planning of cancer research in this field, however, consists to a considerable extent in listing the basic questions of cell-biology.[59] The definition of and approach to these questions are not a function of strategic planning aimed at the goal of combatting cancer, but rather of the internal dynamic of the discipline. The rhetoric of direction employed in the NCP is misleading on this issue: cell-biology is not a type of cancer-specific basic research. The external problem lacks selectivity compared with the internal dynamic of the discipline. Direction becomes promotion.

Under such conditions, the only possible reference to external problems is through transfer research – with the exception of the possibility of carrying out functionalized exploration with the complex cancer systems themselves. Transfer research is a form of orientation which does not impinge on the research fronts of the paradigmatic disciplinary development, but is rather subordinate to them. It takes up the results formulated for simple model systems in basic research, and adapts them to the more complex circumstances of external problem areas. Much of what is customarily regarded as "applied research" or "technical development" falls into this category of orientation: examples would be the systematic scaling-up of laboratory experiments into industrial processes or the development of a metabolic effect tested with animals into a medicament which can be given to humans.

Transfer research is not dependent on the completion of the theoretical dynamic of the paradigmatic phase. It transfers the results of basic research (elements of theory, experimental effects, methods) as soon as they become available. The formation of theory and the application of theory in the paradigmatic phase are two concurrent, but separate lines of development. Their synthesis is the specific form of the orientation of science in the post-paradigmatic phase.

1.3. *Orientation of Science in the Post-paradigmatic Phase*

External orientation at the level of theory-development becomes possible once theoretical maturity has been achieved. The research fronts now lie in the application of the general concepts of the discipline to more complex systems. This "application" cannot be compared with the transfer of the results of basic research in the paradigmatic phase, nor is it a mere "exercise in deduction".[60] We observed above that fundamental theories cannot be applied to the complex systems for which they are valid in the same way that the laws of gravitation can be used to calculate the orbits of satellites.

The application of general theories is a process involving the development of special theories. In contrast to the paradigmatic phase, however, the direction of theoretical development in the post-paradigmatic phase is not indicated by internal regulatives, but can be determined by external goals. In the following we show that mature theories can form the basis on which it is possible to allocate specific theoretical fronts of a discipline to external problems and how the external goal acts to regulate the development of theory.

The first of these points can be illustrated by looking at the relationship between the relevant external goal and theoretical basic research, on the one hand in cancer research and on the other in fusion research. In both cases there is agreement that advances in basic research are the prerequisite of any breakthrough in the solution to the external problem: advances in cell-biology for cancer research, and in plasma physics for controlled fusion. However, since cell-biology has not yet attained the post-paradigmatic phase, it is impossible to define a specific cancer front in the discipline – as we saw previously. Consequently, the planning proposal of the 1974 National Cancer Program "Strategic Plan" in practice merely amounts to the promotion, and not direction of basic research.[61]

The position is quite different in the case of the relationship between plasma physics and fusion research.[62] The fundamental principles of plasma physics had already been established when the problem of the fusion reactor was first posed. The fundamentals were laid down by classical mechanics, electrodynamics, and more specific special theories such as magnetohydrodynamics. This enabled the specific phenomena and problems in plasma physics to be allotted to the technical problems defined by the external goal. This goal demanded: controlled fusion, a positive energy balance and continuous discharge; fusion was not to occur in an explosive instant, and was not to consume more energy than it released. The corresponding scientific problems in plasma physics are, amongst others, containment of the plasma, heating, conditions for the stability of plasmas at high temperatures, and more specifically, stability in toroidal configurations which are regarded as necessary for a positive energy balance. It is not crucial that these specific scientific strategies will eventually lead to a solution of the problem; what is crucial is that the explanation of particular plasma phenomena which are central to the aim of nuclear fusion can be defined as a research front of plasma physics.

The *Physics in Perspective* report cited "controlled fusion" as one programme element of basic research in plasma physics.

The understanding of these phenomena and work towards the solution of the applied goal represents the most advanced application and understanding of plasmas and of the physics of plasmas.[63]

In order to demonstrate that fusion-oriented plasma physics involves *theory* development prompted by external goals, we will analyze how general theories are progressively specialized in their application to the problem of controlled fusion.

The most general theory of plasma phenomena is contained in the *general equations of motion* of classical mechanics, electrodynamics and hydrodynamics. These equations in turn represent concretizations of even more general principles: principles of conservation of certain variables (energy, momentum, mass) and of symmetry for specific systems, e.g., matter in a defined physical state. "In principle" all plasma phenomena are describable through the motion of any particle in the field of other particles. However, the theory cannot be applied. In real systems it is impossible either to determine the intial conditions, position and velocity of the particles, or to solve the number of equations involved (10^{20}). The fundamentals require reformulation before they can be applied to any concrete object-area. Heuristic models are developed which cover specific types of phenomena, and *new special equations of motion* are derived (magnetohydroynamic equations, multi-fluid equations). This process of specialization can be externally influenced. The goal of controlled fusion defines specific phenomena as the relevant object field, and indicates the conditions which have to be considered when theories are reformulated; for example, the condition found only in the non-natural circumstances of a fusion reactor that the plasma must have a toroidal configuration.

These new systems of equations themselves are also frequently not solvable, and cannot yet be used to predict or explain specific phenomena which lie within their domain. *Assumptions and methods of approximation* (models, methods of calculation) therefore have to be developed which allow the special equations of motion to be solved for the areas in which they are to be applied.

It is only below this level of theoretical specialization that the application of a theory becomes an antithesis to the development of a theory. Only then may the application of a theory take the form that given initial conditions are inserted into the theory to calculate the consequences – or, conversely, that the theory can be used to identify the initial conditions that have to be realized in order to achieve a given goal. However, external goals already regulate the process of theoretical specialization by defining research problems and stimulating the production of conceptual extensions of the general theories for which there is no internal scientific necessity.

The problem of controlled fusion defines a research front in plasma physics and establishes a direction of development which would have no compelling necessity according to an internal programme of the field as defined, for instance, in purely academic plasma physics. On the contrary, such a direction would be highly improbable in view of the extreme complexity of the problems generated. The discovery of new effects in fusion physics such as fusion and the instabilities of plasmas has, however, renewed the dynamics of the general theoretical fronts of plasma physics.[64]

The external orientation of the development of post-paradigmatic scientific fields should not be regarded as "applied research" if this is seen as representing an antithesis to theory-formation and "basic research". Far from being a contradiction, in post-paradigmatic science the application of theory implies the progressive specialization of theory. Each special theory has a dual face: it will be an application as far as its predecessors are concerned, and fundamental in relation to its successors. Although the specializations are of diminishing scientific significance, any clear dividing-line between "basic" and "applied" science would be arbitrary in this context. Any specialization of theory is basic research in comparison to the outright application of a theory and its apparatus of calculation for a given set of initial conditions. On the other hand, "application" in the sense of orientation to external goals is already possible within the process of theoretical specialization. Such a process can justly be termed "oriented basic research".[65]

We formulated the phase-model of the development of research fields taking theory-formation in sub-fields of the classical scientific disciplines as examples (mechanics, organic chemistry or genetics). Theories from these fields are basic for theoretical concepts of more specific objects, since their own objects correspond to relatively general properties in nature.

However, the model does not have to be confined to the development of theories dealing with general objects. The application of general theories is itself a process of theory-formation, and it also exhibits the features of a paradigmatic phase – irrespective of whether the application takes place subject to external goal-settings. The emphasis on theory tends to lead to internal problem-generation. For example, the formulation of the specific concepts of agricultural chemistry can be seen as the paradigmatization of the traditional discipline of husbandry; similarly, the boundary-layer theory represents a paradigmatization of technical hydraulics. From the perspective of the general theories, the search for these paradigms represents oriented specialization (finalization); from the perspective of the technical problem-area, it represents a theorizing of areas previously tackled using empirical-explorative methods.

This section has presented a number of types of the orientation of science which represent various options for the translation of external goals into scientific strategies. These options are conditioned by the stages of development of the disciplines concerned. They differ according to the specificity of the relation to the external problem, and the coincidence or competition with internally defined research fronts in the discipline. This implies differences for the relative prestige of oriented research, the resistance of disciplinary scientists to external direction, and the suitable forms for this direction (for instance, funding through contracts or grants). We do not deal here with the consequent problems for the direction of science.[66]

2. THEORETICAL VERSUS EMPIRICAL STRATEGIES FOR THE SOLUTION OF PROBLEMS

Finalization is only one of several possible forms in which the orientation of science may take place. Science policy also has the option of pursuing an external goal with empirical – rather than theoretical – means, using functionalized research. Is finalization, that is, theory development, the superior strategy? Two reasons suggest that it is: firstly, it is superior to empirical methods as a technique for solving problems; secondly, it not only offers means for achieving goals, but also for subjecting goals to criticism.

2.1. *The Technical Potential of Theory*

Empirical procedures yield singular solutions which are valid under particular, empirical circumstances. Prognoses for other situations can only be obtained by extrapolation. This is unreliable when a departure from the original boundary conditions is desired. The problems that arise when laboratory effects are translated into large-scale technical processes are one example. In chemical engineering, for instance, processes in chemical reactors can be represented in mathematical models. The models are phenomenological descriptions: although they frequently yield reasonable prediction for parameters within the range of data that served to build the model, they usually fail when extrapolation to different orders of magnitude is required. The scale-up of the tasks of chemical engineering therefore remains dependent on trial-and-error procedures. In order to anticipate the non-linear changes that may occur in such processes and to predict the limits of current techniques, theoretical explanation is needed. Only causal theory can provide fundamental solutions that are valid for varying sets of conditions. By generating an understanding of the mechanisms which apply in the problem area, it allows the strategic construction of solutions. It produces structural technical innovations on which it would be hard to hit merely through empirical scrutiny. And it offers opportunities for recognising and controlling the side-effects of techniques. The theoretical solution to a problem is therefore in principle more general, more certain and more efficient than the empirical solution.

Thomas characterizes it as "high technology" and distinguishes it from the "half-way technology" of empirical methods.[67]

Taking the degree of scientific systematisation of the problem area as the benchmark, Thomas distinguishes various levels of medical technology. The extensive field where the doctor cares for the patient without being able to intervene into the disease in a systematically controlled way is designated "non-technology". Techniques such as organ-transplantation or radio-therapy are termed "half-way technology". They are based on functionalized explorative strategies rather than a causal understanding of how diseases arise: they represent,

> efforts to compensate for the incapacitating effects of certain diseases whose course we are unable to do very much about. It is a technology designed to make up for disease, or to postpone death.[68]

The term "high technology" is reserved for those methods which tackle the causes of disease based on theoretical explanations of disease mechanisms, and which aim at a reversion or prevention of damage to the patient. Thomas cites as an example the fight against infantile paralysis through immunisation. This has become possible through the virological explanation of the cause of the disease and the breeding of mutant viruses which can be applied as an anti-serum.

In post-paradigmatic scientific development, the application of disciplinary theory to complex systems in the field can be achieved, and hence the transition from empirical to theoretical strategies for technical development becomes feasible. One case in point is the construction of drugs for combatting infectious diseases mentioned above which was made possible by the application of the biochemistry of the bacterial cell. The application of solid-state physics is also expected to produce a revolution in materials' technology.

Physicists are just beginning to acquire the capability of designing a material for a particular purpose or of precisely predicting properties of a new material. Further development of theoretical techniques ... hold the promise of quantitative materials' design and engineering rather than rationalization after the fact.[69]

Finalized science and the development of "high technology" converge.[70] For science policy, the external orientation of theoretical development has therefore to be considered as the technically superior form in which to deal with problems. Nevertheless, theory formation is not always a necessary condition for solving problems. Epidemic diseases, for example, have been successfully dealt with using empirical procedures such as the draining of swamps, or vaccination with dead bacteria – the latter without an understanding of the mechanisms of the immune response.[71]

2.2. *Orientation Towards Goals and Criticism of Goals*

The reality of the direction of science is not one of excessive theory development in relation to external goals but of a preponderance of empirical strategies, of the increasing functionalization of research. The theoretical treatment of complex problems on the one hand is often difficult and exacting; on the other hand, the scope of empirical methods has been steadily extended through the developments of the sciences – the modern tools of mathematics and data-processing, and existing theoretical explanations which allow more directed, and hence more promising, trial-and-error strategies. However, the most active forces promoting the tendency towards the undistracted instrumentalization of science for the solution of narrowly defined tasks are short-term economic considerations and political opportunism.

Functionalization is not only technically inferior to finalization as a means for attaining objectives; it is also politically inferior as a means for criticizing objectives. Science not only represents a potential for producing the means for reaching goals, but also provides arguments for criticizing these goals. The specification of the social goals towards which science is to be oriented rests on interests which, for their part, are a function of existing knowledge. For example, the decision to develop the fusion reactor is rooted in the interest of access to continuously available, national energy sources. It is informed by existing knowledge as to the preconditions and consequences of the desired goal, the assessment of possible energy sources, and energy demands. There are two ways to criticise the goal: doubt can be cast on whether the underlying interest is legitimate. And doubt can be cast on whether the empirical knowledge upon which the interest relies when operationalizing the goal is appropriate.

In the present context, our point is that even legitimate goals and interests are conditional upon knowledge and can be put in question through the growth of empirical knowledge. Within the framework of functionalized empirical science, however, it is not very likely that criticism of goals will emerge from research. The objectives stated determine very strictly the questions which may be scientifically asked. Moreover, functionalized research is usually practised in institutional contexts (for instance in contract research) in which all forms of non-functional investigation are suppressed precisely in order to protect the goals from scientific criticism and delegitimation.[72]

In contrast, finalization favours the development of the critical potential of science. The desire to address problems through theoretical practice presupposes some institutional freedom of action and relief from immediate practical pressures. It cannot tolerate the type of prescription of aims found in industrial contract research. It implies a comprehensive analysis of all facets of the problem, the search for causes, and general explanations of the problem field. Such an approach raises the likelihood not only of providing instrumental knowledge for pre-stated, positive goals, but also of changing the knowledge which underpins the definition of these goals. Chosen goals may become unattractive or unacceptable in view of the technical, economic, ecological or social consequences of their realization. The theory may lead to the conclusion that the chosen objective is unable to satisfy the interests upon which its specification rests, or is less able to satisfy them than other alternatives.

The case-study of agricultrual chemistry provides an example of theory's capacity to mount a critique of goals. Traditional doctrines of agriculture had set the goal of maximising the yield of the soil – defined in terms of the individual farmer's income. The theoretical analysis of the problem of increasing yields delegitimized this goal. By explaining the basis of soil-fertility and the conditions under which it could be reproduced, the theory revealed that previous methods for raising yields in fact implied the depradation of the soil and would eventually lead to the collapse of agricultural production. Increasing the efficiency of existing procedures of exploitation could not, therefore, be an acceptable objective in either technical or normative terms.

An expansion of knowledge forcing a revision of goals and opening up new possibilities for action may also emerge from areas other than oriented science. However, among the forms of oriented science, theoretical development for external goals (finalization) represents a particular potential for critical reflection on its objectives.

3. THE SOCIAL INTEGRATION OF SCIENCE

3.1. *Finalization and Autonomy*

Finalization is a link between theoretical and social goal-setting. This entails some revision of the separation of the categories of science and society, as implied by the notion of an autonomous knowledge of nature pursuing its own internal logic. Such revision does not, of course, mean that practical judgments can substitute for truth-criteria. The role of external goals as guidelines for theoretical development is not to replace objective validity with social desirability, but rather to replace internal scientific criteria of relevance by social criteria.

The object-field and problems of research are determined by pragmatic interests in new technology rather than by philosophical interests in the theory of nature. The ideal of maximum attainable precision gives way to demands for reasonable precision in view of the practical purpose. Instead of maximising the simplicity and elegance of mathematical formulations, finalization requires formulations of results which are pragmatically operational.

The distinction between criteria of truth and criteria of relevance becomes difficult to sustain as soon as we begin to look at the conditions under which scientisits have historically allowed an explanation, an assertion, or an hypothesis to "count". Such a decision implies valuations of heuristic power, conformity with explanatory ideals, simplicity and so on. Consonance with these internal scientific values can, in analytical terms, be seen as a condition of relevance of scientific knowledge, as an indicator of truth, which is additional to truth-conditions (such as a correct prognosis, for example), but which does not replace them.

"Finalization" implies a retreat from the concept of autonomy envisaged in the idea of a self-regulating "republic of science". Nevertheless, the model of scientific development which we have suggested does legitimate the claim for the institutional autonomy of science in certain areas.

For the scientific community, autonomy does not figure simply as a liberal right to ward off state intervention in individual freedom: scientists claim full self-management, and at the same time full social financing. Such a claim goes beyond the autonomy of work usually granted to doctors, or even mechanics. These too fend off intervention into the technical knowledge over which the occupational group has control. However, the object of their activities and the supply of resources are regulated through exchange relations, and therefore in the final analysis by particular client demands.

A number of reasons are advanced to justify the particular claim to autonomy made by scientists; for example, that there is generalized social interest in new knowledge in general; or that the production of new knowledge is a creative activity which cannot, by definition, be planned. Above all it is asserted that the content and direction of scientific development are not a matter of the subjective judgement of scientists, however gifted they may be, but follow their own compelling logic. And it is assumed that the social system of science possesses norms and contents which act to ensure that scientists will in fact deal with those questions suggested as important and relevant by that logic.[73] Given the last two conditions, social financing of science is sufficient to guarantee the satisfaction of the social interest in science. Regulative interventions are not only superfluous, but also damaging.

If the demand for autonomy is related to our model of the development of scientific disciplines, autonomy emerges as an institutional model tailored for research fields during the paradigmatic phase. Assuming that the development of theories is a social goal at all, this goal can only be pursued in line with the internal conditions of science itself. Incompatibility with orientation to external goals and hence institutional autonomy are the consequence of the cognitive autonomy of scientific development in this phase. As we saw above, this does not exclude appointing scientists from these research fields to problems which can be dealt with through transfer research or functional empirical strategies. Since, however, such oriented research remains separated from the real dynamics of the discipline it is regarded as "second-rate" research, and it is difficult to recruit scientists who have the opportunity to work on the current research front of the discipline.[74]

The particular structure of science dynamics which legitimates autonomy does not characterize every science from the beginning, and nor does it remain constant throughout the process of scientific development. In their early phases many disciplines are indistinguishably intermeshed with technological areas. Many of the origins of chemical exploration can be traced back to the laboratories of eighteenth and nineteenth century pharmacists. Metallurgical workshops provided the starting point for the materials sciences. Internal criteria of relevance which mark out research priorities according to purely scientific interests only emerge once disciplinary development has got underway, for example, by suggesting research techniques and experimental systems which are both easily accessible and produce quick results. The consequence is the development of an internal dynamic which differentiates exploration performed for theoretical interests from technical experimentalism, directed at external goals.

This internal dynamic reaches its apogee during the paradigmatic phase. The autonomy of theoretical development in the paradigmatic phase is, however, subject to a dialectic: it is dismantled through the very success of the science itself. In the post-paradigmatic phase, internal regulatives of science become compatible with external ones. The greater the number of stable theories obtained, the greater the potential for an orientation of further theoretical development to external goals. Cognitive autonomy, in the sense of a compelling internal logic of development, and the consequent incompatibility with external orientation, is more the characteristic of an immature science than of a mature one.

The change in science dynamics in the post-paradigmatic phase of development undermines the legitimation of science's claim to institutional autonomy and changes the function of the existing freedoms of research. As a rule, the "republic of science" does not experience any difficulties in generating further problems which allow research to be continued for fields whose internal programme has already been completed. Occasionally, scientists simply continue the previous research programmes, producing more of the same with a few minor variations. The selection of problems follows methodological or instrumental fashions. It is here that subjective preferences and professional interests begin to guide research without being controlled by any clear internal regulatives. Usually, however, processes of external orientation begin to make themselves felt even in the sphere of academic research. Research fields are developed which are held to be of practical relevance in the judgement of the "scientific community", or in line with its perceptions of social need.[75]

Even when it is not enforced through the cognitive structure of the science, autonomy can be a thoroughly sensible form of social organization since it allows innovation and improvisation free of immediate practical demands, and forms a counterweight to the subjection of research to the particularistic interests of ruling political and social groups. The demand for autonomy may, therefore, be justified on political grounds – which would, of course, hold good not only for post-paradigmatic basic research, but also for technological development. However, it cannot call upon the immanent logic of scientific development: there is no compelling reason why the selection of the direction of research in mature sciences should be left to scientists. Our analysis sought to show that with respect to mature disciplines, the social claim to direct the innovative activity of science in accordance with social needs is both legitimate and feasible.

In actual practice, the development of post-paradigmatic science is widely externally steered, since much of it is carried out in research environments determined by industry – in company laboratories or state programmes for the promotion of growth-technologies or arms production. Objectives of research are subordinated to the imperatives of the economic system, and only contingently relate to social goals defined in terms of the general interest. Nevertheless, in the public debate on the direction of science the influence of the economy (when raised at all) is often depicted as being unproblematic and "appropriate" – not only by economic interests with a manifest direct stake in the matter, but also by representatives of the "scientific community". Antagonism is normally directed at those tendencies pressing for a *political* planning of scientific development – on the one hand, because of the implied demand for the democratic legitimation of goals, and on the other, because of the supposed dangers of bureaucratic control over research.

3.2. *Disciplinary Communities and "Task Communities"*

The emergence of finalized fields of research implies a departure from the traditional structures of disciplinary scientific development not only in cognitive, but also in institutional terms. Scientific communities arise which have ceased to be centred around disciplinary research programmes, and which instead focus on problem areas. The technical sciences exhibit a number of forerunners of such task communities. Although fields such as mechanical engineering, energy technology, the chemistry of synthetic materials and so on are also organised within universities, the prevailing mechanisms for finance, the involvement of technical scientists in expert roles, the orientation of the professional associations (such as the *Verein Deutscher Ingenieure*) and the function of research in vocational training nevertheless ensure a constant interaction with industry and the state as the major clients formulating demand for technology; this, naturally, militates against an exclusively academic orientation of the technical sciences.[76] A similar organization can be discerned in the natural sciences in cancer research. Cancer research is an aggregate of disciplinary strategies (biochemistry, genetics, cell-biology, radiology, clinical research) with diverging approaches, orientations and ideologies.[77] Most "cancer researchers" assign themselves to disciplinary "scientific communities" with specific criteria for relevance and reputation of work, and with career patterns for which research on the cancer problem remains somewhat incidental. Nevertheless, there are forms of community formation in cancer research, conferences, journals, scientific societies which indicate the beginnings of the institutionalization of a scientific specialty.

One precondition for the emergence of such problem-oriented scientific communities is without doubt the appropriate organization of the problem-solving potentials through science policy. This does not necessarily have to take the form of comprehensive planning, as the in the American cancer programme, nor express itself in the establishment of big science institutions, as in the early German cancer research policy. It is probably sufficient that the problem is officially formulated in the political arena and that a funding umbrella is provided to bring together the scientists concerned.[78]

The flow of communication that results from the problem perspective shared by the different scientists constitutes a particular form of scientific public. For those involved, it creates additional reference figures outside the academic culture of their own disciplines. It thus stabilises the problem-orientation of the work. It confers reputation on research because of its social relevance. In a certain certain sense this community superintends the interdisciplinary character of problem-oriented research. In the field of cancer research, for instance, unlike technical industrial projects, the various disciplinary approaches are not coordinated within a division of labour; they run parallel to each other, and the transfer of results between them, the discussion of their respective relevance, and the combination of strategies have to be organised. Inasmuch as this organisation is not achieved through administrative means, it requires a non-disciplinary scientific community in which these problems can be articulated and dealt with. Finally, such a task-community can be expected to produce critical evaluations of the goals of orientation in the light of ongoing research. At the least, it provides a forum for rendering conflicts over the relevance of specific goals and strategies to the general problem visible. In the field of cancer research, for example, this can be seen in the discussion of the role of preventitive versus curative strategies, the significance of basic research versus clinical research, and organization within integrated cancer centres.

The need for suitable social structures able to stabilise the relation between oriented research and the real pragmatic issues implied in the external goal is particularly clear in finalized research fields. Although "finalization" represents an external orientation of scientific development from the standpoint of the basic disciplines, seen from the standpoint of pragmatic action it represents the transition to theoretical strategies which are relatively distant from the requirements of practice. If pursued through theory-formation, oriented research is exposed to the risk of subsumption within a specific disciplinary matrix, and the rules which this matrix sets out for selecting questions that may be dealt with in the theoretical tradition of the field. The problem-area will then become the domain of a disciplinary community which develops it in accordance with existing academic regulatives.

This danger has been claimed to exist in the case of chemical engineering science in Germany. This field emerged as a theorization (at the level of the molecular concepts of physics/chemistry) of phenomena encountered in industrial chemical production. However, its development is not guided by the needs of practice but by criteria from the academic culture of science (for example, using the most sophisticated mathematical methods, constructing simple model systems).[79] Another instance of the academicization of problem-oriented research is that of fusion research in Germany. The transition from research into relatively general questions in plasma physics to large-scale reactor-specific experiments which are less "interesting" in academic terms could only be carried out through administrative pressure exercized by the promoting political institutions.[80]

The emergence and functioning of task communities in the transition area between the traditional disciplines of the natural sciences and technological fields have scarcely been investigated. It is clear, however, that they point to a new social organisation of science in which research is tied neither to the regulatives of a disciplinary matrix nor to specific political or industrial projects, but is able to investigate the implications of a pre-given problem in relative independence.

The pursuit of social goals through the orientation of *theory* is a means of avoiding the instrumentalization of science. The institutionalization of task communities is a means of preventing the academicization of oriented research.

NOTES

1 See Böhme *et al.* (1976) and Böhme (1980).

2 Hund (1972), pp. 394 and 390. *Cf.* Weisskopf (1974), p. 482: "We therefore now understand the essence of molecules, marcromolecules, solid bodies, gases and liquids".

3 Stent (1969), p. 66. Today this assessment is confined to the molecular genetics of the single (bacterial) cell; see Hohlfeld, 'Cancer Research' in this volume.

4 "Intended application" is a term used by Sneed (*cf.* Stegmüller, 1973). The domain of the intended application of a theory is the – relatively unspecific – area of phenomena for which a particular theory is supposed to be valid. Originally the domain of very high speeds would have numbered among the intended applications of Newtonian mechanics, for example.

[5] Truesdell (1976), pp. 127 f.
[6] Kuhn (1962).
[7] Andersson (1976*b*), p. 69.
[8] Heisenberg (1971), pp. 87–94.
[9] Heisenberg (1971), p. 93.
[10] Heisenberg (1971), p. 91. Heisenberg's references are not clear. On the one hand the areas listed are designated as disciplines, and on the other as theories. It is understandable, although not necessary, for him to take Maxwell's theory and special relativity together; however, his formulation does not make it clear whether he only considers classical thermodynamics plus statistical mechanics as a closed theory. It is also not clear why he adds atomic physics and chemistry to quantum mechanics, since these ought to be treated more as spheres of intended application of the closed theory of quantum mechanics. Von Weizsäcker's formulation is more precise: his examples of closed theories are classical mechanics, classical electrodynamics, special relativity and quantum mechanics, (von Weizsäcker, 1980 pp. 155–156).
[11] Bohr (1931).
[12] Von Weizsäcker (1980), pp. 156 f. He defines a closed theory as a theory, "that cannot be improved upon by means of small changes".
[13] Von Weizsäcker (1980), Section II.3, see also (1979).
[14] Von Weizsäcker (1980), p. 156.
[15] Popper (1972*a*), p. 317.
[16] For Popper, Kuhn's normal scientist, "is a person one ought to be sorry for", he has been taught in a "dogmatic spirit" and is a "prisoner" of a theoretical framework (Popper, 1970, pp. 52 and 56).
[17] We return below to why we feel that strategies of justification related to transcendentalism are a promising approach. This involves arguments which bring theory-formation and empirical basis together in a constitutive relation (see Section I.2.2.).
[18] Radnitzky (1974*a*), pp. 42 f.
[19] Thus, it is classical mechanics, not quantum mechanics, which constitutes the foundation for the theories of plasma physics and fluid mechanics.
[20] Andersson (1976*b*), p. 72.
[21] Znaniecki (1968).
[22] Popper (1972*a*), p. 281.
[23] Andersson (1976*b*), p. 71.
[24] Popper (1971), p. 103.
[25] *Cf*. Andersson (1976*a*).
[26] See Böhme, 'Autonomization and Finalization' in this volume.
[27] The phenomenon of the stability of theories (see section I.1.2 above) consequently necessitates a number of corrections to Kuhn's model of the development of science. See Stegmüller (1973), pp. 244 ff. for the necessary corrections as far as Kuhn's notion of paradigm change is concerned.
[28] Popper's perception of science as "proceeding to theories of ever richer content, of a ever higher degree of universality, and of a higher degree of precision" (1972*b*, p. 193) is an illustration of his misapprehension of this dimension of development.
[29] Stegmüller (1973), pp. 218 ff.
[30] Weisskopf (1965), p. 24.
[31] The problematic nature of this distinction can be seen in the difficulties encountered

in fitting molecular biology into the schema of intensive/extensive research. Whereas Weisskopf would "perhaps" put it under extensive research (1965, p. 24), Brooks treats it as intensive research, in company with high-energy physics (1967, p. 26).

[32] This is how Brooks characterizes extensive research (1967, p. 26). The thesis of the finalization approach is that the discovery of the "how" of application is a relevant part of the theoretical development of science.

[33] Stegmüller analogously distinguishes two forms of progress within the framework of "normal science", that is, retaining one and the same theory: firstly, the discovery of new special laws (in Sneedian terms, core-extensions), which enable increasingly substantive statements to be made about the theory's sphere of application; and secondly, the discovery of new elements in the theory's field of application (1973, p. 214).

[34] It does not seem plausible to share Andersson's contention that the development of fundamental theories themselves and their embedding in more general theories is a research front in which a "constantly growing number of fundamental problems" will be generated (1976*b*, p. 72). The number of problems actually tends to fall in this direction of development. This is a product of the fusion of theories and disciplines, where the search is for fewer, but more general principles. According to Brooks (1967, p. 26) scientists on this front, "seek to ask and to answer a very small number of very fundamental questions."

[35] Böhme, van den Daele and Krohn (1976), p. 307.

[36] Radnitzky (1974*b*), p. 802.

[37] Agassi (1966), p. 348. One could similarly extend the Hempel-Oppenheim schema for explanation into a concept of theory-application. Accordingly application = deduction, namely of facts or empirical laws from general laws using antecedent conditions.

[38] Sneed (1971); Stegmüller (1973), pp. 130 ff.

[39] *Cf.* Stegmüller (1973), pp. 123 f.

[40] *Cf.* Stegmüller (1973), pp. 130 and 192.

[41] For hydrodynamics, see Böhme, 'Autonomization and Finalization' in this volume; for plasma physics see Küppers (1979); for the application of quantum physics see *Physics in Perspective*, Vol. I, pp. 336 f.

[42] See Stegmüller (1973), pp. 120 ff. for a fuller discussion of this distinction.

[43] See Böhme, 'Autonomization and Finalization' in this volume.

[44] *Cf.* Stegmüller (1973), pp. 122 ff. and 135.

[45] This is a quasi-transcendental argument which provides the necessary link, noted above, between empirical access and theory-formation. It is only "quasi" transcendental because it rests on the validity of certain empirical laws.

[46] See Krohn and Schäfer, 'Agricultural Chemistry' in this volume.

[47] This external reference is taken up in section II.3 below.

[48] Theoretical maturity is certainly not a sufficient condition for orientation of a discipline to external goals. Tried and tested instruments, developed experimental techniques and mathematical procedures are frequently just as crucial as prerequisites. It is occasionally regarded as questionable whether the theoretical maturity of a field is the necessary precondition for oriented theory-formation at all. Symes (1976) doubts this, for example, by reference to the fact that in the case of agricultural chemistry a special theory was developed on the basis of relatively meagre general chemical theory-formation.

[49] *Cf.* Küppers (1979).

[50] Buchholz (1979*a*).

[51] If the construction of a fertilizer is regarded as the central task of agricultural chemistry, then the most important question is evidently that of the chemical specification of the materials which had to be substituted within a cycle. However, questions of the dosage to be applied cannot be answered without a law of mass action.

[52] See Hohlfeld, 'Cancer research' in this volume.

[53] See NCP (1974), Digest VI.

[54] See van den Daele, Krohn and Weingart (1977) on the role of scientists in the formulation of science policy programmes.

[55] See Hohlfeld, 'Cancer Research' in this volume.

[56] On the relation between the explorative phase and "pre-paradigmatic" science in Kuhn's phase-model see Böhme and van den Daele (1977).

[57] We have also used the term functionalization to denote that tendency to raise strategic accessibility instead of causal explanation into the aim of science: *cf.* Böhme, van den Daele and Krohn (1976), p. 319.

[58] Of course, not all objects of nature meet this requirement. The complex phenomena of turbulent flow are still not readily accessible to scientific study, which accounts for the fact that combustion technology (construction of chimneys and furnaces) has scarcely got beyond a handicraft skill. *Cf.* Küppers (1978).

[59] See NCP (1974), Digest III.

[60] Agassi (1966), p. 348.

[61] "The very foundation of cell-biology, molecular biology and immunology must be strengthened" (NCP, 1974, Digest III-3).

[62] We follow Küppers' approach here (1979).

[63] *Physics in Perspective*, Vol. I, p. 359.

[64] *Cf.* Küppers (1979).

[65] Further examples of this form of orientation are: theoretical chemical engineering as a research front of physics and chemistry (*cf.* Buchholz, 1979*b*), the theory of powered-flight as a research front of fluid mechanics (*cf.* Böhme in this volume), the theory of alloys and polymers as research fronts in solid-state physics and chemistry. Anti-metabolite research represents an example of a research front in biology induced through external goals. By using the biochemistry of the metabolism of bacterial cells it is possible to formulate a principle of drug construction: the development of metabolic building blocks which structurally resemble natural metabolites, and the displacement of the latter in the reaction with the enzymes of the bacteria (see Handler, 1970, p. 636).

[66] See van den Daele, Krohn and Weingart (1977).

[67] Thomas (1972).

[68] Thomas (1972), p. 12.

[69] *Physics in Perspective*, Vol. 2, p. 457.

[70] *Cf.* Böhme, van den Daele and Krohn in this volume.

[71] Evidently, the success of these technologies induced Thomas to classify them as "high technology" despite the fact that solutions were obtained at a non-theoretical level. (*Cf.* Fact Sheet 1974.) This makes his classification of technologies unclear. The distinction between "half-way" and "high technology" may correlate with the probabilities for successful problem-solving, but should, however, merely be used to indicate the scientific level of a technique, not its success.

[72] A report published by the US Food and Drug Administration illustrates the consequences of such a policy: "The most often repeated comment from dissatisfied scientists was that the FDA constantly interferes with medium and long range projects, making it almost impossible to complete them. At least part of the reason, they assert, is that certain results might embarrass the agency either by showing a lack of scientific support for an already established regulatory policy or, conversely, establishing scientific evidence of a hazard which there is no legal authority to bring under control" (*Nature* **231**, 1971, p. 279).
[73] Luhmann (1970), Storer (1966).
[74] *Cf.* Buchholz (1979*b*) on the resistance of molecular biologists to participation in biotechnical research.
[75] *Cf.* the processes of a social self-orientation of scientific disciplines in the case of the transfer of plasma physicists (who mostly worked on astro-physical problems) to fusion research (Küppers, 1979). The effects of the variations in the academic culture of disciplines can be seen in the orientation of research. For example, there appears to be a much greater preparedeness on the part of chemists to take up externally relevant questions than is found with physicists – a function of the former's traditional connections with industry (*cf.* Buchholz, 1979*b*).
[76] See Manegold (1970) for the history of the academic organisation of technology, following the example of the Humboldt University.
[77] See Elliot (1973), and Hohlfeld, 'Cancer research' in this volume.
[78] Usually, a mixed group of representatives of the disciplines and from politics are involved in the formulation of the programme (see van den Daele, Krohn and Weingart, 1977). Nevertheless, as yet there has been no integration of those to whom the research is directed, that is the patients in the case of cancer research, in the process of the steering of science.
[79] Buchholz (1979*b*), p. 95.
[80] Küppers (1979), p. 318.

BIBLIOGRAPHY

Agassi, J.: 1966, 'The confusion between science and technology in the standard philosophy of science', *Technology and Culture* 7, 348–366.

Andersson, G.: 1976*a*, Nichts von der Geschichte gelernt', in *Information Philosophie*, December/November 1976, p. 2.

Andersson, G.: 1976*b*, 'Freiheit oder Finalisierung der Forschung?', in Hübner, K. *et al.* (eds): 1976, *Die politische Herausforderung der Wissenschaft*, Hoffmann und Campe, Hamburg, pp. 66–76.

Böhme, G.: 1980, 'On the possibility of closed theories', *Stud. Hist. Phil. Scie.* **11**, 163–172.

Böhme, G. and van den Daele, W.: 1977, 'Erfahrung als Programm – Über Strukturen vorparadigmatischer Wissenschaft', in Böhme, G. *et al.*: 1977, *Experimentelle Philosophie*.

Böhme, G., van den Daele, W., and Krohn, W.: 1976, 'Finalization in science', *Social Science Information* **15**, 307–330 (originally published under the title 'Finalisierung der Wissenschaft', *Zeitschrift für Soziologie* **2** (1973), 128–144).

Böhme, G., van den Daele, W., and Krohn, W.: 1977, *Experimentelle Philosophie. Ursprünge autonomer Wissenschaftsentwicklung*, Suhrkamp, Frankfurt.

Bohr, N.: 1931, *Atomtheorie und Naturbeschreibung*, Springer, Berlin (published under the title *Atomic Theory and and the Description of Nature*, Cambridge and London, 1961).

Brooks, H.: 1967, 'Applied research. Definitions, concepts, themes' in National Academy of Sciences: 1967, *Applied Science and Technological Progress*, Washington, pp. 21–55.

Buchholz, K.: 1979*a*, 'Verfahrenstechnik (chemical engineering) – its development, present state and structure', *Social Studies of Science* **9**, 33–62.

Buchholz, K.: 1979*b*, 'Die gezielte Förderung und Entwicklung der Biotechnologie', in van den Daele, W., Krohn, W., and Weingart, P. (eds.): 1979, pp. 64–116.

van den Daele, W. and Krohn, W.: 1975, 'Theorie und Strategie. Zur Steuerbarkeit wissenschaftlicher Entwicklung', in Weingart, P. (ed.): 1975, *Wissenschaftsforschung*, Campus, Frankfurt, pp. 213–254.

van den Daele, W., Weingart, P., and Krohn, W.: 1977, 'The political direction of scientific development', in Mendelsohn, E., Weingart, P., and Whitley, R. D. (eds.) *The Social Production of Scientific Knowledge*, Reidel, Dordrecht, pp. 219–242.

Elliot, P.: 1973, 'Professional ideology and social situation', *The Sociological Review* **21**, 211 ff.

Fact Sheet: 1974, *Advances in Cancer Treatment 1974*, Special Communication from the Director, National Cancer Program.

Handler, P. (ed.): 1970, *Biology and the Future of Man*, Oxford University Press, New York.

Heisenberg, W.: 1971, 'Der Begriff "abgeschlossene Theorie" in der modernen Naturwissenschaft', in Heisenberg, W., *Schritte über Grenzen*, Piper, Munich, pp. 87–94.

Hund, F.: 1972, *Geschichte der physikalischen Begriffe*, Bibliographisches Institut, Mannheim.

Kuhn, T.: 1962, *The Structure of Scientific Revolutions*, University of Chicago Press, Chicago and London.

Küppers, G.: 1978, 'On the relation between technology and science – goals of knowledge and dynamics of theories. The example of combustion technology, thermodynamics and fluid mechanics', in Krohn, W., Layton, E., and Weingart, P. (eds.): 1978, *The Dynamics of Science and Technology*, Sociology of the Sciences Yearbook II, Reidel, Dordrecht, pp. 113–133.

Küppers, G.: 1979, 'Fusionsforschung – zur Zielorientierung im Bereich der Grundlagenforschung', in van den Daele, W., Krohn, W., and Weingart, P. (eds.): 1979, *Geplante Forschung*, Suhrkamp, Frankfurt.

Luhmann, N.: 1970, 'Selbststeuerung der Wissenschaft', in Luhmann, N.: 1970 *Soziologische Aufklärung*, Westdeutscher Verlag, Cologne.

Manegold, K. -H.: 1970, *Universität, Technische Hochschule und Industrie. Ein Beitrag zur Emanzipation der Technik im 19. Jahrhundert*, Ducker & Humboldt, Berlin.

National Cancer Program (NCP): 1974, *Digest of Scientific Recommendations for the National Cancer Program Plan*, DHEW Publication No. (NHI), 74–570.

Physics in Perspective: 1972, Vol. 1, Vol. 2, Part A, *The Core Subfields of Physics*, National Academy of Sciences, Washington DC.

Polanyi, M.: 1962, 'The Republic of Science', *Minerva* **1**, 54–73.

Popper, K.: 1970, 'Normal science and its dangers', in Lakatos, I. and Musgrave, A.: 1970, *Criticism and the Growth of Knowledge*, Cambridge University Press, Cambridge, pp. 51–58.

Popper, K.: 1971, 'Die Logik der Sozialwissenschaften', in Adorno, T. *et al. Der Positivismusstreit in der deutschen Soziologie*, Luchterhand, Berlin, Neuwied, pp. 103–123.

Popper, K.: 1972*a*, *The Logic of Scientific Discovery*, 6th impression, Hutchinson, London.

Popper, K.: 1972*b*, 'The aim of science', in *Objective Knowledge*, Clarendon Press, Oxford, pp. 191–205.

Radnitzky, G.: 1974*a*, *Preconceptions in Research: A Study*, Chaucer, London.

Radnitzky, G.: 1974*b*, 'Was kann Wissenschaftsforschung zur theoretischen Grundlegung der Wissenschaftspolitik beitragen?', *Kölner Zeitschrift für Soziologie und Sozialpsychologie* **26**, 801–818.

Sneed, J. D.: 1971, *The Logical Structure of Mathematical Physics*, Reidel, Dordrecht.

Stegmüller, W.: 1973, *Theorienstrukturn und Theoriedynamik*, Springer, Berlin.

Stent, G.: 1969, *The Coming of the Golden Age*, Natural History, New York.

Storer, N.: 1966, *The Social System of Science*, Holt, New York.

Symes, J.: 1976, 'Policy and maturity in science' in *Social Science Information* **15**, 337–347.

Thomas, L.: 1972, *Aspects of Biomedical Science Policy*, National Academy of Sciences, Washington DC.

Truesdell, C.: 1976, 'History of classical mechanics', Parts I and II, in *Die Naturwissenschaften* **63**, 53–62, 119–130.

Weingart, P.: 1978, 'The relation of science and technology. A sociological explanation', in Krohn, W., Layton, E., and Weingart, P. (eds.): 1978, *The Dynamics of Science and Technology*, Reidel Dordrecht.

Weisskopf, V.: 1965, 'In defense of high energy physics', in Yuan, L. (ed.): 1965, *The Nature of Matter*, BNL 888 (T-360).

Weisskopf, V.: 1974, 'Zukunftsperspektiven der Wissenschaft', *Physikalische Blätter* **30**, 481–489.

von Weizsäcker, C. F.: 1979, 'The preconditions of experience and the unity of physics', in Bieri, P., Horstmann, R., and Kräger, L. (eds.): 1979, *Transcendental Arguments and Science*, Reidel, Dordrecht, pp. 123–158.

von Weizsäcker, C. F.: 1980, *The Unity of Nature*, Farrar, Straus, Giroux, New York.

Znaniecki, F.: 1968, *The Social Role of the Man of Knowledge*, Harper Torchbooks, New York.

GERNOT BÖHME, WOLFGANG VAN DEN DAELE AND WOLFGANG KROHN

THE SCIENTIFICATION OF TECHNOLOGY

1. INTRODUCTION

This paper is an historical analysis in support of a thesis which underlies, directly or indirectly, all the papers in this volume. This thesis asserts the unification of science and technology in the course of the development of modern natural science.

The analysis proceeds in the following steps:

– The Renaissance saw the establishment of a number of research fields which lay *between* scholastic natural philosophy and traditional technology (architecture, navigation, surgery). Both scholars and advanced craftsmen were involved. Although these fields witnessed a number of considerable practical successes, these never produced a conceptually elaborated scientific technology or technological science.

– During the period of the scientific revolution a number of these research fields nevertheless became paradigmatic for the aims of the New Science (Baconism). However, the first examples of technically oriented natural sciences were practically unusable (mechanics, optics, hydraulics). Theoretically and mathematically they were too abstract, and in pre-1800 technological practice there was no demand for the degree of precision and purity of the scientific instruments and materials of chemical experimentation.

– The repeatedly asserted affinity of science and technology was not created in practice until the Industrial Revolution began to make itself felt at the beginning of the nineteenth century. New technological demands (precision, efficiency, scaling-up) called for the application of scientific methods, instruments and explanations. The most important result of this development is the derivation of the "principles of application" which have the specific function of mediating fundamentals and applications.

– Finally, the twentieth century witnesses a number of examples of fully scientificated technologies, which demonstrate that the original project has been realised. This does not mean, however, that the totality of contemporary science can be fitted to this pattern.

G. Böhme et al., Finalization in Science, ed. by W. Schäfer, 173–205.

2. THE UNIFICATION OF SCIENCE AND TECHNOLOGY IN THE RENAISSANCE

2.1. *The Emergence of the Science-Technology Relationship*

In medieval times the construction of cathedrals was assigned to architectural workshops. For the organization and the techniques of construction the architects and craftsmen relied on traditional knowledge which was often kept secret by the rules of the guilds and which was only slightly modified with the acquisition of new knowledge. The construction of the largest cathedral of late medieval times, the cathedral of Milan, begun in 1386, raised unexpected problems of mathematics and statics[1]. The recently acquired economic and political status of Milan called for the largest building of the period, but the city desired that the design should not follow northern European rules of construction. Within the Roman tradition of Lombardian aesthetics the northern Gothic style was considered too arching and the supporting system of pillars and flying buttresses was felt to be confusing. Further, this style was determined by a well-established principle of construction which stipulated that the height of the church be equal to its width. The Milanese workshop, however, decided to work from an equilateral triangle as cross section. They hoped that the less arching elevation of this design would allow for the elimination of the confusing system of support, even if the size of the cathedral exceeded the Gothic constructions.

Northern Gothic architecture represented the most advanced knowledge and experience of the time. As the records of the Milanese workshop tell us, its members were not aware that their decision to deviate from this experience would lead them into serious problems. First, it would surpass the ability of the architects to determine arithmetically the height of an equilateral triangle. Second, even if an arithmetical determination were at hand, it would be impossible to proceed with the construction according to the Gothic rules of simple yardstick measurement. Working from the *ad triangulum* model, architects would be unable to calculate in advance the length to which workmen should cut the subsections for the main pillars. The council of the town summoned a mathematician who lived nearby. Fortunately, after some calculations he could offer a satisfactory solution[2].

The mathematical problem was followed by a problem of statics. It would be a great risk to carry out a project of this size in disregard of the Gothic techniques of supporting buttresses. The council of the town again became suspicious and called in scientific experts, this time foreigners from the north. The expert opinion of one of them, the French engineer Jean Mignot, was a scathing criticism: according to the geometric theories of Gothic statics, he observed, the whole edifice would not be solid. The theory required much stronger piers to support the system, while to make the project feasible in accordance with the geometric method it would also be necessary to elevate the main aisle. The Milanese builders and architects, he said, were not prepared to deal with the scaling-up problems arising from an extreme enlargement of the design of Lombardian provincial churches. Mignot's opinion regarding their ideas on statics is recorded:

> Proposals were made in a fashion more willful than sound; and what is worse, it was objected that a science of geometry should not have a place in these matters since science is one thing and art another. Said Master Jean [Mignot] said that art without science is nothing (*ars sine scientia nihil est*)[3].

In opposition, the local master-builders appealed to the Lombardian tradition of architecture. They claimed that they would not find insuperable difficulties in transposing Lombardian techniques of support to more monumental and vast projects. The practice-oriented reply to Mignot ended with the statement, '*scientia sine arte nihil est*'. The local guilds won their argument with the foreign experts, although their concepts were not well-supported either theoretically or by experience. The only argument which speaks in favor of the concept of '*scientia sine arte nihil est*' is that although it was built contrary to the Gothic theory, the cathedral of Milan still stands[4].

This dispute between the experienced practitioners and the calculating theoreticians presupposes that, in principle, there exist two different approaches to solving the same problem. This is the first time that science and technology can be seen to compete in being respectively the better means to fulfilling a certain end[5]. From the Greeks down to the early Renaissance, the tension of the relation between theory and practice was primarily related to the moral question of which life style – the theoretical (contemplative) or the life of labor – was more delightful in the eyes of God. In the Milan controversy neither life styles nor world views are questioned but rather the practical utility of theoretical knowledge and the deficiencies of traditional knowledge for solving new problems. The spokesman of the faction which supported theory was not a philosopher but an engineer. And it was to be engineers, artists and mathematical practitioners who were to play a decisive role in the development and social acceptance of this new type of utility-oriented theory.

2.2. *Early Technical and Natural Sciences*

These new theories – or at least new theoretical claims – cannot unequivocally be classified, either under the rubric of the arts or under that of the sciences. This is only in part due to the fact that in the Renaissance period *scientia* and *ars* meant something different than what they mean in modern times: it is just as difficult to assign these theories to the by now familiar subdivisions of the engineering sciences and the fundamental sciences. The reason for this ambiguity is rather that these theories violate the traditional boundary between the natural and the artificial, between understanding phenomenona (theoretical consideration) and constructing artefacts (poietic practice).

Leonardo da Vinci (1452–1519) was one of the first scientists to combine the technical construction of a desired reality with the discovery of the structures of a given reality. Thus Leonardo came upon new ideas in hydrophysics both by solving problems which had arisen with the new projects for regulating water-flows (such as construction of canals, regulation of rivers, artificial irrigation), and by his observations of meteorological or hydrological processes in nature (for instance the formation of clouds and whirlpools)[6]. Leonardo's discoveries were as much laws of nature (*ragioni*) as rules of operation (*regole*)[7].

In the mid-sixteenth century, Niccolo Tartaglia (1499?–1577), a mathematician as well as an engineer, was working on a theory of ballistics which combined a natural force (gravity) with an artificial (the impulse of the projectile) and led to a unified geometrical representation for both. Tartaglia published his theory in a book which bore the programmatic title, *Nova Scientia*. It was addressed to "any speculative mathematician, artillerist and others". It would be difficult to decide whether this book deals with technical or natural science[8].

The association of natural philosophy and technology and the transformation of both into a new type of science occurs in the fifteenth and sixteenth centuries in many fields of the then relevant sciences. Paracelsus (1494–1541), Ambroise Paré (1510–1590), and Andreas Vesalius (1514–1565) revolutionized the human sciences based on natural philosophy as well as on the medieval art of medicine by laying the foundations of the disciplines of pharmacology and surgery[9]. William Gilbert (1544–1603) framed a theory of magnetism intended to encompass earth magnetism and artificially induced ferromagnetism as well[10]. Peter Apian (1501–1552), Gerhard Mercator (1512–1594), and others worked on theories and processes intended to coordinate terrestrial orientation and astronomy[11].

2.3. *The Knowledge of Possible Nature – The Knowledge of the Nature of the Possible*

Within the science of the late Renaissance there are innumerable examples of scientists engaged in connecting science and technology. One may very well ask whether these attempts were accompanied by the development of a new conception of nature, interrelating its explanation and control. The scientists and philosophers claimed, at least, to have done this.

In his *Discourse on Method*, a later and thus a more reflexive exposition than those of the Renaissance scholars, Descartes (1596–1650) gave account of this new concept of nature and of natural knowledge:

> But as soon as I had acquired some general notions concerning Physics and as, beginning to make use of them in various special difficulties, I observed to what point they might lead us and how much they differ from the principles of which we have made use up to the present time, I believed that I could not keep them concealed without greatly sinning against the law. . . . For they caused me to see that it is possible to attain knowledge which is very useful in life,

> and that instead of that speculative philosophy which is taught in the Schools, we may find a practical philosophy by means of which, knowing the force and the action of fire, water, air, the stars, heavens and all other bodies that environ us as distinctly as we know the different crafts of our artisans, we can in the same way employ them in all those uses to which they are adapted, and thus render ourselves the masters and possessors of nature. This is not merely to be desired with a view to the invention of an infinity of arts and crafts . . . but also principally because it brings about the preservation of health . . . .[12]

To achieve these goals, it is necessary to discover "the principles, or first cause of all that is or can be in the world".[13] Recasting nature to something that is conceived of being possible but does not exist as yet – this is technology. If, therefore, the concept of nature is not confined to that which is objectively given, but covers that which is objectively possible, the concept of nature will include the procedures and products of technology. The purpose of natural knowledge is not only to discover the facts but also to construct 'arte-facts' according to rules which delineate the realm of the possible states of nature. In the context of medieval craftsmanship, rules were simply instructions for making something. For Descartes they have become simultaneously both the laws which were put into nature and also those which establish its system of possible operations.[14] Some notes of Leonardo indicate that this concept of nature already underlay the investigations he made some one hundred years earlier. About the utility of his results, Leonardo writes:

> These rules enable you to discern the true from the false, and thus to set before yourselves only *things possible* . . . .[15]

> If you were to ask me what do your rules bring about and of what use are they? I should answer you that they prevent the inventors and investigators from promising themselves and unto others *things that are impossible* . . . .[16]

Finally, a consideration of Leonardo regarding the concept of nature as the reality which can be constructed with some particular end in view:

> You who speculate on the nature of things, I praise you not for knowing the processes which nature ordinarily effects of herself, but rejoice if so be that you know the issue of such things as your mind conceives.[17]

With this conception the phenomenalism of Aristotelian natural knowledge is replaced by a constructive point of view: *knowledge* of nature has become identical with its experimental and deductive *construction*.

Cassirer has described the new thought underlying this program as the transition from substantive to relational thinking, a process which received its main impulse from the artists and engineers of the Renaissance. The parallelism of the artistic formation of nature and its geometrical description gave primacy to the formal regularities in nature rather than to its ontological qualities.[18]

Piaget characterizes this transition from substantive to relational thought as a decentering process in which it is learned that the framework through which we perceive nature is constructed according to cognitive operations.[19] Kant stated that this revolution of our 'Denkungsart' was brought about by the 'realization' [Einfall] "that we only learn from nature, what our mind superimposes on it".[20] This nexus of cognition and operation provides a more profound insight into the interdependency of the knowing subject and the objects to be known than did either the ideas of traditional natural philosophy or those of advanced craftsmanship alone.

2.4. *The Historical Evolution of the Interrelation of Technology and Science*

Historically, this unity has been built up only step by step and only in particular sciences and technical fields. On the one hand, this has been due to the fact that the social acceptance of the new sciences depended on external chances of selection which arose only gradually as capitalist society developed. Further, the application to the different areas of nature of the epistemic principle that knowing and operation are interdependent was hampered by empirical problems which could only be resolved successively.

In what follows the process of the intellectual realization and social acceptance of the program of the new sciences is subdivided into three phases:

The first phase (ca. 1660–1750) begins in the Period of the Restoration in England and the spread of Absolutism on the Continent. From the vantage point of the history of science it is marked by an institutional and cognitive differentiation of the areas of science and technology. Nonetheless, scientific orientation to techniques remains important in two respects. First, it gives rise to a special development of technology, i.e., the technology of scientific instruments and procedures. Secondly, the principle that to know something means to know how it operates is worked out as a universal pattern of explanation in the form of the mechanistic image of the world.

The second phase starts with the industrial revolution and approximately covers the nineteenth century. The spread of capitalism and its internal dynamics makes technical invention a constituent element of economic reproduction. The dynamic momentum given to technology produces a demand for science, bringing about the first examples of a process that may be called the 'scientification' of technology. At the same time a wider scope is given to the economic utilization of those technologies which were originally developed as scientific instruments and procedures.

In the third phase, the reciprocity in demands and supplies between science and technologies becomes systemized and strategically planned. This process begins in the latter half of the nineteenth century and comes to play an increasingly significant role in science policy and planning of research in the twentieth century. Scientific advance becomes goal-oriented and technologies are planned in conformity with theories.

3. THE SEPARATE DEVELOPMENT OF NATURAL SCIENCE AND TECHNOLOGY

Given the background of the first section, it is not surprising that the institutionalization of the 'new sciences' in the seventeenth century comprised both their tradition of natural philosophy (cosmology, theory of matter and force, etc.) and their technological traditions (mechanics, time keeping, navigation, mining, etc.). The charters of most of the scientific institutions following the foundation of the Royal Society and the

Académie des Sciences propose scientific research which is to lead to results both new and useful.

It is, however, from a sociological point of view, surprising that the institutionalized juncture of novelty and usefulness did not last long.[21] Admittedly, innovation in industry and social structure was by no means the foremost goal of the absolutist governments, but still, innovation played a fairly important role in the domains of warfare, foreign trade, luxury production, standardization and measurement (as a basis for taxation), and transportation.[22]

The conjunction of book learning and craft techniques in the Renaissance had arisen from the breaking down of cognitive and institutional barriers and had been brought about by artists, engineers and individual scholars. Yet the result was not a unified 'technical science', but rather on the one hand, a new technology incorporating scientific method, and on the other, a new natural science which analyzed nature by means of technical apparatus and interpreted it according to the model of such apparatus. Though the ideas of the scientists were intimately tied to technical problems, they themselves were not guided by a 'technical interest in knowledge'.[23] Instead, notwithstanding the proximity of both fields, different goals of investigation brought about a new separation of science from technology. While the engineer designed apparatus and improved their functions as the technical realization of a practical intent, the scientist tried to understand how it functioned, he was directed towards theory. Although scientists often enough began their research in connection with technology, as the investigations proceeded these took on a life of their own, independent of their technical starting points. The scientists' concern was "the true natural philosophy".[24]

The technology of pumps, for example, concerned many scientists in the seventeenth century. The phenomenon that vacuum pumps only lift water to a certain height and its explanation by Viviani that the atmospheric pressure impedes to lift it higher led Torricelli to the construction of an instrument for the measurement of the weight of the air, the barometer. This instrument, then, was used by Pascal to test the existence of a vacuum – and the ultimate result was a general theory of pneumatics.

Similar, the trajectories of projectiles were no doubt an extremely interesting subject for scientists occupied with the problems of mechanics, yet their main concern was the theory of impetus, that is, the question of how a body is kept in motion without a moving agent. In this case, theoretical interest gave rise to a development in the field of mechanics independent of developments in the field of artillery.

The scientific interest in understanding natural and artificial events acted as a source of internal problem generation which made science autonomous of production-oriented technology. Already at the turn of the seventeenth century interest in theory had come to outweigh interest in research oriented to practical needs, even in the Royal Society and in the *Academie des Sciences*, whose founding charters had stipulated that their business was to promote the trades and the crafts by means of the new science. With the Royal Society this change can be traced to the time when Newton became a Fellow.[25]

As scientists came to act independently in their focus on the developments of theories and as their contributions to advances in production, warfare and navigation became fewer, the systematic improvement of these fields devolved to the engineer and engineering became institutionalized as well. The absolutist states organized engineering strictly to meet the needs of transportation, mining, the military, and the navy. This led to the establishment of the various technical corps and schools in France, the *École des Ponts et Chaussées* (1750, the *Corps des Ponts et Chaussées* had been established earlier, in 1720), the *École du Corps des Ingénieurs des Mines* (1778), the *École Royale Militaire* (1753), the *École du Corps Royal du Génie* (1665, for the education of architects), the artillery schools and the naval colleges. In England similar institutions were the Mechanics Institutes, and in Germany the Mining Academies in Berlin (1775), in Freiberg (1765), and the Schools of Agronomy. These schools trained engineers and their curriculum comprised the fundamentals of mathematics and the natural sciences. They were not always the centers of the new sciences, yet, it was within this frame that a turning point occurred which marked a rapprochement with the institutions of science in the nineteenth century.

4. THE ORIENTATION OF SCIENCE TOWARD TECHNOLOGY

Although from the seventeenth to the nineteenth century the new sciences had developed both cognitively and institutionally apart from technology, they nonetheless remained technology-oriented. One reason for this seems to be that for modern science the experience of nature is mediated by instruments which are perfected in conformity with scientific standards. These standards are set up without reference to the usefulness of these instruments apart from their scientific purposes. Another primary reason for the lasting technology-orientation of science is that theory development initially evolved within the framework of the mechanistic worldview. The theories of natural science are instrumental theories, partly because they emanate directly from analyses of apparatus, and partly because their model for the interpretation of nature is the 'big clock'. Thus the internal dynamics of scientific progress involves a technical dynamic.[26]

4.1. *Technological Developments Internal to Science*

The modern experience of nature is an experience of regularities and of apparatus as the models of nature. The experience produced according to specific rules, with data reproducible by anyone who observes these rules, is more significant than the experience of events perceived individually. The data for science is not that given directly to the senses but is its objectified occurrence, that is, its effect through and on apparatus. Both reproduction and objectification of events were already established by craftsmen and engineers; in connection with science, however, they undergo a greater degree of specification and standardization. And it is this which determines the development of scientific instruments and procedures.[27]

Scientific instruments were either designed for scientific purposes as such or, where they already existed in connection with technical-practical needs, they were further developed for scientific purposes and in conformity with scientific standards. The last was the case, for instance, with the clock and with the pump. The development of the clock in its uses internal to science became dissociated from its function in the presentation of the celestial movements and was determined by the idea of the absolute isochrony of periodic processes. This was done by introducing the pendulum.[28] The development of the pump followed a similar pattern. Naturally, pumps had already been employed for technical and practical purposes. But their perfection in the form of the vacuum pump sprang from the scientific interest in the phenomena revealed by the barometer, and was governed by the idea of the perfect vacuum. Philosophically, space without any matter seemed to be impossible; experimentally it was approximately feasible.

A concise description of the scientific standards that actuate the development of apparatus and new procedures is facilitated by subdividing them through a distinction borrowed from chemistry into *analytic* and *synthetic* apparatus and procedures. The term analytic refers to instruments of measurement such as the clock, the thermometer, the calorimeter, the barometer, the voltmeter and the amperemeter, and to processes such as Liebig's analysis of the elements or the detection of grape sugar by Fehling's solution. The term synthetic refers to apparatus such as the vacuum pump, the Voltaic cell and the electromagnet, and to processes such as the synthetic production of alizarin. In general, any experimental device which serves to manifest an effect or a substance can be considered such a synthetic apparatus or process.

The scientific standards guiding the development of the *analytic instruments* are the increase of *precision*, the *unequivocal* relation of data to variables, the *universal comparability* of weights and measures, and the *generalization* of measurements over the whole range of a parametric scale. Increase in the *precision* of scientific instruments of measurement by far exceeded what was practical for production technology or what the engineers could calculate mathematically.[29] It is even more significant that the standards of precision became separated from the standards given by the original paradigm of nature. Scientific standards of precision were defined in accordance with theoretical ideals. Thus, even in antiquity the technical standards defining the vertical and the horizontal were transcended by Euclidian geometry. In the same way, the standards of celestial motions for the construction of clocks (the sun's daily zenith, the lunar month, the solar year) were replaced by the ideal of absolute isochrony. Thermometrics at first defined its scales using the hottest and coldest days of the year (Fahrenheit) or the states of liquid water (Celsius). These scales in turn were replaced by Kelvin's absolute scale, through which it became possible to indicate the adequacy of any of the substances previously employed to function as a temperature standard.

The requirement that instruments give *unequivocal reading* was also unrelated to technical purposes outside of science. The farmer, the metallurgist and the brewer needed indicators of those changes occurring in the processes they were working with, but as a rule they were guided by impressions which from the vantage point of science were produced by many interacting variables. Scientific instruments, on the other hand, are intended to respond to one and only one variable. Thus, after the invention of the thermometer by Galileo, Drebbel, Fludd and Santorio, other scientists tried to make it as independent as possible of air pressure and the material qualities of the substances used.[30]

Until the emergence of mass production and world-wide markets, the *universality* of data was not of great practical value. To be sure, attempts to unify weights and measures had been made in various countries throughout the eighteenth century, but not until the nineteenth century were general conventions on weights and measures established. For science, on the other hand, the universal comparability of data was a standard from the very outset. Nor can universal comparability be attained simply by international agreements on units of measurement, as its realization also implies overcoming a variety of technical problems.[31]

In the same way, *generalization* is not a standard of immediate technical interest. Generalization means the establishment of procedures of measurement covering the whole range of a parametric scale. The potter has one method for determining temperature, and the beerbrewer a different one. Correlation of their scales is a matter of indifference to both of them. Scientists, however, search for a theoretical concept of heat for which the different procedures of measurement are but partial interpretations.

When in the nineteenth century production started making severe demands on purity of materials, precision of machine parts, and universality of measures and weights all over the world market, the analytic apparatus and procedures for testing materials and the control of processes were available as a result of a development internal to science and could be modified for industrial purposes.

The development of *synthetic apparatus and procedures* are guided by different standards. Empirical investigations within the framework of modern natural science thus presuppose that phenomena can be isolated, are reproducible, and are disposable regardless of regional or seasonal factors. Thus, though electricity is made manifest by rubbing amber against a cat's fur or by a flash of lightning, for the advancement of the science it was important to isolate electrical phenomena from related ones like magnetism and to make electricity freely available, as was accomplished by Hauksbee's Electrical Machine, and later by the Leyden jar and Volta's electric pile. Analogous examples are to be found in chemistry. A condition for analyzing the structures and properties of a substance is to isolate it, and to produce it in sufficient quantity and purity. Synthetic apparatus and procedures serve to isolate and present substances and effects, and where they exist only in insufficient strength or volume for investigation, to store, accumulate and reinforce them.

When a need for these substances and effects emerges in the economy or in the industry of warfare, techniques developed within science itself become a part of production technology.

4.2. *The Technology-Orientation in the Theory of Mechanics and in the Mechanistic World Image*

Science and technology did not only diverge in the development of apparatus and procedures; they also differed in their theoretical concepts. Even though mechanistic theories were supposed to cover, at least 'in principle', technical effects, in their formation it was necessary to develop concepts which had no bearing on technological purposes extraneous to science. The most conspicuous example is the idealization of bodies to 'mass points' in mechanics after Newton. The formation of scientific theories took a different direction in reference to the ideal of producing useful knowledge. Since it was either mathematically too complex or not worth its efforts to 'deduce' technical devices from mechanistic theories, little contribution could be made to technological advance.

In the period of the Renaissance, technology and theoretical knowledge were still connected only on the basis of Aristotelian categories. In interpreting their discoveries, the practitioners of the experimental method still had to appeal to the 'natural place' of bodies and to teleological forces moving them. The unity of technology and science was practised in experimental research, but its theoretical interpretation lacked consistency. Only in Galileo's new physics was it possible to conceive physical (natural) movements as technical (artificial) movements and vice versa.

Descartes' mechanistic philosophy realized this unity, although his theories in physics and biology were rationalistic constructs rather than explanatory frameworks guided by experience.[32] Descartes' model of nature was a technical one – change of motion is caused solely by push, pull and pressure – i.e., he explained natural by artificial motion presupposing the conservation of a constant amount of motion. In this way, the forces which were known as operative in mechanistic devices were held to be the only possible forces existing in nature. But, even disregarding its inherent difficulties, this argument does not provide the whole answer. To explain nature from the principles of technical mechanics ultimately requires that mechanics be explained from nature.

If mechanics is supposed to be not only a simulation model of nature, but also its real representation, then the mechanical forces must be derived from some universal principles of movement as such. Newton's theory of motion and force was highly successful in formulating this common basis of science and technology.[33] Nonetheless, a comparison of this scientific mechanics with the technical mechanics of the so-called five simple machines–wheel and axle, the lever, the pulley, the wedge, and the screw – shows that although Newtonian mechanics explains the functioning of these machines, the machines themselves no longer figure in it. The mathematical derivation from the Newtonian principles entails non-trivial processes of integration. For this reason, the engineers continued working with the 'already integrated' simple machines. This relationship is found also in later fields of research originating within the frame of mechanistic approaches as, for instance, was the case with electrical engineering and Maxwell's electromagnetic theory of light. The technological study and use of condensers, resistors, valves, and so on was on the whole independent from their theoretical foundation. However, although the mechanistic theories did not lead to useful applications, their formation was oriented toward the explanation of technological problems like torsion, elasticity, fluid resistance. Moreover, many of the early theories of modern science came into existence as theories of specific instruments. This is true in physical optics with the theory of the microscope and the telescope, in pneumatics with the theory of the pump and of the barometer, and in thermodynamics with the theory of the steam engine. Theories explaining instruments enable one to distinguish contingent from systematic limitations of the instrument's performance and efficiency. This distinction is then in many cases the condition for using apparatus economically or for the construction of technical alternatives to them.

4.3. *Scientific Supply to Production*

The development of technology internal to science in the late seventeenth and eighteenth centuries consisted in the perfecting of instruments and procedures and in their theoretical explanation. This 'scientification' of technology was not directly associated with the sphere of production nor did there exist a considerable demand for scientific technology in production.

However, given the economic dynamics which obtained after the industrial revolution in the nineteenth century, science and productive technology came closer together again. With increasing economic opportunities for technological inventions, the technological achievements internal to science became relevant to industry. Furthermore, scientists started devising economically exploitable technologies and experimental inquiry was no longer exclusively devoted to such goals as the construction of ideal regularity, the identification of elementary components or the testing of fundamental principles.

In the nineteenth century, the scientific supply to production consisted primarily of the synthetic procedures and instruments originally developed for scientific purposes. The development of the telegraph may serve as an illustration. It was based on scientific procedures and instruments which were not modelled on the techniques of craftsmanship, but which led to effects which technical-practical strategies could simply not have produced.

In 1820 the Danish professor of physics, Christian Oersted, discovered the magnetic effect of currents. The technical potential of this discovery for the transmission of signals was recognized immediately and the role of scientists in the development of the electromagnetic telegraph was considerable. Contributions were made by Schilling, a Russian State Councillor, and by the physicists Henry (Princeton), Jacobi (Petersburg), Gauss and Weber (Göttingen). In 1837 Cooke and Wheatstone built the first practicable telegraph, which was put into operation by the London and Birmingham Railway. Cooke himself had not received a scientific education. He made a living from the sale of self-constructed anatomic models. When he started his work on telegraphs, he encountered electromagnetic problems, and to solve them joined forces with Wheatstone, who was then a professor at King's College in London. Electric telegraphy was the first of a vast series of new technologies resulting from the direct exploitation of the principles and scientific discoveries of electricity. Other examples of this kind were Marconi's wireless telegraph, which was derived from the results of the work of Maxwell and Hertz,[34] and Bell's telephone, which proceeded not only from the foundations of electromagnetic induction as laid by Oersted, Lenz and Faraday but also from Helmholtz' fundamental studies of physiological acoustics. In all these cases science had already assumed the role of supplier to production.

5. THE ORIENTATION OF TECHNOLOGY TOWARD SCIENCE

So far we have analyzed the orientation toward technology inherent in the development of the theoretical sciences. We shall now examine tendencies to reconcile science and technology which were fostered by advances in those technologies not tied to scientific purposes. The technological advances central to the scientific revolution were developed without recourse to scientific theories and without significant transfer of the experimental results of academic or university science. This is true of early labor saving machinery (the loom), power engines (improved hydraulic motor), transport (the locomotive), and metallurgy (cast iron).

The relationship between science and technology was visible only in their partially congruent methods of research. Experimental variation, measurement and quantification, and geometric description had become procedures which promoted the development of technology. Many of the achievements of the industrial revolution stem from such a methodological development of technology.[35] On the other hand, under Absolutism, governmental regulations as well as the guild's control of production inhibited innovation and consequently slowed down the rate of technological progress relative to the special development of scientific instruments. The industrial revolution, based on a few innovations with key significance, in turn set new incentives for technological research and established a demand for scientific knowledge and methods in technology.

The following four points indicate the growing need for technological help in industry, which led to a demand for science:

1. *Additional raw materials.* In many fields a shortage of raw materials limited industrial expansion. The availability of saltpeter, soda, and steel determined the volume of technology-related outputs in the chemical and the engineering industries. Efforts were made to overcome these limits by means of scientific analysis, frequently through attempts at the synthetic production of the respective raw materials.

2. *Increasing efficiency.* Low efficiency can render a technology economically useless. Thus the early Newcomen steam engine, because it consumed enormous quantities of coal, could only be used in mining areas. The same applied to the locomotives used at the start of the nineteenth century, which only had a speed of a few kilometers per hour and thus could not be used for other purposes than for transport in mining. Optimizing the efficiency of the steam engine became an object of science. Carnot, the French engineer and physicist, theoretically established the principle of the thermal engine and projected the maximum efficiency achievable in each case.

3. *Scaling-up of critical orders of technical dimensions.* The increase in the dimensions, the mass, the pressures, the temperatures, and the like, of technical procedures led to mainly non-linear changes in stability conditions, in quality of materials, and in modes of operation, which could no longer be achieved on the basis of existing technology.

4. *Greater precision of technical procedures.* Precision comprises guarantees of the purity of products and materials (thus the availability of pure yeast for proper fermentation was a major problem for brewery

techniques) of the accuracy and constancy of the size and material qualities of machine parts, and finally, of the accurate control of industrial processes and the correction of disturbing effects.

These problems cannot always be resolved through step-by-step improvements on existing procedures. The substitution for natural of synthetic fertilizers, the development of a separate condenser for the steam engine, the replacement of the bucket-wheel by the turbine in the use of water power, and the like, often start with or even require the adoption of principles other than those which underlie the established procedures. If social conditions are static, the limitations of improving traditional techniques often constitute the limits of technological progress as such. Technological progress exhausts itself in the slow perfecting of already known principles, as with the water wheel and the windmill. Only under dynamic social conditions do these limits constitute a challenge and generate a demand for science.

To develop technology beyond these limits, scientific analysis is, if not a necessary, at any rate in the long run a superior strategy. On the one hand, theory provides technical principles which cannot be discovered on the basis of extrapolations from known processes. Theory can play the role of an heuristics of invention.[36] On the other hand, scientific research can open the 'black box' of functional interrelations by subjecting the factors involved to causal and elementary analyses. In this way, both the theoretical optimum of a technical process, and the limiting conditions of the materials and procedures actually applied, become determinable. Thus, theory offers technical prognoses, in some cases even for non-linear modifications to which the rule-of-thumb method is inapplicable.

Two examples may illustrate how scientific analysis served to solve problems of technological progress. The first instance is the improvement of steel production by the Gilchrist–Thomas Basic Process (1871–1879), invented by the cousins Percy Gilchrist and Sidney Gilchrist Thomas. In 1856 Bessemer's converter had signalled the decisive technical break to cheaper and speedier production of steel. Bessemer's process, however, had failed to remove the phosphorous which most English and European ores contained. The problem of making steel from phosphoric ores was tackled by Thomas. He had worked as a junior police court clerk in the London docks and continued his studies at Birkbeck College (a kind of 'open university' for artisans and others in full-time employment) in the evenings, concentrating on chemistry. He read the scientific and technical literature, and after four years of scientific analysis and experimenting he found the solution of the problem: it was that of absorbing the oxidized phosphorous or phosphoric acid in a basic lining made from magnesium limestone.[37]

The second example is that of the "laws of the construction of locomotives" by Ferdinand Redtenbacher (1855). The building of the locomotive and the working out of construction problems, associated primarily with Robert Stephenson, who insured that the engine's weight rested equally on all wheels so that they would not spin and developed a direct drive between the cylinder piston and the driving wheels, was not the result of any notable contribution on the part of science. In the introduction of his 'laws', Redtenbacher stated that practice "with its sound instincts and perceptions" had reached a stage close to perfection (p. III). As with other major constructional engineering techniques, all that was left for science to do was to keep running after the technical dynamic with theoretical explanations in hand. But there was a point where the building of locomotives was connected with science, i.e., in the problem of the steadiness of locomotives at high speed. The unbalanced reciprocating weights of pistons, crossheads and connecting rods, which were originally set to cancel out only straight fore-and-aft forces, at high speeds led to dangerous oscillations which could accidentally throw the locomotive off the tracks.[38] The engineers tried to improve matters through a more equal distribution of the weights on the locomotives (e.g., by moving the cylinders rearwards), but it was difficult to design the dimensions and the setting of the reciprocating weights. Redtenbacher, who was a Professor at the *Polytechnische Schule* in Karlsruhe and one of the founders of scientific constructional engineering, attempted a fundamental solution of the problem by means of a theory of disturbing effects in express locomotives. He defined the various forms of the disturbing effects –jerking, rolling, nosing, surging, pitching – and the forces causing them. He worked out a set of differential equations determining these movements and stated the methods for solving them, and further defined both the conditions and the limiting factors for eliminating the disturbing effects of the imbalance of reciprocating weights. It is not certain whether

Redtenbacher's formulas came to have any importance for practice. Possibly it was only the theory of balancing of Le Châtelier (1849), who was more closely associated with locomotive building than was Redtenbacher, which actually came into application. Nonetheless, it was a *theory* of the disturbing effects which ultimately resolved the problem of dynamic balancing by making it possible to predict with some confidence the results of the said effects. This theory is even today an essential element in mechanical engineering.

'Applied science', theoretical and experimental research aiming to solve given technical problems, is a characteristic form of the interaction between scientific and technological development in the nineteenth century. On the one hand, this interaction was rooted in the demands for technical advances prompted by the industrial dynamic of the age, and on the other, in the increasing potentialities and proficiency of scientific output.[39]

In the writings of the scientists of the nineteenth century, the growing significance of science for technology is frequently seen not as the result of a *demand* from technology for scientific problem-solving, but rather as the result of the *supply* of new techniques following upon the autonomous development of theoretical science. The industrial expansion experienced by society indeed had the effect that in the nineteenth century the technical potential of scientific discoveries was increasingly applied in technology. We have illustrated this point by the telegraph and other techniques in the field of electricity which exemplify a technical dynamic made possible by developments in science. This gave rise to the general argument that pure or fundamental research is the basis of the technical dynamic. With this argument the scientists sought to protect their recently gained professional autonomy against the demand for a science oriented toward technical applications. Pasteur said that there was no category of science which could be called an 'applied science'. "There is science and there are the applications of science joined together as is the fruit to the tree on which it grows".[40] Yet Pasteur himself had spent many years of his life in the search for scientific solutions to special practical problems, and his central preoccupation was always with the discovery of the causes of a problem and the scientific explanation of the factors involved.

Except for the ideological reason just mentioned, it is hard to understand why this type of externally oriented research directed at the medical-technical solution of urgent problems should not bear the name of applied science. Pasteur's medically-oriented microbiological research, and similarly Liebig's agricultural chemistry, are both closely related to the 'demand model' of science-oriented technology which has been exemplified by the Thomas converter and the theory of locomotives.

6. THEORETICAL TECHNOLOGY

The relationship between scientific progress and technological innovations in the nineteenth century was still unsystematic. This applies both in the case of scientific inputs into new technologies and in that of the demand for scientific solutions to given technical problems. The relationship between science and technology was determined by contingent factors such as personal contacts among scientists and practitioners, the technical and economic interests motivating individual scientists, and the degree to which technicians had had access to scientific education.[41]

At the end of the nineteenth century and increasingly in the twentieth century the technical dynamic actuated by capitalism and by war leads to an expansion of industrial research and development, to independent institutions in applied research, and to a shift of technical education from training at the shop to academic learning. The connection between science and technological progress is in many aspects an offspring of the nineteenth century 'supply model'. New technologies emerge as the 'fall-out' of basic research. The structure which characterizes the new level of the interaction between science and technology in the twentieth century, however, is the ramification of science into special technological theories, or, from the vantage point of technology, the pursuit of technical goals through the construction of theories.

Natural knowledge and technology are no longer only methodologically equivalent, nor are they linked only by the epistemic structure of operational knowledge. They now tend towards unification at the level of theory. In the twentieth century the development of useful technology through scientific theory building becomes possible and can be planned strategically. Its basis is the successful formulation of theories for a series of object areas. In contrast to the conditions from which science started in the nineteenth century, science at present disposes not only of successful theories of mechanics, but also of reliable electrodynamic and chemical theories: for such phenomena as moving bodies, heat, light, electricity, science has achieved solutions of most of the fundamental problems, and these solutions can serve as the foundation of special theoretical models, explaining complex technical phenomena. The process of interconnecting technical phenomena with a fundamental theory by means of special models has two complementary aspects. It includes both the formulation of theories for technical structures and the concretization of general scientific theories. The examples of the sciences of chemical engineering and fluid mechanics will serve to illustrate both these aspects.

Toward the end of the nineteenth century the engineering design and optimation of chemical-technical production processes were in the main still based on personal experience and trial-and-error variations. By the mid-twentieth century, however, they are the subject matter of a theoretical science: the science of chemical engineering.[42] The development of chemical engineering as a scientific discipline covers three distinct stages. The first, which lasted until World War I, is marked by the aggregation of the disciplinary competencies of the engineer and the chemist in the area of chemical engineering. (In Germany this took place through the cooperation of traditionally trained engineers and chemists, and in the U.S.A. and Great Britain through the formation of the profession of the 'chemical engineer'.) New processes were increasingly being developed experimentally in engineering research laboratories. But translating the results of this research into practice was still a matter of chance and 'finger knowledge'. The disciplinary methods of chemistry and engineering were not yet fused, and therefore a direct scientific treatment of the complex operations in the 'real' technical processes was not feasible.

The second stage of 'scientification' – especially in the U.S.A. after World War I – consists in classifying production processes on the basis of unit operations. The units of technical operations such as distillation, filtration, mixing, crushing, grinding, and the like were differentiated from each other and organized according to practical requirements. The operations were dealt with by means of scientific methods, but these methods were not directed towards theoretical analysis on the level of fundamental sciences such as molecular chemistry, thermodynamics or hydrodynamics. By about 1930, chemical engineering had been established as a scientific discipline with its own methods, standard solutions and conception within the framework of phenomenological theory.

The third stage, beginning in the thirties and accelerating after the Second World War, entails the integration of chemical engineering into the framework of physical and chemical theories. The characteristic of this third stage is that the real processes in chemical reactors are more and more successfully made comprehensible by means of microtheoretical concepts within the framework of thermodynamics, hydrodynamics, aerodynamics, kinetics, and other theories. Work on chemical engineering problems becomes for scientists a possible avenue for the continuation of physico-chemical research.[43] They develop special concepts and mathematical models which interconnect the more complex technical processes and operations ('macrophenomena') with the idealized objects of fundamental science. Thus, for example, 'macrokinetics' is developed to understand chemical reactions which are influenced by physical transport phenomena (transport of heat, energy and matter).

In the ideal case the theoretical treatment of chemical engineering makes possible the theoretical calculation of the conditions of technical apparatus and of process operations. It also allows engineers to design productive means with a minimum of uncertain extrapolation on the basis of existing processes or trial-and-error experimentation.

The second example of a 'theoretical technology' is fluid mechanics, with its various special models such as that of the friction of bearings and of lubrication, or that contained in the theory of the airfoil.[44] As was the case with chemical engineering in the third stage of its development, fluid mechanics is not only 'in principle' related to specific technical devices, (as was Newtonian theory, for instance, to the windmill) but its special theoretical models also instruct the engineer how to construct a particular device: they make possible the calculation of the required technical solution, though its feasibility may remain a problem. These theories make it possible, by varying certain parameters and holding others constant, to select optimal solutions. They also predict what will occur if thresholds of a certain order of magnitude are surpassed, and they identify the causes of the limitations to which a particular technique is subject.

As opposed to chemical engineering, theories of fluid dynamics may be called theoretical technologies: they did not emerge from the systematization of the corresponding technologies. In the nineteenth century theoretical and technical hydraulics evolved in parallel with almost no connection to each other. This was due to the fact that the Navier-Stokes' equations, which 'in principle' had made classical hydrodynamics encompass all the processes of fluid mechanics, were solvable only in a very few instances which had a bearing on technology. The transition to technological theories in the field of the mechanical effects of fluids was the result not of systematizing technology, but of theoretical efforts. It was Prandtl's concept of the boundary layer which enabled the transition to the 'real' technical processes to be made. To put it briefly, the boundary layer concept allowed fluids on the whole to be treated as viscosity-free, but required that regard be paid to the effects of viscosity at the boundary layer. The boundary layer concept made it possible to specify general hydrodynamic theory into distinct theories, which covered particular types of technical operation such as lubrication, the airfoil, and the propeller.

The technology of aeronautics was from the start a scientific technology, not only in the methodological sense that the first designers of aircraft made scientific experiments (the brothers Wright even had a wind tunnel) but also in the sense that theory played a crucial role in the development of this technology. Kutta's theory of the airfoil wing, by virtue of which one could calculate the lift of the wing, under idealized conditions, did not have any practical bearing on the first flights made. But Prandtl's theory of the airfoil for real aircraft wings (which, unlike Kutta's, were of 'finite length') made the theory practicable. Step by step, on the basis of this theory it became possible to compute the lift-to-resistance ratio for airfoil wings, to construct wings with a calculated lift distribution, and to determine the lift distribution for a given wing. Further, it became possible to predict the functional relation of lift to the angle of attack and later on to explain the dangerous phenomenon of stalling in flight. Investigations into the stability of the boundary layer made it possible to delay stalling in flight through the design of more appropriate wings and the influence of the boundary layer by means of blasts of air, suction and artificial turbulence.

This, then, is a theory whose propositions always have a direct technical relevance and which is also open to further development in line with the state of technology. The use of biplanes and propellers led to the elaboration of the multi-airfoil theory, in that higher air speeds first required a theory for swept-back wings, and then later a special airfoil theory which would take account of the problems of the sound barrier, and finally a theory for the field of supersonic flight.

These two examples present theoretical techniques which are both based on valid, fundamental physical theories, and are both guided by goals that are not already given within the framework of these fundamental theories.

A different kind of connection to the natural sciences can be found in more formal theories of technology, as for example in systems theory or in information theory (cybernetics). These utilize mathematics and logic to understand the formal structures of artificial information-processing systems. Originally developed as theories of technical systems, they were later applied also to natural structures.

At the outset information theory confronted the task of developing techniques for the transmission of messages, inventing various systems of coding, and transmitting messages without redundancy or distortion. Certain quantitative concepts such as information and redundancy were formed, and certain relations were established which made possible the theoretical solution of the optimisation problems. Information theory was later integrated into a general theory of information-processing systems. This theory now is applied not only in computer science but also in cybernetical models of biochemical, neurophysiological, ecological, physiological, and communicational structures. Technological progress has reached a stage at which its theoretical capacity yields special theories of natural processes as a sort of by-product.

The cases of fluid mechanics and chemical engineering have indicated how general scientific theories of nature such as hydrodynamics or classical mechanics became interconnected with attempts to solve technical problems by means of theoretical models such as the boundary layer or macrokinetics. In the case of information theory, the reverse also holds true: a general technological theory is here applied to natural phenomena. For example, the model of the double helix interconnects general information theory with genetics and the model of the metabolism of the cell interlinks information theory with the 'life' of the cell.

Both the development of sciences into special theories of technology and the development of technologies into special theories of natural structures show that after four hundred years of the development of science and technology their unification is no longer merely a philosophical project.

NOTES

1. The history of the construction of the Cathedral of Milan, including all the construction and design problems, has been described in detail by Frankl (1945), Panofsky (1945) and Ackerman (1949).
2. The otherwise unknown mathematician Stornaloco gave the following solution: The width of the cathedral or base of the bilateral triangle is 96 brachia (54 m). This measure was already established with the beginning of the construction in 1386. This yields the height $h = 96/2 \times \sqrt{3} = 83.134\ldots$ Rounding it off to 84, Stornaloco could subdivide the cross-section into rectangular squares with lengths of 16×14 br. This subdivision made it possible to figure out all the relevant heights of the buildings by means of basic arithmetic, as is shown by the following diagram (from Ackerman, 1949, p. 89):

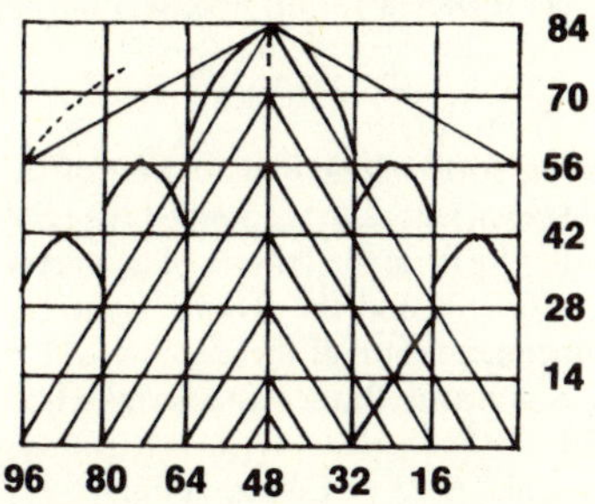

Panofsky (1965, p. 63) explains the advantages of Stornaloco's solution:

> The problem was to reduce an irrational value containing the factor $\sqrt{3}$, to a reasonably accurate expression by means of a formula that had to fulfill three conditions. First it had to achieve its purpose by means of simple multiplication and division instead of by means of extraction of roots. Second, it had to be applicable without involving an extensive use of vulgar fractions of which the Middle Ages, not yet familiar with the decimal system, were understandably afraid. Third, it had to meet the requirements of architects who measured the sides of their triangles in large units ('unitates') and wished to have the results, viz., the approximate values of the heights of these triangles, in small units ('quantitates'), one unitas, we remember, comprising 8 quantitates.

Looking at Stornaloco's formula from this point of view, we discover that it is very practical indeed. It operates like a slot machine: we throw in unitates, pull the lever and receive a reasonably accurate result in quantitates. For all we have to do is to multiply the given number of unitates (expressing the side of the triangle) by 100, multiply by 700, divide the result by 1010 – and we obtain a number which, when divided by 10, expresses with reasonable accuracy the height of the triangle in quantitates. Calling the number of unitates *n*, we can rewrite the formula in general terms:

$$10h \, \frac{n100 \cdot 700}{1010} .$$

If *n* is 6, that is to say if the side of the triangle is 6 unitates or 48 quantitates, 10*h* is, as we have already seen, 415 (85/100), and it is at once evident that *h* itself is a little more than 41½ or, as Stornaloco would say, a little less than 42. If *n* is 12, that is to say if the side of the triangle is 12 unitates or 96 quantitates,

$$10h \, \frac{12 \cdot 100 \cdot 700}{1010} = 831 \, \frac{69}{101}$$

and it is equally evident that, in this case, *h* itself is a little more than 83, or, to speak with Stornaloco, a little less than 84. If *n* is 4, the result for 10*h* is 277 (27/101), and for *h* itself a little more than 27½ or a little less than 28; and so forth. We now perceive the advantage of writing out the formula 10*h* instead of for *h*. For in doing so, we automatically obtain what would be called today the first decimal point of the final value, the 0.5 after 41, the 0.1 after 83, the 0.7 after 27, etc.; so that the fractional rest can be neglected for all practical purposes.

3. Ackerman (1949) has reprinted the respective parts of the records of the Milan architectural workshop. See p. 109.
4. This phrase is Ackerman's, p. 104. The explanation of the Milanese success is to be found in the fact that Medieval statics overdesigned the reinforcements of the construction, and, in fact, worked more from the criteria of geometry than from those of statics.
5. The conflict between the Gothic and the Romanesque tradition took place in other towns as well. A short documentation of the disputes at Florence in the second half of the Fourteenth century is given by Prager and Scaglia (1970), p. 2ff.
6. See Gille (1966), p. 179ff.
7. For example For. III 43 v. and C.23 v. (Ed. Zamboni, p. 33). For further interpretation see Krohn (1977), p. 95f.
8. For a more detailed description, see Olschki (1927), pp. 76f. Tartaglia uses another significant phrase, the 'practical speculation'. See for example the title of the ninth book of his *quesiti et inventioni* (1554) and the title of the sixth part of the *trattato de numeri et misurie* (1560).
9. For Paracelsus see Pagel (1958); for Paré see Zilsel (1945); for Vesalius see Farrington (1938).
10. For Gilbert see Zilsel (1941).
11. See Averdunk and Müller-Reinhard (1914).
12. Descartes (1967), p. 119.
13. ibid., p. 121.

14. For the compounding of the idea of the law of nature with technical rules by Descartes, Bacon, Newton, Hooke and others see Zilsel (1942), p. 86.
15. Leonardo, C. A. 119 v.a. (Ed. Zamboni, p. 13). Our italics.
16. Leonardo, C. A. 337 r. 6 (Ed. Zamboni, p. 17). Our italics.
17. Leonardo, G. 47 r, (Ed. Zamboni, p. 27). One might consider whether the above quotation refers to the *social* purpose of discovery. It is more probable, however, that Leonardo's position was to seek the *causa finalis* more in artificial things than in things natural, as indicated by the following remark: "You investigators therefore should not trust yourselves to the authors who by employing only their imagination have wished to make themselves interpreters between nature and man, but only [to the guidance] of those who have exercised their intellects not with the signs of nature but with the results of their experiments". I 102 r (Ed. Zamboni, p. 29). All translations into English of Leonardo's notes as cited in this paper are taken from Leonardo, Ed. by MacCurdy (1938).
18. Cassirer (1927) p. 161ff.
19. Piaget (1975), p. 82. For a more fully elaborated interpretation of the Scientific revolution of the Renaissance in the light of Piaget's theory of cognitive development, see Krohn, (1977).
20. Kant, *Critique of pure reason*, B XIV.
21. For an analysis of the social structures that underlay the founding of the Royal Academy see van den Daele (1977).
22. See Heckscher (1932), Vol. I, 165ff.
23. See Habermas (1968). Habermas overstates the significance of the "functional circuit of instrumental-rational action" for early modern science. The sweeping analogy between instrumental action and modern scientific ideals of explanation does not reach the concrete level of theory formation.
24. See Boas-Hall (1974).
25. See Mason (1961), p. 311; Clark (1970).
26. The fact that many and particularly the more general theories of modern science are theories about the functions of instruments leads to metatheoretical consequences which are outlined in Janich (1978).
27. It is very important to realize that the development of this technology particularly useful for scientific purposes was independent of the technology employed in the productive forces of the economy, though, of course, there existed a small market for scientific instruments.
28. The precision of the pendulum clock was useless for solving the most urgent problem of timekeeping of that period: timekeeping at sea.
29. See Koyré (1948).
30. See Middleton (1966). p. 28ff. and 48ff.
31. In the making of thermometers it was first proposed that all scientists should procure their thermometers from one and the same instrument maker. The problem of universality was later resolved fairly simply by establishing two fixed points. But even this solution presupposes the use of pure mercury and the regularity of the capillary tube.
32. Among the best examples of Descartes' deductive approach to experience is the *Traité*

de L'homme. Here even the interaction between the determination of the will by the soul and the activities of the body is mechanistically constructed.

33. See Dijksterhuis (1956) p. 534 and 547f. On the other hand, Newton was not able to work out his mechanistic program for all levels of nature. Material qualities and chemical processes were not covered by the "principia mathematica" and the general principle of gravitation was not in line with mechanistic technology, which excluded *actio in distans*.
34. See Aitken (1978).
35. The role of science in the industrial revolution is emphasized, for instance, by Schofield (1963) and Musson and Robinson (1969).
36. The non-scientific resources of invention, of course, remain extremely important in the nineteenth century. There are examples showing that the problems of scaling-up, of rendering procedures more precise, and of controlling disturbing effects, are resolved more or less satisfactorily by the empirical method, occasionally independent of already existing offers from theoretical science. Thus Stephenson optimized the use of steam power in the steam engine by practical inventions, and without any recourse to Carnot's theory of the motive power of heat. Similarly, in 1856 Bessemer patented a new principle of steel production which he had discovered through systematic experiments based on a theoretical explanation. But at the same time the same principle was discovered by the American Kelly in empirical-practical experiments. A parallel example is provided by Layton (1978). But for the production of synthetic dyes, by the end of the nineteenth century the techniques and concepts of organic chemistry were an indispensable condition.
37. See J. Bernal (1963), 107 pp.
38. The problem of the steadiness of locomotives at high speed was highly relevant to save Stephenson's main-line long-boiler engine of 1841, especially after a long-boiler engaged in a demonstration before the Gauge Commissioners in 1846 left the road and turned over (Snell (1971), p. 100).
39. Obviously there was a demand for science to solve technical problems even before the nineteenth century. Thus, for instance, when in pre-revolutionary France the production of saltpeter, which was the major constituent of gunpowder, had diminished, a prize contest with a view to an improvement of saltpeter production was offered under the auspices of the French Academy. For Lavoisier, who had formulated the conditions for the prize competition, it was clear that the replacement of the existing procedure – the collection of saltpeter was practiced generally by scraping walls and sweeping stables – was primarily a scientific problem. "The chemists have not established an entirely satisfactory theory of the principles of nitric acid, on its true origin and the mode of its formation" (Multhauf 1971, 166). But the program did not bring a solution, as the analytic capacities of eighteenth-century chemistry were not equal to the task. On the other hand by 1870 Thomas could proceed on the basis of the much more sophisticated techniques of elementary analysis and of the theory of acids and bases.
40. Pasteur, *Oeuvres* V, 215.
41. The contingency and broad variety of the relations between science and technology is amply demonstrated by a series of case studies in this field. See for example Mayr (1975) on the invention of the high speed steam engine, and Mayr's essay (1971) on the

development of the technology of speed regulation. See also Layton (1978). More classificatory and comparative ideas are developed in Layton (1971) and in Layton's and Mayr's essays in *Technology and Culture*, Vol. 17, Nov., Oct., 1976, which is dedicated to the historical analysis of science and technology and is focused on the second half of the nineteenth century.

42. The presentation of chemical engineering is based on a study by Buchholz (1974).
43. Eucken and Kirschbaum, both pioneers of the theoretization of chemical engineering, pointed out in 1934 that "in the construction of apparatus a vast and fruitful research field lies barren in that it is possible to clarify the outstanding question through systematic research". Cited after Buchholz (1974), 10.
44. The presentation of fluid mechanics draws on 'Autonomization and Finalization', in this volume.

BIBLIOGRAPHY

Ackerman, J. S., 'Ars sine scientia nihil est'. Gothic Theory of Architecture at the Cathedral of Milan, in *The Art Bulletin* **XXXI** (1), March 1949, p. 84f.

Aitken, H. G. J., 'Science, Technology and Economics: the invention of radio as a case study', in Krohn, Layton, and Weingart (eds.), *The Dynamics of Science and Technology*, Dordrecht: Reidel, 1978.

Artz, F. B.,*The Development of Technical Education in France 1500–1850*, London: MIT Press, 1966.

Averdunk, H. and Müller-Reinhard, J., *Gerhard Mercator und die Geographen unter seinen Nachkommen. Petermanns Mitteilungen*, Ergänzungsheft, Nr. 182, 1914.

Bernal, J. D., *Science and Industry in the Nineteenth Century*, Bloomington: Indiana University Press, 1963.

Bernal, J. D., *Wissenschaft* (*Science in History*). Reinbeck: Rowohlt, 1970.

Boas-Hall, M., *Science in the Early Royal Society*. Paper presented to the Summer Meeting of the British Society for the History of Science, Leeds, 2–4 July 1974.

Böhme, G., Van den Daele, W., and Krohn, W., *Experimentelle Philosophie*, Frankfurt: Suhrkamp, 1977.

Buchholz, K., *Zu Stand und Entwicklung der Verfahrenstechnik*, Ms., Frankfurt, 1974.

Cassirer, E., *Individuum und Kosmos in der Philosophie der Renaissance*, Leipzig and Berlin, 1927. New printing: Darmstadt: Wissenschaftliche Buchgesellschaft, 1974.

Clark, G., *Science and Social Welfare in the Age of Newton*, Oxford: Clarendon Press, 1970 (1937).

Van den Daele, W., 'The Social Construction of Science', in Mendelsohn, Weingart, Whitley (eds.), *The Social Production of Scientific Knowledge*, Dordrecht: Reidel, 1977.

Deffossez, L., Les savants du XVIIe siecle et la mesure du temps', Lausanne: *Ed. du Journal Swisse d'Horlogerie et de Bijouterie*, 1946.

Descartes, R., *The Philosophical Works* (translated by E. S. Haldane and G. R. T. Ross), Vol. I, Cambridge, Mass.: U. P., 1967 (1911).

Dijksterhuis, E. J., *Die Mechanisierung des Weltbildes*, Berlin, Heidelberg: Springer, 1956.

Farrington, B., 'Vesalio and the Rain of Ancient Medicine', In: *Modern Quarterly*, Vol. I, 1938, p. 23 ff.

Frankl, P., 'The Secret of the Medieval Masons', in *The Art Bulletin*, XXVII (1), March 1945, 46f.

Gille, B., *Engineers of the Renaissance*, Cambridge, Mass.: MIT Press, 1966.

Habermas, J.: 1972, *Knowledge and Human Interests*, London: Heinemann.

Hall, A. R., 'Merton Revisited or Science and Society in the Seventeenth Century', in *History of Science*, Vol. II, 1963.

Heckscher, E., *Der Merkantilismus*, 2 vol. (translated from Swedish by G. Mackenroth) Jena: Fischer, 1932.

Krohn, W., 'Die "Neue Wissenschaft" der Renaissance', in Böhme *et al., 1977*, p. 13ff.

Koyré, A., 'Du monde de l'à peu près à l'univers de la précision', in *Critique* **4** (28), pp. 809–882.

Layton, E., 'Millwrights and Engineers', in Krohn, Layton, and Weingart (eds.), *The Dynamics of Science and Technology*, Dordrecht: Reidel, 1978.

Layton, E., 'Mirror-Image Twins: The Communities of Science and Technology in 19th Century America', in *Technology and Culture* **12**, 1971, 562ff.

Leonardo da Vinci, *Philosophische Tagebücher*, edited and translated by Giuseppe Zamboni, Hamburg: Rowohlt, 1955.

Leonardo da Vinci, *The Notebooks*. Engl. (translation by E. MacCurdy), New York: Reynal Hitchkock, 1938.

Mason, S. F., *A History of the Sciences*. Rev. ed., London: Macmillan 1971.

Mayr, O., 'Victorian Physicist and Speed Regulation: An Encounter between Science and Technology', in *Notes and Records of the Royal Society of London* **26** (2), 1971, p. 205ff.

Mayr, O., 'Yankee Practise and Engineering Theory: Charles T. Porter and the Dynamics of the High-Speed Steam Engine, in *Technology and Culture* **16** (4), 1975, 570ff.

Middleton, W. E. K., *A History of the Thermometer*, Baltimore, Maryland : Johns Hopkins, 1966.

Mittelstrass, J., *Neuzeit und Aufklärung*, Berlin, New York: De Gruyter, 1970.

Multhauf, R., 'The French Crash Program for Saltpeter Production, 1776–1794', in *Technology and Culture* **12,** 1971, 163–181.

Musson, A. E. and Robinson, E., *Science and Technology in the Industrial Revolution*, Toronto: University of Toronto Press, 1969.

Olschki, L., *Geschichte der neusprachlichen wissenschaftlichen Literatur*. 3 vls., Leipzig and Halle 1919–1927. Reprint, Vaduz: Kraus 1965.

Pagel, W., *Paracelsus: An Introduction to Philosophical Medicine*, New York: S. Karger, 1958.

Panofsky, E., 'An Explanation of Stornaloco's Formula', in *The Art Bulletin*, **XXVII** (1), March 1945, 61f.

Pasteur, L., *Oeuvres completes*, Paris: Masson, 1922–1939.

Piaget, J., *Die Entwicklung des Erkennens II. Das physikalische Denken*, Stuttgart: Klett, 1975.

Prager, F. and Scaglia, G., *Brunelleschi: Studies of his Technology and Inventions*, Cambridge, Mass.: MIT Press, 1970.

Redtenbacher, F., *Die Gesetze des Lokomotivenbaus*, Mannheim: Bassermann, 1855.

Scheler, M., *Die Wissensformen und die Gesellschaft*, Bern, Munich: Franke, 2. Aufl., 1960.

Schofield, R. S., *The Lunar Society of Birmingham. A Social History of Provincial Science in Eighteenth-Century England*, Oxford: Clarendon Press, 1963.

Smith, C. S., *A Historical View of One Area of Applied Science – Metallurgy*, in National

Academy of the Sciences, *Applied Science and Technological Progress*, 1967.
Snell, J. B., *Mechanical Engineering: Railway*, London: Longman, 1971.
von Weizsäcker, C. F.: 1980, *The Unity of Nature*, New York: Farrar, Straus and Giroux.
Zilsel, E., 'The Origins of W. Gilbert's Scientific Method', in *Journal of the History of Ideas*, **II**, 1941, S. 1 ff.
Zilsel, E., 'The Genesis of the Concept of Physical Law', in *Philosophical Review*, **LI**, 1942, S. 345ff.
Zilsel, E., 'The Genesis of the concept of Scientific Progress', in *Journal of the History of Ideas*, **4**, 1945, S., p. 325ff.

WOLF SCHÄFER

NORMATIVE FINALIZATION

> The radicals were wrong; but they are beginning to look less stupidly wrong than they did once.[1]

1. INTRODUCTION

The study of science is still in its early stages, that is to say, without a paradigm. It is a field rich in hypotheses but poor in effective results. Philosophy of science often enough lacks empirical backing, and many case studies in the history of science are not compatible with the philosophy of science. In fact, when we ourselves have tried to check our theoretical approach against insights from the history and sociology of science, we found that, in switching to and fro between theory and empirical data, we tended to lose track of one or the other of these aspects. Moreover, there is an additional danger: as the field of critical science studies becomes professionalized, as it inevitably will, it may become depoliticized.

My presentation here is aimed against the gradual levelling of both the theoretical and political claims raised by our group's approach to science studies. I will focus on the political, that is, the normative issues. I consider these issues important because of the fact that the contemporary advancement of knowledge no longer automatically has emancipatory effects. The destructive capacity of modern weapons systems, or the risks of genetic engineering, require normative regulation. The concept of normative finalization seeks to take account of this.

2. THE THEORETICAL BASIS OF RESEARCH IN FINALIZATION

2.1. *Basic Hypotheses Reconsidered*

The basic assumption of the "finalists" is not in fact the notion of "finalization in science" itself, but the particular scope within science for alternative developments, and the receptiveness of science to social or political goals.

G. Böhme et al., Finalization in Science, ed. by W. Schäfer, 207–231.

We claimed that, "as science progresses, alternative paths for its further development constantly appear. Clearly these alternatives can be characterized by internal scientific criteria . . . Yet they are determined by factors external to science."[2]

The externalist concept of alternatives in science can be taken as a fundamental assumption since it applies to *all* scientific progress. This assumption gives rise to two lines of argument: firstly, the dynamics of science as a whole (general dynamics of science) and secondly, the dynamics within particular disciplines (special dynamics of science).

We formulated two theses which related to the general dynamics of science: that of *de facto* Darwinism in the history of science, and that of the trend toward finalization in current science. The former implies that until late in the nineteenth century, "it is not adequate to understand the development of science either as an approximation to 'true knowledge' or as the unfolding of some true essence of science; nor . . . as a conscious goal-oriented endeavour of human practice."[3] Or to put it more positively: the evolution of science in its Darwinistic period takes those paths which the socio-cultural environment allows it to take. Scientific evolution at this stage does not yet progress towards a predetermined goal but is more a departure from a starting point. The finalization thesis, on the other hand, points out that modern science can produce "conditions necessary for a research planning based as much on the interests of science as on the needs of society."[4]

As far as the *special dynamics of science* are concerned, we developed a three-phase model of the growth of theories with reference to particular disciplines. In the model the development of a discipline is considered to pass through (a) a pre-paradigmatic, or explorative phase, (b) a phase of paradigm emergence, or autonomous phase, and (c) a post-paradigmatic phase, in which finalization becomes possible.[5]

(a) The pre-paradigmatic phase of the special dynamics of science is determined by the growth of experience through the use of apparatus and instruments, through the systematic ordering of empirical data, and through the formation of heuristically meaningful analogies and models of reasoning. Although the explorative phase therefore possesses certain structures it still lacks a comprehensive theory. Tools are transformed into instruments, phenomena into effects, and ideas into concepts. The unity of pre-scientific experience rooted in everyday life is shattered, and a field of scientific inquiry is opened up.

(b) In the subsequent phase of paradigm articulation, the search for the fundamental theory of an experimentally structured field becomes dominant.

This theory must be capable of resolving all the fundamental problems of that field, thereby reconstituting the fragments of experience into a theoretical unity. We found that science follows an autonomous path of development in this phase, and regard this autonomy as being essential on cognitive grounds. The various theoretical approaches, each with their own internal problems and debates, give rise to a clear-cut research front. The individual scientist has, by then, come to realize that the most important and interesting problems of the field are those that will provide it with the theoretical foundations it has, up to then, lacked.

(c) The knowledge accumulated in the preceding phase of cognitive autonomy makes it possible to formulate the fundamental theory of a particular research field. Scientification of the field having therefore achieved a stage of theoretical maturity, further theoretical development can be linked with specific social interests. Thus, the strategic direction of fundamental research begins with the post-paradigmatic phase. The cognitive resistance to external direction (which marks autonomous fundamental research) becomes obsolete once the fundamental theory has "in principle" resolved all the problems of that field of research.

Accordingly, the three-phase model of the special dynamics of science offers an explanation of the internal logic of scientific development. This model, in turn, allows us to answer the question as to why a particular field reaches the point of finalization, at which external goals can be realized consciously as guidelines for further theory development.

The concept of the *general dynamics of science* presupposes some kind of transition from the "Darwinist" to the "finalization" stage of scientific progress. This changeover sheds light on two epochs in scientific history, and two sets of alternatives open to science. We assume that the "replacement of Darwinism in the development of science"[6] took place in the nineteenth century. The feature which distinguishes "Darwinistic" and "finalist" alternatives in science is that in the former, useful knowledge is produced *spontaneously* (*naturwüchsig*), while in the latter, it is produced *strategically*. "The history of science may be characterized as Darwinistic insofar as decisions on alternatives in science have occurred which can be accounted for neither by the deliberate strategies of scientists nor perhaps by cogent criteria of scientific rationality (truth), but have to be explained as resulting from influences in the environment of science."[7]

Finalist alternatives in science invert the sequence postulated by classical evolutionary theory. Variation no longer precedes the selection of new mutants but results from prior selection.

> As soon as the internal mechanisms of scientific progress . . . become transparent, they can be strategically employed. Theories which so far, under suitable conditions of the socio-cultural environment have, so to speak, grown "naturally" can now be designed as strategies. "Normal science" then will no longer be confined to exhausting the respective field of knowledge in accordance with scientific viewpoints, but, with the help of a kind of cartography of the philosophy of science, will be able to traverse the terrain in the direction of the goals set by policy.[8]

Since we have assumed that the proliferation of alternatives in science is determined by external goals, we have to clarify our position on externalism. We believe that external factors regulate the *direction* of scientific development. This concept of externalism, however, concerns only the *reconstruction* of particular events and developments in the history of science; it does not impinge on the truth-value of scientific propositions. A stronger externalist position, which we have at times voiced but never really worked out, would extend the influence of external factors to the *theoretical* development of science. Exposing the hard core of science to socio-cultural influences would mean that cognitive structures could be explained by historical and sociological analysis. Figures 1 and 2 indicate the connection between the two externalist approaches – the weaker and stronger positions – and our two lines of argument.

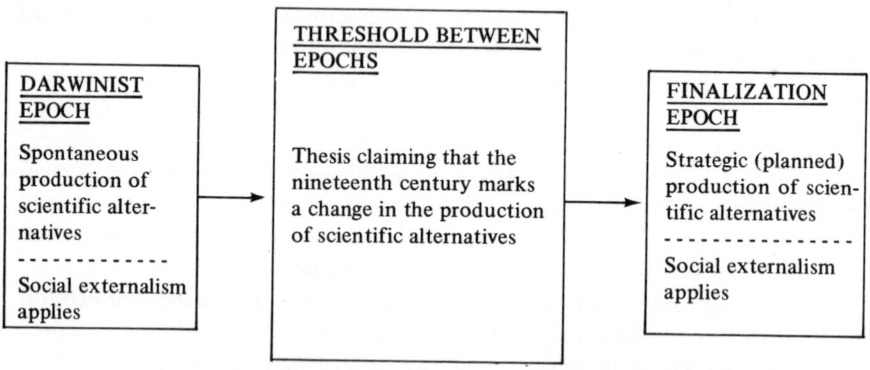

Fig. 1. The general dynamics of science (historical dynamics).

To avoid possible misunderstandings arising from the use of the attributes "weak" and "strong" (which might imply that externalism is of interest only in its strong variant), I would prefer to use the terms *social* and *cognitive* externalism.[9]

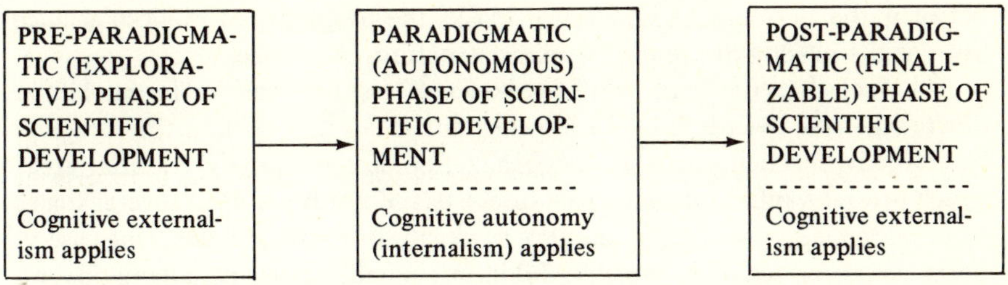

Fig. 2. The special dynamics of science (structural dynamics).

Social externalism is predicated on the premiss that the history of science and social history are interrelated. The historian of science will therefore endeavour to trace the ways in which socio-cultural factors have influenced the direction of scientific progress, and in particular, the changes of direction they have induced. The philosopher of science who has opted for cognitive externalism will have to defend the thesis that social factors influence not only the genesis but also the truth of objective knowledge. Adopting this thesis implies shifting the focus of science studies from the context of discovery (which is the object of enquiry of social externalism) to the context of justification (which is the object of enquiry of cognitive externalism), because the intention of cognitive externalism is to work out the social definition of the truth-content of propositions. However, it should be stressed that *if* the claim for cognitive externalism is made, the only areas in which it can contend for validity in our model are the pre- and post-paradigmatic phases of the special dynamics of science, as indicated in Figure 2.

Finally, I want to clarify the connection between our two lines of argument. It is possible to correlate the explorative phase of science and the spontaneous production of alternatives in science on the one hand, and on the other, fundamental research open to finalization and the planned production of alternatives in science. But it would be wrong to match the periodization of general science dynamics with the three-phase schema of special science dynamics. The argument behind general science dynamics is an evolutionary one and points to the emergence of a new epoch of science during the nineteenth century. By contrast, the argument behind special science dynamics is a structural one and presupposes a sequence of three successive stages of theory development. This sequence is defined not in historical but in systematic terms. Thus, the historical transition referred to in Figure 1 cannot be used to periodize the special dynamics of science. The threshold between the two epochs of science masks the beginning of modern science but not the passing of pre-paradigmatic structures in science.

The fact that finalization of fundamental research can be found in the nineteenth and twentieth centuries does not mean that *all* scientific progress has already reached the state of finalization. Indeed, many research areas which are relevant to current social interests and needs combine several stages of development, revealing a complex aggregate of research fields. One case in point is cancer research, which exhibits pre-paradigmatic, paradigmatic, and post-paradigmatic structures of scientific development.[10] Consequently, the application of the three-phase model to contemporary science in no way legitimizes what "anti-finalists" suppose, namely the totalitarian "control of science, including even attempts to block the advancement of certain fundamental research areas."[11] Rather it enables us to identify the different levels of theory development, which in turn imply different means of orientation and control. So that, for example, one would keep science policy from interfering with an autonomous phase.

Before ending this section, it might be worth noting that until now we have never made an explicit distinction between general and special science dynamics. However, the differentiation between the two lines of argument should now make it evident that, within the frame of general dynamics, we are operating with a macro-theoretical two-phase model (Darwinist epoch and finalization epoch), while within that of special science dynamics we are using a micro-theoretical three-phase model (pre-paradigmatic, paradigmatic and post-paradigmatic phases).

2.2. *Normative Claims Revised*

In the past we linked the finalization thesis with normative claims. "Where natural science is normative, the point of reference of scientific generality should be universality in society, not in nature. Thus generality in the form of unlimited reproducibility would yield to the generality of social consensus."[12]

While offering some orientation, this position is still doubtful. It correctly points to the inadequacy of "unlimited reproducibility", as well as to the urgency of political debate about the future development of science; however, the thesis itself is too radical. It seeks to shatter the *cognitive* framework of natural science. The effect of such a revolution would be to subordinate experimentally-derived natural laws to discursively formed goals of social development. The purpose of this revolution would be to provide a social definition of scientific truth. This, however, is not yet accomplished.

The thesis is also unclear because it is concerned with the future. There is no way of knowing whether scientific progress and scientific truth will ever take the generality of *social consensus* as their point of reference. The future path of a science open to finalization would become even more uncertain if its guidelines were to be dictated by the particular interests of power groups. Science can take either an emancipatory or a repressive course. That is why we have always stressed that finalization subject to such particular control would lead to an asocial science.

Finalization means precisely receptiveness to theoretical internalization of external goals. Within this structure science can in principle maximize both its conformity to power and its services for particularistic social interests without degenerating itself as a useful science. The possible achievements of such a science go even further: by internalizing the stabilization of power as a goal into its theory it renders domination itself scientific . . . Should this kind of scientificity come to dominate the science system then the emancipatory phase of science has been but an episode.[13]

The domination of scientific truth by the generality of social consensus is problematic, whereas the normative claim for a socially oriented science is not, as long as this claim is not associated with cognitive externalism. If we are willing to put cognitive externalism aside, we can develop the conceptual framework of "normatively oriented science"[14] without first having to substantiate the proposition that "theoretical internalization of external goals" does take place.

2.3. *The Notion of a Normative Finalized Science*

The idea of a "normatively finalized science" was first outlined in 1973. Since then we have not systematically elaborated or revised this idea. Neither do I intend to discuss whether the generality of consensus about the objectives and goals of society should replace the classical generality of scientific laws as a point of reference. Instead, I am concerned with the *convergence of two types of discourse*: theoretical discourse about objective knowledge and practical discourse about social interests.[15] With reference only to this concern, I will briefly examine the question of the "limitation of the claims to the general validity of scientific laws".[16]

Our group started from the premiss that the universal validity of classical physics had been relativized. The first limitation of the concept that events can be reproduced without limit comes with the discovery that nature – both animate and inanimate – is historical. Thermodynamics and Darwinism have both brought out the temporal dimension of the world and its historical species. The second limitation arises from the concern with complex phenomena. Meteorology and the scientification of wind and weather (to serve weather forecasts and in the long-run climate technology) both seek to give increasingly accurate definitions of the singular event instead of producing general idealizations which are not concerned with concrete events. The third limitation of unlimited reproducibility results from the historical consequences of degeneralizing scientific knowledge. The process of degeneralization is induced by the extensive use of scientific knowledge to dominate and control nature. Thus, the once valid proposition that DDT acts as an insecticide was established by laboratory experiments: but repeated millions of times in nature, these led to the selection of resistant insect strains, thereby invalidating the proposition.

The internal relativization of the ideal of natural science can be interpreted constructively if the above-mentioned limitations are not seen as destroying the traditional conception of science, but, instead, as an advance towards an empirically more meaningful conception of science which includes,

– the temporality of nature,
– the singularity of phenomena, and
– the transformation of natural objects and fundamental life-relationships through the technical application of scientific knowledge.

Furthermore, this new conception of science would have to reflect the "unity of man and nature" on a theoretical level as the "circle of knowledge",[17] and, since empirical sciences produce the experiences they make, on a practical level as the impact of knowledge. Such a conception of knowledge would be capable of integrating both the human and the natural history of nature.

Developments in ecology have made it evident that this new conception of science is not merely the product of a brainstorm. Ecological thinking reveals, on the one hand, the peculiarities and variability of natural reproduction cycles, and, on the other, that scientific and technological interventions within nature fulfil purposes different to those of nature itself. In ecology the desired states of nature and the social purposes of society are interrelated. Hence, we proposed that ecology should serve as the prototype of a normatized natural science which, unlike traditional natural science, combines objective *and* normative elements consciously and intentionally.

The modern interest in ecology as a science is a reaction to the results of the "construction of nature" by the scientific-technical interventions of man. Ecological analysis has to define its subject matter as a reproduction process (system). This implies on one hand conceding that nature has norms (goals) of its own – the distinction between "is" and "ought" is appropriate even within the frame of natural science itself. On the other hand, however, since man himself and nature constitute interdependent systems for which stability and reproducibility are required, it seems to be impossible to determine even those natural norms without a recourse to some human norms for the intervention within nature. The reproduction of nature is no longer determinable solely in terms of natural variables. The definition of the nominal values of natural systems will increasingly be mixed together with "human determinants" of nature, i.e., with forms and purposes of man's development of nature. Ecology thereby develops from an objective natural science in the traditional sense to a science with normative, strategic elements, supplying evolutionary concepts of nature. It thus becomes open to the analysis of processes by which nature reproduces itself in terms of alternative goals.[18]

Our concept of a normative finalized science has been shaped by ecology. We hold that the *rational* coordination of natural norms and social interests within finalized science has become not only possible, but also necessary. Technological intervention within nature has reached global dimensions and is gaining in intensity, running the risk both of injuring the purposes of nature as much as the interests of society. Consequently, natural systems should be "exploited" in a way that does not overlook or, even worse, disobey the laws which govern the reproduction of these systems, and that does not come into conflict with the vital interests of human society.[19]

For us, the minimum normative requirement for an orientation of those sciences open to finalization would be that it does not damage the natural environment, as the basis for life, or violate fundamental humanistic goals of social development. Finalization without such a minimum of normative regulation leaves the social orientation of scientific progress at the mercy of the established powers in politics, industry and the military. We therefore have to confront the problem of how to guard strategically developed theories against particular interests. Since we considered that particular interests could only be mediated rationally through universalistic discourse, we postulated "rationally generated social consensus" as a norm (although not a very realistic one). However, relating the laws of nature to the needs of society means correlating *true knowledge* and *rational interests* in such a way that our scientific discoveries are governed by universal social norms and that our social purposes develop on the basis of objective knowing.

Our neo-conservative critics have rejected this view instead of offering a different answer because there are particular questions they do not want to see raised. Although the development of our research may branch in different directions, there are nonetheless a number of fundamental questions which are simply too relevant to be suppressed.

3. RETRACTION OF BASIC HYPOTHESES AND SUPPRESSION OF NORMATIVE CLAIMS

3.1. *Theoretical Retraction of Basic Hypotheses*

Our approach to science studies, as drawn together above, initially promised to be far-reaching. It indicated "alternative lines of development" in science, and contended that "certain lines of development may die out" or "become 'dead branches' of the history of science."[20] It engendered a "system in which existing sciences could be classified and from which missing disciplines could also be identified";[21] it made clear that it "is imperative today that we take command of the development of science."[22] Our progress in achieving these objectives has so far been slow. We have not yet identified the lines of development which may die out or "be reactivated later"[23] as alternative developments. Nor have we, as yet, managed to clarify the empirical conditions of the "paradigm change of science itself."[24] I attribute this to the fact that our plans were ambitious, although not illegitimate.

In natural science the existence of true states of fact is normally taken for granted, irrespective of whether they are actually discovered or not. Efforts to penetrate such a rigorous concept of truth have repeatedly proved unsuccessful. The problems which the concept of cognitive externalism gives rise to are but part of a whole tradition of promising attempts to conceptualize a New Natural Science. The history of this tradition stretches at least from the early works of Marx to the later works of Marcuse. The question is: should one reduce or retract this theoretical framework in the light of the conceptual difficulties which are inherent in this approach?

Wolfgang van den Daele has drawn a number of ambivalent "consequences" from his analysis of the rise of modern science (*cf.* van den Daele, 1977*a*, pp. 166–172, and 1977*b*, pp. 45–48). Pointing to the historic compromise which led to the political selection of positive science, he argues against an internalist interpretation of this process. Yet his arguments against the demands for an alternative form of science rest on the cognitive principles of positive science. Finally, while wanting the social role of science to be modified, he no longer questions positive science itself, since, as he puts it, "the limits for a social reconstruction of science lie first in the society and not in science."[25]

The extent to which van den Daele has reduced our position becomes clear when he says that the originally planned investigation of alternative developments in science is not possible.

Current theories in philosophy of science are incapable of systematically analyzing the alternatives to positive science. They are founded in contemporary science. Current philosophy of science is capable of reconstructing existing science but not of constructing a new science *a priori*. What is scientifically possible is determined by science, not by reflection on science. It appears that the only method of proving that an alternative science is possible is actually to make it.[26]

Admittedly, the strongest proof of the possibility of something is its existence: the construction of a new science would therefore undoubtedly constitute striking evidence. However, is the technical demand actually "to make it" really justified in this instance? Can one draw up and manufacture an alternative science, in the same way as an alternative energy system? Do we even have to present the new science to contemporary science in the form of a prototype for approval? In my opinion, it cannot be accomplished as easily as that. An alternative science cannot be "invented", either by scientists or philosophers of science; van den Daele is clearly correct on this score. This is not what is being maintained, however. What is being maintained is that there is a difference between positivist and critical science studies. Positivist science studies excludes our theoretical and political imagination and keeps to what actually is – the existence of positive science. Critical science studies makes use of our theoretical and political imagination and also investigates what actually is – namely the historical and current tendencies of scientific development which can no longer be fitted into the conceptual Procrustean bed of positive science, such as normative ecology.

3.2. *Political Suppression of Normative Claims*

In addition to the danger of theoretical retraction, our project was threatened by external repression. Prominent German opponents of finalization have endeavoured to outlaw the "Starnberg approach" to science studies as "neomarxist".[27] They have proved unsuccessful on the theoretical level: however, it would be mistaken to underestimate the social influence of such an ideological campaign against the social orientation of scientific progress.[28]

The "paradigm change of science itself" has not yet reached public consciousness because, as far as the history of science goes, the possibility of the strategic production of scientific alternatives is relatively novel. Antifinalists are in tune with a state of consciousness which, for the time being, is representative. They can attack the demand for a fundamental change in the relationship of science and society by appealing both to the traditional self-understanding of scientists, and to the current conception of science. To show that the antifinalist polemics lack legitimacy, it must be made evident that the new possibilities (and risks) inherent in finalization cannot be seized using the classical conception of science, valid until the mid-nineteenth century.

The arguments which led me to oppose the antifinalist ideology of science are: firstly, fundamental research needs to be goal-oriented; secondly, scientists need to be willing to adopt partisan positions; and thirdly, theoretical and practical discourse need to converge (see 'The Finalization Debate' in this volume).

The three essential points can be summarized as follows.

(a) The anti-finalists objected that any kind of goal-orientation of fundamental research would signify the end of free science with "disastrous consequences for the advancement of knowledge". While admitting that goal-oriented basic science might bring "short-run successes", it was argued that it would, however, "in the end destroy the fertile humus for application-related research and technology."[29] This vision of the future of science assumes that only "research without a goal yields the greatest degree of goal-realisation."[30] I do not share this view, and would answer the anti-finalists by saying that external goals become the guidelines of theory construction in the post-paradigmatic phase *anyway*.

(b) The line of attack regarding the "modern cult of partisanship",[31] which the anti-finalists seem to think they can most effectively oppose by being addicted to it ran as follows: partisanship implies the "divorce of political legitimacy from truth";[32] Max Weber was quoted;[33] it was said that there exists "an objective truth which is independent of the knowing subject, his wishes, preferences and interests," and that "the cognitive value of scientific findings cannot be determined on the basis of evaluations that are extraneous to scientific knowledge, and that it is not up to scientific knowledge to make value judgements regarding its object of inquiry."[34]

I cannot accept these arguments. The traditional categories that distinguish scientific knowledge and social interests in no way rule out that one and the same person may form descriptive *and* evaluative judgements concerning one and the same subject matter. Besides, the analytical distinction between these types of discourse tends to obscure rather than clarify the issue since it leaves unanswered the issue of whether theoretical and practical interests have equal truth-claims, that is to say,whether both practical interests and theoretical knowledge are equally value-related.

(c) The anti-finalists' claim that a partisan science can only result in "totalitarian confusion"[35] extends their critique beyond the individual scientist's partisanship to include the project of a partisan science. For them, the convergence between science and society constitutes a "utopian fusion of theory and practice" which does not bring with it "peace and reasonable freedom, but politico-pedagogical and perhaps even worse forms of terror."[36] Admittedly, our assertion that "the perspective of the finalization of science embodies a growing coincidence of theoretical aims and social norms"[37] might be accused of aiming at such a "utopian fusion". Since we had in mind no irrational amalgamation of theoretical and practical discourse, I propose to use the term *growing convergence* instead of "growing coincidence" between the two types of discourse. This should avoid further misinterpretation.

Theoretical and practical discourse are linked in a special way by the notion of the growing convergence between theory and practice. Growing convergence between these discourses means interrelating science and society so that scientific knowledge *rationalizes* conflicts involving social interests, while conversely, social interests *orientate* the pursuit of scientific knowledge. The specific rationality of either practical or theoretical discourse is not in doubt. But, if practical and theoretical reflection are kept separate and left to their respective "experts", each discourse risks being completed without the benefit of the other. Instead of fighting "totalitarian confusion", concerned anti-finalists should realize that the danger is this one-eyed rationality, as recently illustrated, for example, by the socio-cultural shortcomings of the "green revolution". One is therefore led to conclude that human action in nature and society will only become responsible and enlightened if the mediation of true knowledge and rational interests is widely supported and actively encouraged.

4. THE SOCIAL RECONSTRUCTION OF SCIENCE

4.1. *A Utopian Perspective for Science*

The use of science and technology to oppress and destroy human beings and to exploit nature no longer belongs to the realm of science fiction. An alternative integration of science and society would have to bring the norms of nature and the interests of human beings into accord, instead of violating them. To advocate such an alternative is to articulate a political interest in the social reconstruction of science. This political assertion has been formulated as a programme for science by Everett Mendelsohn.

> A reconstructed science would value truth, but also compassion. It would have an inbuilt ethic that would defend both being and living; that is, knowledge would be non-violent, non-coercive, non-exploitative, non-manipulative; knowledge that would set out to pursue harmony with nature rather than dominion over nature; knowledge that would include subjective experience, that is, interior knowledge; that would renounce finally the Faustian quest to achieve the limits of the universe or total knowledge; that would work to construct models that would be more explanatory and more inclusive – science practised among and derived from the public. What if we were to say that we would not undertake to develop what could not be understood and publicly absorbed, that we were intent on building a science not confined to academies and institutions. This is only the beginning of a list. It would have to go beyond this in new directions. But this certainly would remake and reshape scientific activity. Change of this sort would represent a revolution – a revolution no less in magnitude than that of the seventeenth century – a revolution in science. But I would say that the crisis of knowledge that we face is no less great today than that which faced the pioneers of the new science of the seventeenth century.[38]

Our own "manifesto" might also start this way. We share Mendelsohn's political commitment and his theoretical perspective. Our research programme questions the pursuit of knowledge for the sake of knowledge. We criticize the "asocial" positivism inherent in the traditional idea of truth; we too are looking for an ecological orientation of scientific and technological progress. In short, we conceive science as a "*collective utility*".[39] And yet, we too have failed to go beyond the opening sentences of this manifesto.

4.2. *Two Strategies for Research*

The social reconstruction of science not only relates, of course, to particular sciences but to science as a whole; our assumption that "a kind of paradigm change of science itself" has taken place refers to science in this global sense. By contrast, when studying the origins of agricultural chemistry or cancer research, we are dealing with concrete sciences which allow detailed descriptions and statements about the particular. Accordingly, the social reconstruction and development of particular sciences within the microtheoretical frame of special science dynamics must be distinguished from the social reconstruction of science as a whole within the macrotheoretical frame of general science dynamics.

Research on the special dynamics of science yields results for a critical social history of science incorporating forgotten or suppressed alternatives.

Such research examines the internal developmental logic and the theory-building within particular disciplines. Its objective is to identify the external factors which influence the history, and, possibly the cognitive structure of a given science. Furthermore, it analyzes the options open to modern research policy through the finalization of science, although it is not restricted to post-paradigmatic theory development. As such, it serves to rationalize the planning of scientific progress. This is the line of research we have so far been mainly concerned with.

Research into the general dynamics of science is concerned with science as a whole and focuses on the structural changes in humanity's interaction with nature. It transcends the historical and theoretical frame of reference of modern science by inserting it into the "human history of nature"[40] which does not relate to the pursuit of knowledge for the sake of knowledge, but to the evolution of technology.

The threshold between the spontaneous and strategic production of alternatives in science indicates that our knowledge of nature evolves towards finalization, and not solely towards an increasingly deeper and more exact knowledge of the universe. Finalization ends autonomous scientific progress for its own sake, and begins to advance human progress by means of science. The highest stage of scientific progress is represented by scientificated technology – hopefully with a human face. From this point of view, finalization in science is revealed as scientification of technology: this step is but the latest stage in the culture–nature interaction which can be said to have begun with the earliest tools. The whole socio-cultural learning process through which humanity actively conceptualizes and transforms nature needs to be framed theoretically. We realize this, and intend to make it a topic of further research.

The social reconstruction of science will be a blind venture as long as we limit our investigation exclusively to either the special or general dynamics of science. The history of nature has to be reconstructed from the vantage point of both these research areas if it is to reveal more than half the field of vision: the general dynamics can shed no light on the internal conditions which determine the social receptivity of science. Conversely, the special dynamics do not encompass the epochal changes in the "natural states of humanity" outlined by Moscovici.

4.3. *Toward a Revolution in the Social Paradigm of Science*

Natural science is a social enterprise, but this does not make it a *social* science.

Science is permanently shaped by external influences, and that is enough to invalidate the internalist notion of an "unsocial" science, but not sufficient to delineate the novelty of a *social natural science*.

Our hypothesis of a "fundamental change in the relationship between science and its social environment"[41] seems to reveal a new quality in the interrelation of scientific progress and socio-cultural factors. In fact, this quality is "the replacement of Darwinism in the development of science", or the transition from spontaneous to strategic production of scientific alternatives. Compared with early modern science, the strategic production of alternatives gives modern science a higher degree of instrumental functionality. Science becomes technologically better but not socially good. The new way in which modern science generates its alternatives does not, therefore, indicate a fundamental change in the relationship between science and society, but rather increases its efficiency as an instrument for manipulation. The relationship between science and society will, presumably, only gain a new quality if normative finalization is put into practice. And yet, giving finalization a normative edge merely means giving modern science back what early modern science lost.

Wolfgang van den Daele's analysis of the social construction of science has revealed the price society had to pay for the new science of experimental philosophy in England and France during the latter half of the seventeenth century. Almost all the social and democratic alternatives envisaged by the new learning were both politically and theoretically suppressed. Proponents of the emancipatory alternatives were denounced as rebels[42] and the Baconian reform projects, which were intended to serve the "commonwealth", were barred from the corpus of positive knowledge.

Christopher Hill has reconstructed this first attempt to produce a socially beneficial science.

> The radicals of the English Revolution made a last attempt to see the universe as a whole, science and society as one: Copernican astronomy had ended the distinction between heavenly and sub-lunary: the radicals aimed at completing this by ending the distinction between specialists and laymen. They wanted to drive scholastic theologians out of the universities, to end the dominance of Latin, Greek and Hebrew; but they did not want science to be handed over to a new set of mumbo-jumbo men. Dee, Bruno, Fludd and many others had aspired to understand the whole universe in all its aspects. Comenius was perhaps the last serious thinker to attempt an all-embracing synthesis and apply it to human life. Winstanley wanted science, philosophy and politics to be taught in every parish by an elected non-specialist, drawing on the pool of scientific and other information which something like Hartlib's Office of Addresses would have furnished. He and the radical scientists wanted science to be applied to the problems of

human life: this was the practical significance of their emphasis on astrology, alchemy and natural magic. Their defeat, however scientifically necessary and desirable, also meant the end of dreams of an all-embracing Weltanschauung accessible to ordinary people.[43]

Up to the mid-seventeenth century the New Learning was connected with moral, political, social and educational goals; its distinctive features were:

– *anti-authoritarianism*, or rejection of the classical authority of Aristotle, Ptolemy, Galen and others.

– *progressiveness*, or the idea of progress as an open future to be attained through critical scrutiny of the old, and through methodological investigation and accumulation of the new.

– *anti-elitism*, or revaluing intellectual and manual labour in favour of the latter and in opposition to scholastic erudition.

– *educational reform*, or reconstructing the scholastic educational institutions in order to provide an education that was practical and that promoted the public good.

– *humanitarian orientation*, or paying attention to the needs of the poor, especially the provision of free medical treatment.

The foundation of the Royal Society in London by Charles II and of the Académie Royale des Sciences in Paris by Louis XIV mark a turning point in the early history of science, opening the way to a new science, stripped of all radical or emancipatory aspirations. A science defining itself as "pure", without admitting that its purification came from above. It gained political legitimacy by virtue of royal edicts at the cost of explicitly renouncing normative claims.[44] The normative (moral, educational, political and social) neutrality which resulted from this historic compromise is now the predominant feature of scientific positivism, objectivism and factualism. It is worth noting that neutral science was necessary, not cognitively but politically, for the social institutionalization of science in the latter half of the seventeenth century.[45]

One can sum up by saying that the selection of a normatively neutral science was not made on cognitive grounds. The contest between the variants of the new sciences was not decided rationally, nor within science, but outside it, with the external influence of political power as the decisive factor. When positive science was institutionalized, the cognitive criteria that could have demonstrated the superiority of normatively neutral science did not exist. Instead, it was the Restoration which, in 1660, ended the association between the New Science and "Moralls and Politicks".

The immediate and unequivocal consequences of the separation of morals and politics from experimental philosophy served the interests of those who held power. "The frame of *Men's bodies* . . . the *Arts* of *Men's Hands* . . . the works of *Nature*"[46] were established as the legitimate objects of scientific enquiry, thereby rendering the New Morals and Politics of the seventeenth century illegitimate. Within these domains controlled by state and church, a niche was created for those wanting to pursue scientific activities without moral or political pretensions, in which the guiding rule was the pursuit of knowledge for the sake of knowledge. Blocking the New Science from research into society, and equating "true" science with normatively neutral science had unpredictable, and hence, unintended effects: namely, to obstruct the scientification of morals and politics in the long run, and to deprive science as well as society, of moral and political control over the advancement of science itself. As a result, subsequent social progress was realized without scientific enlightenment and the ensuing scientific progress was achieved without social evaluation.

The first impulses for the scientification of morals and politics can be found in the declaration of the True Levellers and in the efforts of astrology and magic.[47] Suspected of being politically radical and charged with being pseudoscientific, they were crushed. Complementary efforts to moralize and politicize experimental philosophy suffered the same fate. Hence, the relationship of science to society could only be changed fundamentally if this paradigmatic suspension of practical rationality were to undergo a crisis, be transformed and replaced. And indeed, this state of crisis has been reached.

Bourgeois-capitalist societation[48] of the new science led experimental philosophy away from its emancipatory goals and established a hazardous relationship between pure natural knowledge and the profitable control of nature. Political discourse, such as Winstanley's[49] about the twin objectives of "creation" – that is "community of mankind" and "community of the earth" – was discarded as an unscientific mode of reasoning. The fact that the early modern claims for a communitarian science of humanity and nature look less stupidly wrong to us is due to an ever-increasing awareness that our exploitation of humanity and nature has its limits. This awareness was aroused in the nineteenth century. The vision of Liebig and Marx enabled them to see the dawning crisis behind the booming expansion of capitalism, science and technology. They illuminated theoretically the worst form of agriculture and industry as "Raubbau" (rapacious exploitation).[50] Each of them envisaged a revolutionary solution to this key problem. Liebig insisted on the ecological reintegration of humanity into nature; Marx advanced socialism as the means for mutual human integration. Since these two programmes still have to be put into practice, let alone unified into a new and better social order, my aim can only be to ask whether a change in the social paradigm of science is conceivable.

The social paradigm of normatively neutral science has enabled the production of true knowledge without any regard for desirable social interests. As yet, it is uncertain whether it will be possible to overcome this positivist constraint of rationality. But, it does seem possible to outline those arguments which speak for the convergence of practical and theoretical discourse and against their separation. These arguments may be distinguished as changes of consciousness in the socio-cultural domain.

(1) *A new conception of history* shook up the old order of things. In the period from 1750 to 1850 our temporal experience was formulated,[51] the human and the natural world were historicized and reordered:[52] Marx and Engels discovered the human history of nature, Darwin and Wallace discovered the natural history of humanity.

(2) *A new conception of nature* made the interconnection between natural and social history intelligible. The expansion of the capitalist mode of production into a world-wide system made clear that the industrial "maîtres et possesseurs de la nature" (Descartes)[53] utilized the natural riches of the earth and the technological potential of science in destructive ways.[54] Aiming to command nature and to bring about the "magna instauratio imperii humani in naturam" (Bacon), they commenced to destroy it. The early modern conception of nature that disregarded its human history was consistent with the little-developed technology which existed up to the mid-nineteenth century. Since that time, it has lost validity. Ecology which initially centred on natural processes, started to pay attention to the "human" factor.[55] Now human ecology, environmental research, and the concept of "ecological socialism"[56] all point to a sociocentric conception of nature.

(3) *A new conception of science* paralleled the reconstructed conception of nature. The temporality of the world, the singularity of phenomena, the "circle of knowledge" and the socio-cultural transformation of nature are its distinctive features; they have already been mentioned. These four novel perspectives reveal the limitations of the classical conception of science, and illustrate the anthropofugal dynamic of early modern science. Counter to this, an "anthropocentric orientation of the theory programme of natural science"[57] is strengthened by the modern tendency towards finalization. This trend raises the problem of the norms which should orientate scientific and technological progress.

(4) *A new conception of society* emerged as a result of the scientification of morals and politics. It started in the early nineteenth century with the development of sociology and the socialist theory of society. The attempt to link the positive knowledge produced by the social sciences with the emancipatory goals of modern social movements (such as the labour movement, or later the ecology and feminist movements) called into question the traditional structures of privilege and power. In our own period, Jürgen Habermas has developed a research programme for a theory of social evolution which draws on this tendency to enlightened politics. His thesis is: "The species learns not only in the dimension of technically useful knowledge decisive for the development of productive forces but also in the dimension of moral-practical consciousness decisive for structures of interaction".[58] This is perhaps too optimistic a proposition, and therefore I would like to give it an outright normative reading:The human species is obviously recognizing that it *should* learn in both dimensions.

We are approaching the theoretical integration of these new structures of consciousness. The goal is known in its outlines, as drawn by Liebig and Marx. The lobotomization of rationality can be seen all the more clearly against the backdrop of these new conceptions of history, nature, science, and society. The tendency toward the convergence of theoretical and practical discourse is undermining the legitimacy of the social paradigm of early modern science. A normatively neutral science, estranged from practical discourse, is no longer rational in this light. Rationality, as conceived by modern consciousness, requires one to interrelate theoretical and practical discourse in order to ocvercome the demarcation lines set up for each of them. The objective is a science that is open to moral and political reflection upon its progress, and a society that promotes scientific criticism of its goals and ends. Normative finalization makes this prospect conceivable; it undertakes to link true knowledge to rational interests, and to revolutionize the social paradigm of early modern science through social natural science.

NOTES

[1] Hill (1975), p. 293.
[2] Böhme *et al.* (1978), p. 72.
[3] Böhme *et al.* (1978), p. 87.
[4] Böhme *et al.* (1976), p. 312.

[5] Received after a lecture on finalization at the Science Center, Harvard: "Murphy's Paradigm: A New Sociology of Science. The authors propose a new sociology of science with three clearly defined phases. One: *Pre-Murphy's Law*. During this period the scientist is trying to come up with a new idea. He even considers his morals but is doubtful about them. The external world is allowed to help in the lab, but often drops test tubes and fudges results. Things are still advancing since the scientist has an idea. Two: *Murphy's Law*. Everything that can go wrong does, but the scientist is happy. He has managed to exclude the external world from his lab, but the external world insists on peeping through the lab window kibbitzing and betting that things will go wrong. They do. Three: *Post-Murphy's Law*. The scientist now feels so confident about his theory that he invites the external world back into the lab. This proves to be a mistake since the external world is rewriting the whole experiment for television. The external world also wants to wear a lab coat. The scientist is thinking of taking up batik or metaphysics. We are ready for phase one again, thereby avoiding finalization, or any other end to the story." The fact that the three-phase model can be expressed in this somewhat light-hearted manner testifies to its essentially heuristic character. Too literal an interpretation of the schema fails to grasp its function as a *guide* for approaching the history of science.

[6] Böhme *et al.* (1976), p. 312.

[7] *ibid.*

[8] *ibid.*

[9] It might be objected that the term "social externalism" is but a pleonasm, as externalism already implies social factors influencing science. Nevertheless, I want to make the following distinction: "social externalism" refers to the influence of social factors on the *history* of science; "cognitive externalism" to their influence on scientific *theory*.

[10] Pre-paradigmatic molecular embryology, for instance, is at present working on the solid ground of macrotheoretical physiology, and, at the same time, employing corroborated microtheoretical results from biochemistry and molecular biology (*cf.* 'Cancer Research' in this volume).

[11] Radnitzky (1976), p. 30.

[12] Böhme *et al.* (1976), p. 325.

[13] Böhme *et al.* (1976), p. 326.

[14] Böhme *et al.* (1976), p. 325.

[15] The concepts of theoretical and practical discourse are used here in the sense referred to by Jürgen Habermas: the purpose of theoretical discourse is to examine the correctness of truth claims, that of practical discourse to examine the correctness of norms.

[16] Böhme *et al.* (1976), p. 321.

[17] von Weizsäcker (1980), p. 83. Von Weizsäcker thinks that one could search for a "higher unity" than the "unity of physics", namely the "unity of man and nature" (1980, p. 6); for "we must comprehend natural science, with all its concepts of nature, as the work of man, and man, with all his cognitive abilities, as a child of nature" (1980, p. 6). This "circle of mutual dependency" is new: "Science would be dishonest if it claimed more than a first glimpse into this state of affairs" (1980, p. 83).

[18] Böhme *et al.* (1976), p. 323 f.

[19] Although there is at present no institutionalized process for the differentiation of particular and universal interests, the increasing threat to humanity's inner and outer nature is obvious. Genetic engineering threatens his/her body; operant conditioning

threatens his/her soul; informatics threatens his/her social life; technology threatens his/her physical environment. This yields one overriding interest about which there must be general agreement: the preservation of the human habitat. This interest may be considered vital in both the literal and the figurative sense.

20 Böhme *et al.* (1978), p. 74.
21 Böhme *et al.* (1978), p. 80.
22 Böhme *et al.* (1978), p. 87.
23 Böhme *et al.* (1978), p. 74.
24 Böhme *et al.* (1976), p. 307.
25 van den Daele (1977*b*), p. 48.
26 van den Daele (1977*a*), p. 169.
27 *Cf.* Hübner *et al.* (1976).
28 The shutting down of the Max-Planck-Institut for the Study of the Conditions of Life in the Scientific-Technical World, Starnberg, was not only accompanied by a press campaign comparable to that of the 1976 Finalization Debate, but also by a not entirely inaudible "sigh of relief" in a number of circles.
29 *Cf.* Andersson (1976), p. 67.
30 *Cf.* Tenbruck (1976), p. 215.
31 Albert (1976), p. 152.
32 Radnitzky (1976), pp. 25 ff.
33 *Cf.* Lübbe (1976), pp. 18 ff. and Albert (1976), pp. 152 ff.
34 Albert (1976), p. 151.
35 Lübbe (1976), p. 20.
36 Lübbe (1976), p. 19.
37 Böhme *et al.* (1973), p. 128.
38 Mendelsohn (1973), p. 52.
39 *Cf.* Böhme *et al.* (1973), p. 143.
40 The term "L'histoire humaine de la nature" comes from Serge Moscovici. Moscovici has correlated the differentiation of inventive labour (from the artisan through the engineer to the scientist) with three successive "natural states" of humanity (the organic, mechanic, and cybernetic). The cybernetic natural state of humanity enables both the scientific invention of labour and the synthetic construction of nature (*cf.* Moscovici, 1968).
41 Böhme *et al.* (1976), p. 307.
42 *Cf.* Hill (1975), pp. 294 f.: "After 1660 everything connected with the political radicals had to be rejected including 'enthusiasm', prophecy, astrology as a rival system of explanation to Christianity, alchemy, and chemical medicine. Proponents of the latter were dismissed as 'fanatics in physic', 'as a sort of men not of academical but mechanic education', supporters of 'the late rebellion', who wanted to open medicine to 'hatters, cobblers and tinkers'."
43 Hill (1975), pp. 296 f.
44 Robert Hooke's "draft Statutes of the Royal Society (1663) stated: 'The Business and Design of the Royal Society is: To improve the knowledge of natural things, and all useful Arts, Manufactures, Mechanics, Practices, Engynes and Inventions by Experiments (not meddling with Divinity, Metaphysics, Moralls, Politicks, Grammar, Rhetoric or Logick)' " (van den Daele, 1977*b*, p. 31).
45 This reading is confirmed by the circumstance that even after 1660 there existed no

clear idea as to which concept of science should count as "true": "The period was undecided. 'There was still room in the third quarter of the seventeenth century for serious disagreement over the choice of the new science . . . ' " (van den Daele, 1977*b*, p. 38).

[46] Spratt (1667), p. 83.

[47] *Cf.* Hill (1975), pp. 107–150, 287 and 290–293.

[48] The use of this term seems historically justified if one bears in mind that following the restoration of the monarchy in 1660 it was not the feudal Cavaliers but the capitalist bourgeoisie who, for the first time, seized its chance in the political arena: "The new Monarchy was 'limited', not feudal, i.e., a constitutional *bourgeois* capitalist monarchy" (Holorenshaw, 1971, p. 17).

[49] See Winstanley (1973), pp. 198 f.: " . . . for if ever the Creation be restored, this is the way which lies in this twofold power: First community of mankind . . . The second is community of the Earth . . . These two communities, or rather one in two branches, is that true levelling which Christ will work at his more glorious appearance . . . " (*A New Year's Gift for the Parliament and Army*, 1650).

[50] Since Liebig is already extensively quoted in the study on agricultural chemistry in this volume I will simply quote Marx: "Capitalist production, therefore, only develops the techniques and the degree of combination of the social process of production by simultaneously undermining the original sources of all wealth – the soil and the worker" (Marx, 1976, p. 638).

[51] The historical consciousness of modern times originated in this period: "The sense of mobility, of variability, of acceleration, the sense of an open future, the revolutionary trends . . . , a modernity that at all times seeks to surpass itself – the sum of this temporal experience of the modern era has been reduced to its simplest expression by virtue of the collective singular which is history" (Koselleck, 1971, p. 7; *cf.* Koselleck, 1975).

[52] The breach which this radical change of experience made in the "Naturalis Historia" has been traced by Wolf Lepenies as a revolutionary transition from a static to a dynamic history of nature (see Lepenies, 1976).

[53] Descartes (1967), p. 119.

[54] Obviously Descartes had in mind "le bien général de tous les hommes". But he failed to observe the contradiction between universal rational claims and private "possessors of nature".

[55] *Cf.* Küppers *et al.* (1978), p. 96: "In its earlier as well as in its later phase, ecology has almost exclusively focused on 'natural' and 'nature-related' eco-systems; neither a practical orientation nor the factor of 'human intervention' are ruled out by this."

[56] Ronge (1978), pp. 188 ff.

[57] Böhme *et al.* (1976), p. 311.

[58] Habermas (1979), p. 148.

BIBLIOGRAPHY

Albert, H.: 1976, 'Die Idee der Wahrheit und der Primat der Politik. Über die Konsequenzen der deutschen Ideologie für die Entwicklung der Wissenschaft', in Hübner, K. *et al.* (eds.): 1976, *Die politische Herausforderung der Wissenschaft*, Hoffmann und Campe, Hamburg, pp. 149–158.

Andersson, G.: 1976, 'Freiheit oder Finalisierung der Wissenschaft?', in Hübner, K. *et al.* (eds.): 1976, *Die politische Herausforderung der Wissenschaft*, pp. 66–76.

Böhme, G., van den Daele, W., and Krohn, W.: 1973, 'Die Finalisierung der Wissenschaft', *Zeitschrift für Soziologie* **2**, 128–144.

Böhme, G., van den Daele, W., and Krohn, W.: 1976, 'Finalization in science', *Social Science Information* **15**, 307–330.

Böhme, G., van den Daele, W., and Krohn, W.: 1978, 'Alternatives in science', *International Journal of Sociology* **8**, 70–94.

van den Daele, W.: 1977*a*, 'Die soziale Konstruktion der Wissenschaft. Institutionalisierung und Definition der positiven Wissenschaft in der zweiten Hälfte des 17. Jahrhunderts', in Böhme, G., van den Daele, W., and Krohn, W. (eds.): *Experimentelle Philosophie*, Suhrkamp, Frankfurt.

van den Daele, W.: 1977*b*, 'The social construction of science. Institutionalization and definition of positive science in the latter half of the seventeenth century', *Sociology of the Sciences* **1**, 27–54.

Descartes, R.: 1967, 'Discourse on Method', in *The Philosophical Works of Descartes*, trans. E. Haldane and G. R. T. Ross, Cambridge University Press, Cambridge.

Habermas, J.: 1979, *Communication and the Evolution of Society*, Beacon Press, Boston.

Hill, C.: 1975, *The World Turned Upside Down. Radical Ideas During the English Revolution*, Penguin, Harmondsworth.

Holorenshaw, H.: 1971, *The Levellers and the English Revolution* (with a foreword by J. Needham), Howard Fertig, New York.

Hübner, K., Lobkowicz, N., Lübbe, H., and Radnitzky, G. (eds.): 1976, *Die politische Herausforderung der Wissenschaft. Gegen eine ideologisch verplante Forschung*, Hoffmann und Campe, Hamburg.

Koselleck, R.: 1971, 'Wozu noch Historie?', *Historische Zeitschrift* **212**, 1–18.

Koselleck, R.: 1975, 'Geschichte', in *Geschichtliche Grundbegriffe. Historisches Lexikon zur politisch-sozialen Sprache in Deutschland*, Vol. 2, Klett, Stuttgart.

Küppers, G., Lundgren, P., and Weingart, P.: 1978, *Umweltforschung – die gesteuerte Wissenschaft?*, Suhrkamp, Frankfurt.

Lepenies, W.: 1976, *Das Ende der Naturgeschichte. Wandel kultureller Selbstverständlichkeiten in den Wissenschaften des 18. und 19. Jahrhunderts*, Hanser, Munich/Vienna.

Lobkowicz, N.: 1976, 'Erkenntnisleitende Interessen', in Hübner *et al.* (eds.): 1976, *Die politische Herausforderung der Wissenschaft*, Hoffmann und Campe, Hamburg, pp. 55–65.

Lübbe, H.: 1976, 'Planung oder Politisierung der Wissenschaft. Zur Kritik einer Kritischen Wissenschaftsphilosophie', in Hübner, K. *et al.* (eds.): 1976, *Die politische Herausforderung der Wissenschaft*, Hoffmann und Campe, Hamburg, pp. 13–23.

Mendelsohn, E.: 1973, 'A human reconstruction of science', *Boston University Journal*, Spring, pp. 45–52.

Marx, K.: 1976, *Capital*, Vol. I, Penguin, Harmondsworth.

Moscovici, S.: 1968, *Essai sur l'histoire humaine de la nature*, Flammarion, Paris.

Pfetsch, F.: 1979, 'The finalization debate in Germany: some comments and explanations', *Social Studies of Science* **9**, 115–124.

Radnitzky, G.: 1976, 'Dogmatik und Skepsis: Folgen der Aufgabe der Wahrheitsidee für Wissenschaft und Politik', in Hübner, K. *et al.* (eds.): 1976, *Die politische Herausforderung der Wissenschaft*, Hoffmann und Campe, Hamburg, pp. 24–51.

Ronge, V.: 1978, *Die Gesellschaft an den Grenzen der Natur*, AJZ-Druck und Verlag, Bielefeld.
Spratt, T.: 1667, *History of the Royal Society*, edited with critical apparatus by Cope, J. and Jones, H. W., Washington University Studies, 1958.
Tenbruck, F.: 1976, 'Wissenschaft und Verwissenschaftlichung', in Hübner, K. *et al.* (eds.): 1976, *Die politische Herausforderung der Wissenschaft*, Hoffmann und Campe, Hamburg, pp. 213–220.
von Weizsäcker, C. F.: 1980, *The Unity of Nature*, Farrar, Straus and Giroux, New York.
Winstanley, G.: 1973, *The Law of Freedom and Other Writings*, Penguin, Harmondsworth.

PART III

PROSPECTS

SUMMARY OF CONTRIBUTIONS

The essays *Science in a Crisis of Legitimation* and *Towards a Social Science of Nature* outline the main themes and objectives of our current research after bringing the studies in finalization to a close.

Science in a Crisis of Legitimation begins with the proposition that the diminishing acceptance of science and public criticisms of scientific progress are occasioned by scientific progress itself. The practical dissolution of the theoretical distinction between the domination of nature and the explanation of nature, the inclusion of human nature in the practice of research and the scientific production of unanticipated opportunities for practical intervention into this sphere have exposed the legitimacy of further scientific progress to a number of critical questions. Should scientific progress be slowed down or actually stopped in the name of traditional values, or should support be lent to the scientific and technical transformation and supersession of these values? The impossibility of providing an empirical answer to this question leads to the conclusion that empirical science research can and should involve itself in the elaboration of a socially acceptable notion of progress by means of studies designed to explore the relation of scientific developments and changes in values, and in particular, at the emergence of new goals and values as a product of science.

Towards a Social Science of Nature proposes that critical science studies should acquire a new object, in which its former confinement to the mutual relationship between science and society, or between science and politics, is to be superseded by a new orientation to the mutual interactions between science, politics, *and nature*. The current state of these interactions is presented via a comparison of traditional and contemporary relationships between science and politics, science and nature, and politics and nature. This historical-theoretical analysis serves as the background for establishing the main task of the project for a Social Science of Nature – namely, the investigation of the normative problematic of a post-industrial politics of nature, that is a politics with and not against nature. A consideration of the current state of normative concepts of nature in both social reality and in the social sciences indicates that this problematic can probably only be adequately formulated within a critical theory of social reproduction which acknowledges not only

G. Böhme et al., Finalization in Science, ed. by W. Schäfer, 235–236.

the differentiated character of knowledge *about* nature but also the antagonistic interests directed *at* nature. This draft for a new orientation of research concludes with some thoughts on the likely theoretical, social, scientific, and cultural-political consequences of a conscious development of the human history of nature.

WOLFGANG VAN DEN DAELE AND WOLFGANG KROHN

SCIENCE IN A CRISIS OF LEGITIMATION

As long as care for examination of the effects of our discoveries on human life is not felt to be as self-evident as for the execution of our experiments, then I believe that we are not yet ready for life in the age of technology.[1]

1. INTRODUCTION

Following Max Weber, legitimations can be regarded as widely-held beliefs through which institutions obtain their justification and, in cases of conflict, a defence against opposition. Science is clearly in need of a defence. The bombs dropped on Hiroshima and Nagasaki made it clear to a number of concerned scientists that, in the final analysis, science could not evade responsibility for the social consequences of the knowledge which it produces.

At the same time, many scientists apparently remain surprised by the fact that science has run into such problems of legitimation. Scientific freedom seems to be inseparable from social freedom. *Objectivity*, *value-freedom* and political *neutrality* ought to be sufficient to secure the general acceptance of science. Gene researchers gathered in Asilomar in 1975 were certainly stunned to discover that, in the view of those legal experts present, there was nothing in the liberal constitution of the USA to prevent the imposition of major restrictions on their activities as a consequence of social unease over the unknown risks involved in gene research.

2. THEORETICAL AND PRACTICAL PREREQUISITES FOR THE LEGITIMATION OF SCIENCE

Modern natural science is characterized by its operational concept of truth. Its central criterion for validity consists in successful experimental manipulation and its explanatory ideal in the presentation of "effective causes". A fact becomes acknowledged as such once it is possible to specify the steps through which it can be repeated. Within the development of natural science,

G. Böhme et al., Finalization in Science, ed. by W. Schäfer, 237–250.

natural law "is increasingly simply a description of the possibility and of the outcome of experiments – a law of our ability to call up phenomena."[2]

This operational concept of truth blurs the distinction between the domination of nature and the explanation of nature, between technology and natural philosophy. Writing at the beginning of the seventeenth century, Bacon made the first explicit reference to the fact that modern natural knowledge was power.

Human knowledge and human power meet in one; for where the cause is not known the effect cannot be produced. Nature to be commanded must be obeyed, and that which in contemplation is as the cause is in operation as the rule.[3]

At the same time, modern natural knowledge is *merely* power. Although it establishes a potential for the purposive domination of nature, it does not provide a legitimation for any specific purpose. The nature which it presupposes is devoid of moral qualities and has lost the dignity of a meaningfully ordered cosmos: it is a "storehouse of matters"[4] open in principle to any demand placed on it by humans.

Natural knowledge is therefore value-free. Kept within the confines of its conception of truth it can neither act to guide human conduct, nor to generate a general philosophy of the world. It nonetheless requires a philosophy within which the value-freedom of knowledge and the accumulation of the means for the domination of nature can be granted moral justification. "Science is value-free, but it is based on values, amongst which, paradoxically, the freedom from value-judgements is one of the first."[5] This conception of the world originated in the early modern period: it endows human subjectivity with the status of the sole moral and cognitive reference point for value and meaning in the world, and simultaneously imposes a categorical division between the human individual as a moral and as a knowing subject and nature – a distinction classically demonstrated in Descartes' notion of *res cogitans* and *res extensa*. Nature can then indeed be treated like a warehouse since, by definition, human beings are not a part of the warehouse but operate it according to their interests.

This presupposition allows the two ideals of the modern period to be reconciled: operational knowledge of nature on one hand, and a human subject capable of reason and moral action on the other. It also allowed Bacon to dismiss as superstitious any suggestion that the "inquisition of nature is in any part interdicted or forbidden."[6]

The social value of natural knowledge consists in its potential usefulness in relation to technically attainable objectives. Ever since the seventeenth century academic institutions, learned societies and universities have been able to trade their utility for a number of institutional privileges, at the core of which lies the almost total exemption of scientists from any accountability for the social application of knowledge, for the appearance of unexpected consequences, and for the risk that merely hypothetical or even possibly incorrect scientific postulates may be used as the basis for guiding action. The institutional differentiation of research as a sphere of social activity was able to meet with general acceptance since, at least, the results of the activity were public – enabling their potential utility to be judged by all – and since, in principle, these results were available for any private or public use. Legal restrictions on use also had to ground themselves in this public judgement.

The divorce between the knowing and moral subject and nature represents the philosophical and theoretical prerequisite for the classical legitimation of the natural knowledge of the modern period; the social and practical prerequisite is represented by the institutional separation of the production of knowledge and its application. Both these presuppositions are placed in question by the progress of the very knowledge which they make possible.

3. FIRST PARADOX OF UTILITY: FROM EXPERIMENTING FOR HUMAN BEINGS TO EXPERIMENTING ON HUMAN BEINGS

The demarcation of the human individual as subject from nature has remained problematic through the philosophy of the modern period. It is true Kant's "Paralogisms" and "Antinomies" of pure reason were seen as proof that the thinking and moral ego not only ought not to be subjected to empirical investigation but *cannot* be so treated. Consciousness cannot be explained from neurology, and moral action – exercised in freedom – from observable behaviour in time and space. His position has, however, not remained fully convincing. Neither epistemology nor moral philosophy are immune to advances in empirical knowledge. And in particular, no interpretation of the thinking or moral ego can be conceived independently of, or even in contradiction to, its material conditions and real manifestations. Hegel opposed Kant with his view of the necessity of the *existence* (Dasein) of freedom and of the person (*Philosophy of Right*, Sections 33 and 51). Correspondingly, the attributes of the individual as a subject which the political constitutions of bourgeois democratic states are intended to protect have acquired an "empirical" and "objective" meaning which can be both defined in terms of actual social action and is enforceable. Values such as "human dignity", "the free development of the individual", "life", "freedom from harm", "equality", "freedom of thought" which are written into basic laws and declarations of human rights refer to real human characteristics and empirically observable actions, and conditions for action.

These real correlates between human subjectivity as thinking and moral ego, and the values related to this subjectivity are, in principle, a possible object for empirical natural research. In becoming so in practice, however, they would be exposed to the danger of alteration and dissolution through the consequent increase in operative knowledge. The possible extent of scientific-technical encroachments on human characteristics can be seen in the accompanying table (Table 1) indicating actual, probable or at least projected developments in biomedicine.[7]

TABLE 1

Human Reproduction
BY 1985
Extensive development of chemical birth, fertility and desire control agents.
Extensive development of chemical abortifacients.
Sperm banks with the capability of storing human sperm more than ten years.
Artifical insemination with options of choice of sex and to some degree choice of qualities.
Artificial inovulation with both natural and foster mothers.
BY 2000
An artificial womb.
The first cloned mammals.
AFTER 2000
"Baby factories".
Cloned people.
Physical Modification
BY 1985
Extensive transplantation of organs, limbs, and glands after the rejection phenomenon is overcome.
Extensive use of artificial organs and limbs.
BY 2000
Organ and tissue regeneration.
Frozen organ banks.
An extensive, although not complete, understanding of human genetics and heredity.
Limited negative eugenics.
AFTER 2000
Total mastery of human genetics and heredity.

TABLE 1 cont'd

Extensive positive and negative eugenics.
Synthetic plants and animals.
"Humanants", "Humanals" and "Plantimals".

Mental Modification

BY 1985
An expanded list of mind-modifying drugs.
Pain relief through electrical simulation of the brain.

BY 2000
Disembodied animal and human brains (cyborgs).
Memory injection, editing and erasure with drugs.
Extensive intelligence and personality reconstruction with drugs.
Mini-computer "boosters" implanted in the brain.

AFTER 2000
Head and possibly brain transplants.
Brain–machine, brain–computer, brain–brain chimeras.
Genetic engineering of mental characteristics.

The Prolongation of Life

BY 1985
Increased power to postpone clinical death with machines and drugs.
The conquering of cancer.
The initial use of hibernation and body-temperature lowering drugs.
The expansion of freezing research.

BY 2000
Most major diseases conquered.
Frozen organ banks.
Discovery of the cause of ageing.
The power to increase life-spans 10 years or more by various techniques.

AFTER 2000
Indefinite postponenment of death, and even immortality, through the development of anti-ageing, freezing, hibernation, suspended animation, and sleep techniques.

The Creation of Life

BY 1985
The production of artificial viruses.

BY 2000
The synthesis of unicellular organisms.
The discovery of extra-terrestrial life.

AFTER 2000
The synthesis of multicellular organisms.

Many items in this catalogue probably defy the moral sense of most people, although we suspect only a few will prove to be beyond scientific realisation. Techniques which were regarded as scientifically or morally inconceivable a few years ago are now almost matters of routine – such as heart-transplants, sex-changes, and child-bearing via "surrogate mothers". Human nature is becoming an attractive and successful research front for an increasingly large number of scientific disciplines such as molecular biology, neurophysiology and experimental psychology.

As a consequence, the moral qualities of the subject, which are still regarded as normatively inviolable, are being increasingly operationally explained and rendered amenable to alteration. Although value-free natural knowledge presupposes a thinking and a moral subject as an independent point of reference, its consequences eventually swallow up this subject within denormatised nature. The advance of research therefore undermines its own legitimation.

4. SECOND PARADOX OF UTILITY: THE PURSUIT OF TECHNOLOGY FOR ITS OWN SAKE

Since the knowledge held by the natural sciences merely extends the means for action, but does not determine goals, it is considered value-free and politically neutral. The belief in this neutrality was a precondition for the general social acceptance of science. That means are neutral is logically correct ("is" does not imply "ought") but sociologically false ("can" usually implies "will"). The non-neutrality of means is revealed by the fact that, historically, societies are more characterized by their means of production and means of power than by their varying goals. This characterization not only affects social structures but also value-systems. Many of the social relations and moral beliefs in agrarian or industrial societies, or even in a possible "nuclear state" (Robert Jungk) are not the product of the society's goals but of the means through which these societies are structured. The neutrality of the application of technologies should not be interpreted as implying that they do not have a normative influence on society, but the direct opposite of this: that their influence is universal and all-embracing. This applies with particular force to societies such as our own, in which sub-systems are institutionalized (economy, armaments) which become the sites for the development and acquisition of new technologies without any specification as to goals, with new areas of application created for these technologies *ex post facto*.

The sciences not only constitute a basis for this technical dynamic, but are increasingly integrated into it: successful scientific disciplines converge with technologies. Considered schematically they switch from a phase with predominantly analytical activities, in which the given reality is theoretically explained in terms of its elements and their interaction, to a phase of predominantly synthetic activities, in which they begin to extend reality. Mechanics, electrodynamics and chemistry already made this transition in the nineteenth century: molecular biology is currently passing through this phase. As long as disciplines are predominantly analytic in terms of their procedures, their findings are usually relatively neutral as far as the social application of knowledge is concerned – they supply theoretical explanations in the form of general laws. However, as soon as they begin to operate synthetically they provide the drafts for the construction of artefacts (arcs, aniline dyes, recombinant DNA and so on) which are not only methodologically but also materially indistinguishable from technologies (*cf.* 'The Scientification of Technology' in this volume). The distance between the production of knowledge and its technical application therefore disappears; synthetic disciplines are the production of technical applications. This loss of distance applies, in particular, to the industrial sciences which currently account for more than half of all research. There is no public discussion of the findings of this type of research, which hides behind the law of patents and is able to evade political control over its application to a considerable degree. The intertwining of interest in application and the development of research characteristic of synthetic disciplines extracts itself from the sphere of political control through the self-financing of research in industry. The development of science as a productive force still contradicts the feelings of many scientists who believe in the pursuit of knowledge for its own sake. But strong objections against this development can scarcely be raised – it is, after all, built into the conception of truth of modern natural knowledge. However, this development renders ineffective value-neutrality and public accessibility as the foundations of the legitimacy of research.

The classical model for the legitimation of scientific progress is nullified by the very success of science itself. The argument that there are no moral problems in intercourse with nature – since nature is merely a means for human purposes and never a moral purpose in itself – cannot be applied if the nature concerned is *human nature*. The argument that the production of knowledge must be distinguished from its application becomes untenable if science itself becomes technology, both in terms of its institutions and its content. Science becomes inevitably drawn into social conflicts: conflicts over the moral limits of the content and methods of research, and conflicts over the social aims and consequences of the application of research. There is no clear model for the regulation of these conflicts. Clearly, the relation between scientific and technical progress, social goals and moral values must find a new equilibrium. The following sections sketch out some of the problems encountered in the pursuit of such an equilibrium.

5. CONTROL OF RESEARCH AND CHANGE IN VALUES: BETWEEN DOGMATISM AND SCIENTISM

The collision between empirical claims to knowledge and moral interpretations within the sphere of human nature can generate two theoretical extreme outcomes. The first, *a crisis of science*: culturally established moral interpretations of human nature are posited in normative terms and preserved as ideals, or as in our penal code, even as fictions. The interpretations are dogmatised as part of the value-system and immunised against possible dissolution through empirical research – if necessary by the prohibition of research. American law, for example, forbids any open or secret empirical examination of the deliberations of juries so as not to jeopardize the moral authority of this institution (18, United States Constitution, Article 1508).

The second, *a crisis of the value-system*: science dissolves those interpretations and beliefs which originally established the basis for its own legitimacy. Operational explanation and manipulation cease to be bound to a system of moral values and instead *replace* this system (scientism). As in earlier disputes over the concept of a divine, meaningful cosmos, or the notion of a vital-force, concepts such as dignity, freedom, self-consciousness and identity are eliminated as "non-operationalizable" and "superfluous" hypotheses. In this case, the sole orientation for human action, and only moral constraint on scientific development, would be provided by the functional imperatives of the social system. The particular contribution of the social sciences to the maintenance of the social system may then lie in the control of deviant behaviour, the improvement of human mental and physical endowments, and the control of motivations.

Both tendencies can be seen in the current conflicts over the moral limits of biomedical research. On the one hand, the conduct of research is subjected to clear legal constraints as far as the individual rights of human subjects are concerned. Encroachments on these rights, such as experiments on humans without their valid approval, cannot be legitimated either through a scientific knowledge-interest or technical usefulness. It is quite possible to argue that any increase in knowledge which might require such an encroachment, such as double-blind experiments, or experiments in psychology involving deception, experiments with children, disabled people or the residents of institutions, simply has to be done without. On the other hand, there are instances where the cultural definitions of what is morally unique about human nature are revised in the light of growing technical capabilities. This was the case with heterologous insemination, in which the doctor is empowered to select the sperm-donor according to such criteria as race, status, background, etc. And it currently takes place in embryo-transfer ("test-tube" babies). It can also be seen in the sanctioning of scientific experiments on living, but non-viable, foetuses (although a number of American states have legally forbidden such experiments).

In our culture the dynamics of the development of means for intervention dominates value patterns and political orientations. This would suggest that the only human characteristics able to continue as elements in the "inviolable dignity" of an individual will be those which have not *yet* been rendered accessible to scientific manipulation. The "breeding of humans" for their chosen properties is unanimously regarded as immoral and in contravention of the Federal German constitution. ("The dignity of man is inviolable. To respect and protect it shall be the duty of all state authority".) However, will this view still prevail once controlled eugenics becomes a possibility? The first justifications for the development and application of the new technology will be medicinal; then parents will appeal to their subjective right to equip their children with "useful" characteristics "made-to-measure"; and finally, the state will discover the need to influence the overall development of the nation's gene-pool, just as it currently seeks to control the level of education. Scientific-technical progress not only creates additional options for human conduct; it also dissolves the value-positions that have functioned to govern which options could or could not be adopted. Whether this dissolution will also apply to the formal, individual right of self-determination, or whether we will inevitably reach a stage in the further advance of knowledge which is "beyond freedom and dignity" (Skinner) may still be undecided. Dissolution of moral evaluation is already visible for elements of human nature over which the individual still exercises some degree of "self" determination (physical make-up, character, temperament, intelligence).

If the advance in knowledge within science and technology is a factor in the change of values, how can its reconcilability with existing morality be ensured? In many instances the only option seems to be to place limits directly on the *production* of knowledge rather than on its *application*.

Decisions about the relationship between science and society are not only made on the basis of existing values – they are simultaneously decisions about the possible changes in these values. The decision always to permit the production of knowledge implies the decision to expose existing morality to a transformation via this knowledge. Such a position corresponds to the dominance of the cognitive factor and to the prospects for change in our culture. These positions could conceivably change, however, and society may attempt to control the influence of the scientific-technical dynamic on the value-system through controlling the expansion of knowledge. Such an attempt cannot be considered as unscientific or irrational: it seeks to establish a new equilibrium between knowledge-interests and moral values. This is a political conflict in which, in the final analysis, the issue is whether moral values are to be protected from a change in social reality, after which the nullification of these values might not be perceived as representing any kind of a loss.

6. ACCOUNTABILITY FOR THE SOCIAL CONSEQUENCES OF SCIENCE

When science converges with technology, research ought to be expected to assume some responsibility for its own consequences. Although scientists can point to the fact that, logically, the distinction between value-neutral, and hence non-accountable, production of knowledge and morally-evaluable application continues to hold even within the field of technology, it is doubtful whether such a response would now find many ready listeners.

A chain of action lies between the production of knowledge and the application of knowledge, and although each link in the chain posits additional conditions for the success of the application, each individual step can still merely be interpreted as the production of the possibilities of action, with no implication for its social implementation. The encroachments into the protected sphere of the individual made through micro-electronics represent a case in point. The discovery of the semi-conductor, or the invention of the micro-chip, cannot be held responsible for these encroachments on privacy; but then neither can the exploration of the technical possibilities for applications in the field of surveillance, the development of relevant technologies – nor even logically, their industrial manufacture or supply to the market. The division of action into sub-steps leads to a "Taylorization of morality". Each step presupposes further steps which can always be pointed to as the *real* area in which morally accountable actions are undertaken – an argument which can be taken to a *reductio ad absurdum*. In fact, the demarcation of accountability within a chain of actions is not determined by the logical distinction between the production and the exploitation of knowledge, but by the assessment of the hazard implied by each individual step.

The augmentation of new possibilities of action through new knowledge cannot therefore be removed from the sphere of accountability. The more the application of knowledge takes place in discrete steps, each of which is organised as a process of research, the more difficult it becomes to defend any particular individual step as the border between the non-accountable production of knowledge and accountable application. There seem to be no fundamental arguments against laying ultimate responsibility for nuclear war and the nuclear threat at the door of basic research in physics.

Responsibility for the consequences of knowledge cannot rest with scientists, however, but with the society, or societies, which institutionalise science. This responsibility encompasses the possibility, where necessary, of controlling the production of knowledge. The difficulty in anticipating the consequences of knowledge can of course, equally well be used as a reason for dispensing with such control as for introducing it as a preventative measure. The political likelihood of exercising such responsibility depends on how important the possession of operational control over nature is felt to be by society.

The differentiation of sub-systems of intrumental rationality – such as the economy, administration, science – tends to endow functional thinking and controllable intervention with the status of the dominant orientations for action, and to replace substantial moral values by the functional exigencies of the system as the reference points for conduct. Conflicts over scientific progress are clear signs that such a change in values cannot take place without frictions. Science has not yet reached a point at which the production of knowledge and new possibilities for intervention are inevitably regarded as justified. Nevertheless, it is conceivable that society may not succeed in the task of exercising moral control over the development of knowledge and its application. In such an instance, unlimited scientific-technical progress would prevail, not, however, because it is justified, but simply because it cannot be prevented.

7. PERSPECTIVES

The foregoing analysis may well give the impression that it is easier to present the paradoxes and cultural contradictions into which modern scientific development has led modern society than it is to indicate some ways out of this predicament. This is in fact our own impression, despite the existence of a number of proposals in the literature. At the political level it has been suggested that the "social contract" between scientists and society be renegotiated, and the limits to the freedom of science and technology, and the human rights affected by the consequences of that freedom, redefined (see Nelkin, 1978). More radical than this is Feyerabend (1978) who advocates a thoroughgoing disestablishment of science, its separation from state support and its reduction to the status of other belief systems. Others propagate the development of a new ethics able to extend value commitments and moral obligations to more and more distant consequences of our actions (Jonas, 1979). Finally, there are those who rest their hopes in the rise of a counterculture, in which science would be replaced by a more integrative wisdom as the paramount type of knowledge in our society (Roszak, 1972). All these proposals are more of an indication of the range of the problem rather than its solution. Pragmatically, we find ourselves in a state of confessed helplessness. Theoretically, the situation demonstrates that we still lack categories with which to understand the cultural changes that are impending.

If our diagnosis of a fundamental legitimation crisis of science is correct, it would seem consistent to work at producing a new basis for its legitimation (unless one considers that a functionalist culture has already simply replaced the old humanistic values, see Luhmann, 1971). However, such a project does not seem to be feasible at the present. On the one hand, such a legitimation cannot be the outcome of theoretical reflection, but must emerge out of the conflicts and controversies between concerned interest groups. And on the other, the aim cannot simply be the redefinition of the relations between existing knowledge and social values. Instead, we have to consider the social value of knowledge itself, and take into account a new knowledge about our values, their preconditions and consequences. What is at stake is a revision of some of the attitudes and orientations that have guided the modernisation of our societies over the last century.

In particular, we have to depart from the concept of scientific progress subscribed to so far. However, a retreat from the idea of progress can scarcely be a total one. It is evident that at present the progress of science and technology may be incompatible with the freedoms of individual and social action. And it is part of our helplessness that we feel that the necessity of progress is deeply engraved in the major institutions of our society, and therefore that all proposals for a more rational use of that progress seem bound to fail from the very outset. But it is likewise evident that historically the pursuit of knowledge has been a means of individual and social liberation. In *principle* science can relieve humanity from both the constraints of nature and of its own history, and increase the options for a rational projection of social and technical evolution. Nobody would today subscribe to this Baconian principle without major qualifications. However, "in principle", it seems to remain a principle.

What can science contribute to the elaboration of a socially acceptable notion of progress? An important topic would be the renewal of conceptual and historical investigations of the relations between scientific development and changes in values. The studies presented in this volume have only been able to touch on these themes. Their aim is to establish that in the course of its internal dynamics, science becomes more and more amenable to external direction. This essay suggests that the notion of direction and the implied idea of a priority of social over scientific goals are only valid in the contexts of the application of science and the scientification of technology. It seems to fail wherever social goals and values demand a revision of that context, and a reorientation of science and technology. In fact, it appears that the precise opposite is the case – that such goals and values become irrelevant and insignificant if they stand in the way of the dynamics of science and technology. However, it is not simply the marginalisation of these goals and values which can be observed; at the same time, new goals and values are created, as it were, as dependent variables of cognitive development, equipped to push the adaptation to and support of this development further forwards. These processes still remain unexplored. The area is seldom investigated – and then, at best, as a marginal contribution to studies on modernization. Nevertheless, they constitute the material which we now need to better understand the changes to which we are currently subjected.

NOTES

[1] C. F. von Weizsäcker (1957), p. 15.
[2] C. F. von Weizsäcker (1960), p. 173.
[3] F. Bacon (1860), p. 47.
[4] F. Bacon (1860), p. 225.
[5] E. Weil (1965), p. 184.
[6] F. Bacon (1860), p. 20.
[7] See Utke (1978), pp. 109–111.

BIBLIOGRAPHY

Bacon, F.: 1860, *The Works of Francis Bacon*, collected and edited by Spedding, J., Ellis, R. E., and Heath, D. D., Vol. IV, London, reprinted Holzmann, Stuttgart-Bad Cannstatt, 1963.

Feyerabend, P.: 1978, *Science in a Free Society*, NLB, London.

Jonas, H.: 1979, *Das Prinzip Verantwortung*, Suhrkamp, Frankfurt.

Jungk, R.: 1979, *The Nuclear State*, Calder, London.

Luhmann, N.: 1971, 'Systemtheoretische Argumentation', in Habermas, J. and Luhmann, N.: 1971, *Theorie der Gesellschaft oder Sozialtechnologie*, Suhrkamp, Frankfurt.

Nelkin, D.: 1978, 'Threats and promises: negotiating the control of research', *Daedalus* **107**, 191–210.

Roszak, T.: 1972, *Where the Wasteland Ends*, Doubleday, New York.

Utke, A. R.: 1977, *Bio Babel*, Knox, Atlanta.

Weil, E.: 1965, 'Science in modern culture, or the meaning of meaninglessness', *Daedalus* **94**, 171–189.

Weizsäcker, C. F.: 1957, *Die Verantwortung der Wissenschaft im Atomzeitalter*, Vandenhoeck & Ruprecht, Göttingen.

Weizsäcker, C. F.: 1960, *Zum Weltbild der Physik*, Hirzel, Stuttgart.

GERNOT BÖHME AND WOLF SCHÄFER

TOWARDS A SOCIAL SCIENCE OF NATURE

1. THE PROJECT OF THE MODERN AGE – RECALLED FOR REFORM

The first attempt has proved a failure: the "Universal and General Reformation of the Whole Wide World" – the humble aim of the Modern Age so aptly formulated by Johann Valentin Andreae [1] – did indeed take place, but in a form which has shattered all the hopes of the Early Modern Age for a union between scientific and technical progress and the universal satisfaction of human needs. Work is now in progress on the second draft. But what the late-twentieth century demands and needs is:

– an end to the scientific-technical construction of rational paradises, and instead the *social reconstruction of the scientific-technical world*;

– an end to attempts to exercise mastery of nature, and instead *the social reform of control over science and technology*;

– an end to the scientific revolutionizing of the world to the exclusion of "Moralls and Politicks",[2] and instead *the renormatization of scientific knowledge*.

Works such as "The Global 2000 Report"[3] and other contemporary scientific dystopias have revealed just how much now hinges on the success of such a programme to reform our increasingly deformed "whole wide world". The ultimate issue today is the possible end of humanity, on an earth, which as far as human beings are concerned, will be permanently destroyed. This is a stark reality: what is less clearly visible is an indication of what might assist and sustain the cultural project of reforming the project of the Modern Age and what might jeopardise it.

2. CRITICAL SCIENCE STUDIES MOVED FORWARD: FROM SCIENCE AND SOCIETY TO THE POLITICS OF NATURE

In past studies we have examined the controversial issue of the politicization of science, and sought to answer the question as to how the influence of society upon the development of science is to be understood. We are now proposing to take a step further forward, whilst at the same time turning the question around. Going a step further means expanding the framework

G. Böhme et al., Finalization in Science, ed. by W. Schäfer, 251–269.

of "science and society" to embrace the additional dimension of nature. Turning the question around means asking how we should understand the combined influence of science and society on nature.[4]

Our claim is that within the relationship between science, society and nature, the influence of culture upon nature has become problematic, and that advanced industrial societies today confront the task of developing a new politics of nature. The adaptation of critical science studies to this question requires going beyond the constraints of the programme of the classical History and Theory of Science – namely, the preoccupation either with the chain of events within the context of discovery, or the chain of propositions within the context of justification – and concentrating with equal intellectual rigour on the social and scientific history of nature. The "Social Science of Nature Project" is designed to study the impact of our instrumental mastery of science on nature, and to assess the role which the social sciences will have to play within a conscious and deliberate politicization of nature: in our view, the main element of this role will be enlightenment as to the normative conceptions of nature. This does not, however, mean that our scientific re-orientation is linked to short-term policy expectations, or hopes; on the contrary, the requirements of the present period of profound cultural crisis are not for stopgap thinking, but for theory and reflection. What we lack is a new "Ortsbestimmung der Gegenwart" – a clear theoretical location of where we stand at the present. Our contribution to this clarification, and to the establishment of the uniqueness of the present position, will be to contrast the traditional relations of science, politics and nature on the one hand, with their current interconnections on the other.

3. SCIENCE, POLITICS, NATURE: TRADITIONAL AND CONTEMPORARY RELATIONS

3.1. *Science and Politics*

(a) The *traditional relation* between science and politics was based on a historic compromise worked out in the mid-seventeenth century. The New Science of "experimental philosophy" relinquished all of the moral, political, educational and social aims established for it by Bacon, Comenius, Winstanley and many others in the early-seventeenth century. The absolutist state rewarded this renunciation of the ideals of science as a radical project for socio-political reform with the offer of wide-ranging privileges for "pure" natural science; the founding of the Royal Society in London (1662) and the

Académie Royale des Sciences in Paris (1666) formed part of this process. Royal support for the New Science was paid for by separating science from politics.

(b) In the *contemporary relationship* between science and politics an increasing politicization and social orientation of science can be discerned. This orientation to social goals or needs within a scientific discipline is characterized by the *finalization* of its theory-formation, and the political and/or industrial overseeing and exploitation of scientific progress. This new tendency in science began to manifest itself around the middle of the nineteenth century – and it was more than coincidental that this was also the period of great advances in industrialization.[5] Currently, all those fields of science open to finalization are systematically pillaged by the research of the "science-based industries" – the most notable example being possibly the lucrative field of the New Biology. The attractive idea of the strategic development of new vaccines, new hormones, new drugs and new crops has led some very eminent scientists to establish their own companies for the commercial exploitation of the discoveries made in their university laboratories, and consequently blending the academic world of "pure" gene-splicing research with the world of "pure" capitalistic enterprise; indeed, not an especially pleasant proof of our theory.

Finalization, however, is not only compatible with the economic interests of industry-guided research, but also with general social goals in the context of public political decisions about science. As a result, the difficulties encountered in political decision-making about science in the general social interest have increasingly less to do with the intrinsic resistance of science to external influences, than with the resistance of particular interests against their social supersesssion. The struggle over the social control of scientific development is an indication of the extent to which contemporary science has lost the protection of the privileged sanctuary it previously enjoyed: instead, it has now become a politically relevant force of the first order. Both the lucrative technological efficiency of finalized science and the increasing costs of ensuring social viability in the face of new technologies, together with the military-industrial menace both to the natural environment and to human life – *in toto*, the development of science from a productive into a destructive force – have delegitimized the classical ideal of a research free from politics and domination, and have consequently triggered the current breaking down of the barriers between science and politics.

3.2. *Science and Nature*

(a) The *traditional relation* between science and nature was determined by the methodological ideals of classical mechanics, which defined it as a relationship of "pure" cognition. The scientist enters into this relationship not as an active element but as a detached mind simply recording what is observed. Nature – or rather the isolated physical facet of it under investigation – is grasped not as an object experienced in human practice, and not as an object of sensory cognition (since sense-perceptions can be ambiguous), but rather as the object of a physical measuring apparatus. The needle of the measuring instrument, which ideally is supposed to function just as "naturally" as nature itself, points precisely to the new qualities, the objectivity, positivity, and realism of early modern science. The object of knowledge is itself understood as a-historical: only those things which can be reproduced at will, at any time and in any location, can count as lawlike and true.

(b) The *contemporary relationship* between science and nature allows a more realistic epistemological assessment of both the research process and the object involved in acquiring scientific knowledge. The ideal of a pure, or undisturbed, cognitive process has been relativized by the discovery of unavoidable limitations on the objectivity of the act of measurement (indeterminacy in quantum theory). The ideal of the full replicability of experimentally secured knowledge has been contradicted by the reactions of natural objects to the technological application of such knowledge (such as the selection of DDT-resistant strains of insects). Finally, nature itself, the supposedly atemporal object of knowledge, has been increasingly historicised through disciplines such as geology, cosmology, thermo-dynamics and Darwinism. All these factors, combined with the increased investigation of complex and variable phenomena (dissipative structures) has led to a revolution within science itself, overthrowing all mechanistic concepts and destroying the classical deterministic worldview.[6]

This revolution in the traditional relationship between science and nature is also associated with the research approach of human ecology, which concerns itself specifically with the real consequences of human activity at the level of natural systems (such as the introduction of fluorocarbons into the ozone layers of the atmosphere). Although older ecological theories did not fail to take note of the numerous effects humans have on nature, they did so with the objectivistic attitude of classical natural science, regarding such effects either as "disturbances" or as "natural", and thereby failing to grasp their real significance. The more recent approach of human ecology, on the other hand, is on the point of making a decisive break with such an attitude, having consciously chosen to make the active human alteration and shaping of nature as the central object of its both theoretically illuminating *and* practically relevant research.

3.3. *Politics and Nature*

(a) The *traditional relation* between politics and nature was originally marked by weak interactions in practice but extravagant fantasies in the realm of theory. Bacon, for example, spoke of "magna instauratio imperii humani in naturam" (the great restoration of human dominion over nature). Myth and utopia, divine salvation and human realization in history are all condensed into a single formula for occidental modernity. On one side of the equation the myth of the Fall from the Garden of Eden functions as the historical presupposition for a future conquest of, and release from, humanity's post-paradisaical natural nature. On the other, the divine charge to the human species to "fill the earth and subdue it, and have dominion over the fish of the sea and over the birds of the air and over every other living thing" (Genesis, 1: 28) is employed as evidence of the expectation that human history will ultimately achieve the goal of recapturing the original human condition – that is, dominion over nature with knowledge as its most powerful weapon.

The Baconian formula turns a real necessity – the technological weakness of early modern science – into an ideological virtue. The technological potential of seventeenth-century science was in no way commensurate to the dimensions of its utopian dreams, dreams which later inspired Descartes to praise humanity as the "masters and possessors of nature". The science and technology of the early modern era, therefore, restricted by their limited real practical usefulness, scarcely interfered with natural processes at all, and had no impact whatsoever on the poor mortal flesh of Adam: nonetheless, they stood under the ideological sway of the highly charged expectations of an imperialistic politics of nature.

(b) The *contemporary relationship* between politics and nature is characterized by bitter disillusionment. The scientific-technical progress of the nineteenth and early twentieth centuries has gradually removed the original sharp disparity between ideological aspirations and the actual capability of humanity to exercise dominion over nature. In particular, the interplay between capital and labour in industry has begun to implement the utopian programme of the early modern era, but in such an excessive and life-threatening form that the unsolved problem of the present period is no longer simply how to rule nature, but how to control this exercise of power. New social movements, reacting against just these consequences of our industrial history of nature have taken up the struggle for a post-industrial "human history of nature".[7] The New Politics of Nature has consequently entered the public arena of social controversy.

The question of how to control the industrial explosion of our metabolic relation with nature confronts late-modern industrial societies with an abundance of tasks and with a particular dilemma. The tasks are the inheritance of the continuous and predatory exploitation and despoliation of the natural conditions which are the prerequisite for life (a despoliation against which Justus von Liebig first began to struggle over a hundred years ago). Bureaucracies for environmental "protection" have been given a mandate if not to change, then at least to "administer" the type of politics of nature characterized by such an economy of despoliation. It may however soon become apparent that the rapid transformation and progressive destruction of external nature, as well as the incipient modifications of human beings' own physical nature, are not the sort of thing which a government agency can simply administer. If the decisive confrontation of late-modern industrial society with itself is not to lead to a catastrophic "final solution", but to the recognition of a new historical task (the reform of modernity), then the traditional politics of nature as a politics *against* nature will have to be re-examined and a new one developed in its stead. This new politics will either be one of dialogue and alliance *with* nature – or it will be useless.

It is however hard to see how the old politics of nature can be overcome without a radical self-transformation of industrial society; and herein lies the dilemma. To overcome the traditional notion of a politics against nature requires revolutionary changes in the paradigm of industrialization – meaning, for advanced industrialized societies, self-transcendence. Yet, if industrial societies insist on defending their historical identity, they will have no choice but to continue with the old politics of nature – and instead of the prospect of self-transcendence, will be confronted with self-destruction. The desire to prevent such an outcome has fuelled, at least in part, two new social movements – the ecology movement and the feminist movement – in which individuals have joined together outside the existing institutional framework to resist the escalating threat posed by the devastating relation of industrial *man* to humanity's inner and outer nature. Such initiatives attempt to develop alternatives for a post-industrial history of both nature and society. Although, therefore, the validity and increasing necessity of such an opposition seems beyond dispute, this does not necessarily mean that the political platforms of this opposition are always, or completely, on the right track. Too weak or too strong a medicine can be just as dangerous as the illness which it seeks, or claims, to cure.

4. DIGRESSION: GOING NATURAL, GOING LUDDITE, AND GOING TECHNOCRAT. THREE WAYS OF GOING ASTRAY

Modern science has produced undreamt-of means not only for the production and maintenance of life, but also for death, and continues to do so ever more "successfully". The resistance which science is currently encountering is therefore not only legitimate, but indeed necessary for survival. Moreover, such resistance has not emerged from thin air, or from the neo-conservative camp of the counter-enlightenment but from the ranks of responsible science itself,[8] and more recently from the new social movements which have broken with the technological optimism of nineteenth-century working class traditions. The utopian vision of a technological realm of freedom, so enthusiastically attributed to the scientific project over three hundred years ago, is currently being transformed across a broad front into mistrust and criticism of science and technology; what this signifies, however, and what it implies for the future is anything but clear. The historical period now coming to a close will be followed by a new historical process, whose outcome – for well-known reasons – cannot be predicted. As a consequence none of the social agents in the immediate present can claim to possess an uncriticizable standpoint; even the new social movements' opposition to science, representing with the best of intentions the idea of a "good science" must recognize that it cannot avoid bringing problematic tendencies into play.

It is one thing to pose the crucial questions, and another to find the correct answers. Even though it may be evident that a science of nature which actually destroys nature cannot be true, it is not clear what ought to replace this science – true only in its own terms, but possibly disastrous for us and almost all other forms of life. What is the alternative to modern science? How can it be achieved? And no less important: in what ways will it definitely *not* be achieved? The more closely one studies the first two of these questions, the more remote any kind of a convincing answer appears to be: it may, therefore, be more promising to consider the third – in what ways does the growing opposition to science actually harm itself? Three tendencies can be distinguished: two come from below, as products of the contemporary social movements, and one initially from above, in the form of the "Limits to Growth" approach. We term the two more grass-roots based movements "Going Natural" and "Going Luddite", and the tendency which is more likely to descend on us from above, "Going Technocrat".

The fallacy of Going Natural lurks in both the ecology and feminist movements. It combines nostalgic and romanticised views of Mother Nature, a pastoral past, and the rejection of any politics of nature beyond pure and simple conservation and protection. As a consequence a number of important contributions have been flawed through insisting on an organicist world-view: Carolyn Merchant, for example, declares in her study *The Death of Nature*, "The world we have lost was organic."[9] However, the "progress" from a physical-mechanical reduction of the world to a biological one carries with it the danger of an updated doctrine of "Blood and Soil", possibly "scientifically" rooted in sociobiology. The metaphysical concept that animate nature is one integrated whole, and that all its different parts have proper functions only within this whole and for the sake of this whole, and which exist only to fulfil prescribed vital functions may represent an organicism relevant to the world of ants and their societies of natural law and order, but is hardly applicable or appropriate to ourselves and our less organic way of life.

The fallacy of Going Luddite, the aggressive alternative to all elegies about our organic paradise lost, aims at a total abolition of "real existing science and technology". This type of radicalism may not yet seem terribly significant, but may represent an instance of where a small and almost negligible error at the outset can lead to unpleasant consequences later. Too "hot" a radicalism threatens to fan popular widespread doubts about science into sheer destruction – the result of which would be a *tabula rasa* in the true sense of the expression. Those whose tables are full and who profit from the material abundance of the industrialised world, yet who at the same time preach that "man should not live by the bread of science" would be well advised to remember that almost everywhere else in the world men and women have much less bread than they would have if they were able to claim their full due from a science, which although radically re-oriented to serve their needs, is not to be totally abolished.

Going Technocrat, the third way of going astray, is certainly not a fallacy which errs in going too far in its critique of science and technology; on the contrary, it doesn't go far enough. The Technocrat stops short of realizing that he is trying to solve a problem of which he is a significant cause. The Technocrat fails to notice that modern science and technology – the Technocrat's favoured means for keeping the world running – have turned into one of the most conspicuous stumbling blocks to genuine progress. The development of science and technology has long since ceased to provide innocent means for the attainment of rational human ends, and has now become an end in itself. The Technocrat who believes that he is following the traditional maxim of "just applying" science is no longer, in fact, operating with value-neutral means but is contributing to the evolution of uncontrolled objectives. Science has ceased to be his servant and has turned into his master. The establishment of this topsy-turvy relationship ultimately rests on the Technocrat's ignorance of the complexity and interdependence in the contemporary relationships between science, politics and nature which we sketched out above and which we summarize below.

5. THE TRIANGLE OF SCIENCE, POLITICS, AND NATURE ESTABLISHED

The foregoing comparison between the traditional and the contemporary relations between science, politics, and nature has revealed the new structural pattern of modernity and its normative implications – in particular the growing politicization of science and nature, together with an increasing responsibility of science and society for the reproduction of the natural conditions for the possibility of life and civilization. Before the middle of the nineteenth century any strong interaction between politics and science, science and nature, and politics and nature was either in principle, or in practice, out of the question, but in subsequent years the relationships between these areas began to undergo a fundamental and still continuing transformation:

(1) The sharp borders dividing science from society and politics are dissolved by an increasing convergence of both fields of action.

(2) The mutual interaction between science and nature, originally suppressed by systematic neglect, attained revolutionary significance both for scientific theory and for practical research.

(3) The early modern rhetoric of mastery of nature was overtaken by the industrial domination of nature, thereby undermining the legitimacy of an imperialist politics of nature.

Polarities which originally involved little or no mutual and almost no conscious interaction have now become highly charged, producing a situation of critical and unstable interdependency within the triangle of science, politics, and nature. The "Social Science of Nature Project" sees itself as confronted with the task of investigating and disclosing the nature of this interdependency through case-studies and theoretical analyses. Building on the historical and theoretical basis of this projected work, we will in turn attempt to develop political perspectives for a fundamental reform of the industrial domination of nature.

6. NORMATIVE CONCEPTS OF NATURE IN SOCIAL REALITY

The postulate of a normative concept of nature seems to be a paradoxical one, not simply because of the traditional dichotomy between *physis* and *nomos*, but also in terms of the standpoint of positivist science.

Such a postulate becomes much more plausible, however, once the element of *social action* is introduced into the argument. Property in land is a case in point. It represents a concept of nature which sees nature through normative eyes. What are the normative implications of *nature considered as property*? In the form of property, nature, or a part of it, becomes a concession or award to some individuals and a prohibition or exclusion – a taboo – for others. Since property always implies a demarcation, either against commonly-held land, or the property of others, it inevitably implies some form of "enclosure". As a consequence, the conceptual treatment of nature as property will generally be paralleled by its practical treatment in the form of the erection of fences, signs and walls. Even in this narrow sense, therefore, "appropriated nature" has a certain concrete superstructure. This superstructure is most clearly visible from the air: it customarily appears as a net or chequeboard imposed on the natural structure of the landscape. The geometrisation of nature is primarily an expression of its social appropriation. Furthermore, if property is considered as inheritable, the conception of nature will also imply some notion of conservation.

Property which is to be transmitted from one generation to the next must be kept valuable: this forms a backdrop, for example, to the tendency observable amongst some environmentalists to legitimate private property as one means for fending off the ecological crisis. Since property, by its very nature, must be conserved, then nature as private property *will* be conserved.[10]

Nevertheless, the plausibility of this example should not divert attention from the general problem, since there are a number of quite different ways in which nature can be integrated into the social system. It is, however, a fact that the concept of property overshadows the other modes of integration in capitalist societies. This is immediately apparent if we turn to the example of *nature within legislation*. The explosive growth in environmental legislation has meant that elements of nature are increasingly becoming subjects of the law. What sort of problems does such a legislative integration of nature create? One extreme possibility would be to treat nature as if it were a person with a claim to a particular set of rights. In a negative sense this was actually practised during the mediaeval period when cats, dogs and other animals could be put on trial for causing damages – and condemned to punishment if found guilty.

Although such a practice is not what we would particularly conceive as a legal integration of nature, current legislation protecting animals points in this direction. All natural, and not only human beings are considered as having rights of their own, and are therefore entitled to protection not simply by virtue of their possible usefulness to people, but in their own right. This attribution of rights does not however mean that animals are held to be juridical persons: for example, they require someone else to claim their rights for them. In this sense they do not differ legally from mentally disabled people, or children. Any difference between human and non-human beings depends essentially on what is considered to be the basis of personality. As the example of mentally disabled people or children shows, this basis is not considered to lie in the possession of property or reason. Following this line of argument, some authors contend that the basis for the equal rights of animals – both human and non-human – resides in their equal capacity to suffer.[11] Although at first sight a reasonable proposition, this argument breaks down when the very existence of animals depends on human beings: in such circumstances it would be more appropriate to emphasise the responsibility of humans for animals.

In contrast to the case of animals, environmental law remains within the limitations of the law of property. The environment is considered as a collection of goods, the use of which might be contentious. The protection of the environment then turns out to be the protection of the potential interests of another group of persons. There has been considerable discussion as to whether these interests have the juridical status of individual rights (to potable water, clean air, etc.) or whether they are public property. However, the more fundamental question is whether the concepts of property are appropriate to such matters at all. This can be clearly seen in the wording adopted by the UN Conference on the Environment (1972) for its "Declaration of Principles": "Man has the fundamental right to freedom, equality and adequate conditions of life, in an environment of a quality which permits a life of dignity and well-being, and bears a solemn responsibility to protect and improve the environment for present and future generations."[12]

The conditions for well-being – that is, for a healthy life – may still be conceivable in terms of an inherited but inalienable property, such as the body. However, the UN deliberately chose to express the conditions of well-being as a project which awaits implementation. The project-character of the UN declaration is particularly clear in the case of those qualities of the environment which facilitate a "life of dignity". What the UN principle claims is some type of adequacy of nature to the idea of humanity. Although this is clearly a cultural – perhaps even an anthropocentric – conception of nature, as such it in fact serves to reinforce our argument: that social or normative concepts of nature can be found within the socio-political reality of the present.

Whereas the first mode of juridical integration of nature into the social system implied a right of nature against humanity, the latter concedes to everyone a right to assert the condition in which nature *ought* to be. Although it is not quite clear to whom such an assertion is to be addressed, it is clear that it is not to nature: rather, it is a claim for nature. The formulation of human rights on nature implies a concept of nature as a cultural product; what is at stake is the *quality* of nature as a social product.

How is nature integrated into the *system of social production*? The concept of nature which corresponds to this system – in contrast to conceptions based on property or law – is that of nature as resource, or considered in a more extended sense as a "productive force". However, there is a fine distinction between "resource" and "productive force": "resource" suggests that nature consists of a mere stock of means to be exploited through production. In contrast, nature as "productive force" carries with it the flavour of nature as a kind of partner in production, and the valuing of its spontaneous activity as a form of co-production. Nature as a resource is therefore a highly selective and differentiating conception of nature: nature is integrated into the social system essentially in response and correspondence with certain human intentions – with varying degrees of appropriateness. There is a clear gap between humanity (and its wishes) and nature: nature is not evaluated in terms of its conformity with humanity, but in terms of its quality as a set of means. One consequence of such a conclusion is that this form of the social integration of nature provides the raw material for a great deal of social controversy, since different sections of society may not necessarily, and do not, agree on how specific elements of nature should be used. They may even disagree on what type of resource a particular element of nature represents: is a remote river valley an agricultural district, a recreation area, a potential reservoir, or a site for a nuclear power station?

The character of normative concepts of nature implies divergence, and social controversy. Moreover, since a critical consideration of the political reality of the environmental crisis precludes the formulation of an integrative or consensual theory of society, it follows that any theory of normative concepts of nature will at the same time constitute a theory of the antagonistic structure of society itself. The incorporation of nature into such a theory would imply the formulation of a theory of the antagonistic reproduction of society. Nature would figure as the divergently conceived basis of this reproduction, and the reproduction of nature itself would also enter as an object of struggle between different social groupings (communities versus companies, for example). Nevertheless, the conceptual difficulties of such a theory become very apparent once one faction in society refuses to accept, for example, that the conception of nature in terms of property is an appropriate strategy for the reproduction of society.

7. NORMATIVE CONCEPTS OF NATURE IN THE SOCIAL SCIENCES

The previous section dealt with a number of normative concepts of nature corresponding to the economic, the juridical and the productive integration of nature. These different types of the integration of nature into social reality are reflected in the different ways in which environmental studies are incorporated into the academic system. In addition to studies within departments of biology, environmental studies are also organised within departments of economics, the arts and law. This forms the background to the following question: what type of concept of nature is appropriate for the social sciences, and what changes must the social sciences undergo to accommodate such a concept?

As a starting point it should be noted that research in some areas of *social science* and *law* is no longer really feasible without some competence in natural science. The legal aspects of the construction and running of complex productive systems (such as nuclear power plants) is so closely bound up with sophisticated technical prescriptions that they cannot be determined without an adequate technical and scientific training. Within the productive economy, management includes management of the environment and its associated competences. Such developments may involve the dubious tendency of transforming legal questions into technical questions – reifying human affairs in objective facts. In institutional terms, the related problems of equipping personnel are normally dealt with by adding on natural science and technical sub-departments to social science, or through interdisciplinarity.

Although the work of natural scientists within the institutions of the social sciences may be very fruitful, the question still exists as to whether the problems of the social integration of nature can be resolved merely by borrowing knowledge from the natural sciences. The example of the problems encountered in formulating a basis for the legal protection of animals provides a demonstration that such a straightforward borrowing cannot work, and that a transformation of natural science itself is necessary. As noted above, this type of legislation implies the notion of protecting animals for their own sake. But is it not also permissible to raise the question of what animals might want for themselves, what life might be like for animals "on their own terms"? Such an approach is not part of the current methodology of science which demands objectivity – and in particular, the treatment of the objects of knowledge as objects, even where these objects are animals and animal behaviour. Although, in principle, the discipline of the study of animal behaviour could serve to formulate an adequate law for the protection of animals, an accommodation between this type of research and the idea of the volition of animals cited above would imply some fundamental changes in scientific method: the object of research would have to be accepted as a subject, and the fundamental distance between human beings and nature suspended. A different normative concept of nature would therefore correspond to a different type of research. The research implications of this position would differ considerably from those consistent with treating animals as a "resource". The generalization of such a position would be tantamount to the thesis of the social constitution of nature as an object of research. The character of nature as an object of research depends on the type of social integration of nature. This comes close to the traditional criticism of science as ideology: Sohn-Rethel, for example, sees science as conceptualising nature in terms of exchange-values; Moscovici shows how concepts of nature correspond to modes of human labour; and philosophers from Scheler to Habermas have treated natural science as representing a moment within the ideology of purposive-rationality. Following these theories we are obliged to concur that modern natural science is already the science of nature integrated into the social system as potential machinery. But when confronted with the widespread contention that science is value-neutral in cognitive terms, such a position must be buttressed by a demonstration that science would look different under a different set of social circumstances. However, our aim here is not to explore the question of why it was thought that natural science dealt directly with nature (a view fundamentally refuted by Kant who showed that what nature is depends on our view of it).[13]

The more pressing question for us is that of the long-term feasibility of simply "borrowing" knowledge from the natural sciences for use in the social sciences without realizing, in addition, that the concept of nature within the natural sciences already corresponds to a particular social integration of nature.

The social sciences themselves must develop scientific knowledge about nature: this does not mean that they should turn their backs on the achievements of natural science, but rather, that they should make explicit the different modalities in which nature is socially integrated. Seen from this perspective, it is possible to conceive of a science of nature as a subject (or even as a person), a science of nature as a resource, a science of nature as a means of production, a science of nature as a condition for human life, and a science of nature as property, or as a common good.

Would such different sciences of nature (in social science) produce different knowledges about different objects? The traditional self-conception of positive science would clearly say no: there is the knowledge of nature on the one hand, and the application of this knowledge on the other. However, political reality has been transmitting a somewhat different message for some time. Scientific expertise from different areas when applied to ecological problems is often incommensurable since the different parties work with different concepts of what a particular part of nature is, measure and evaluate different parameters, and even employ quite different methods. The effects of partiality on knowledge are only too obvious in such a context. The problem conceptualized at the level of epistemology and the philosophy of science shows that these problems are necessary and not merely contingent products of dealing with nature in ways which touch upon different aspects of its integration into society. We therefore have to study what the effects of the social integration of nature on the knowledge of nature are, or might be expected to be.

We have already noted the most obvious aspect of these effects in terms of the selection of variables. This aspect relates to the treatment of nature as environment – that is, nature not as a complex of laws but as the surface of the globe, the ecosphere. Nature considered in this light is constituted through the variables used to describe it.

The second aspect involves levels of production: that is, the systematic aspect of nature, or parts of nature, understood as systems in constant reproduction, in which human action plays a role. The problem of reproduction embraces the question of whether a system is reproduced at the same level, as measured by one of the system's parameters, and, more crucially, whether it is reproduced as the "same". The latter would refer, for example, to the issue of change in one of the central pattern variables of nature – such as the loss of regenerative capacity, the sudden "death" of a body of water. The effects of the social integration of nature on knowledge can be seen here in the fact that each type of integration would consider certain levels of reproduction as "healthy" or "natural" or desirable on certain grounds. Concepts of evolution or regeneration rest on such theoretically articulated interests. This can also be expressed by saying that each social integration of nature investigates nature against the background of a paradigm (for example, the paradigm of the oyster bank in traditional ecology).

The third aspect of effects is located within the unifying concepts of the object of research itself. Clearly, the character of research cannot be indifferent to whether the research is directed at investigating a certain part of nature as a water-resource or as a potential recreation area. This involves choice of parameters and levels of reproduction and in turn indicates the different disciplines of social research into nature, such as resource management, hydraulic engineering or landscape planning. These disciplines accumulate already differing stocks of knowledge about nature, which can then be called up into the discussion as support for the various antagonistic forces in society in the political struggle over what a particular part of nature is to become.

8. RESEARCH PROSPECTS

(1) From the point of view of the philosophy of the social and natural sciences, the current interconnections between science, politics, and nature lead into a new territory by virtue of the fact that the constructive variation of natural cycles and processes – the increasing intensity of human interaction with nature – is bound up with normative questions. The integration of natural scientific knowledge with social norms cannot, we believe, be achieved through the mere conjunction of natural scientific and social scientific "data". Rather, we expect that the integration of natural and social scientific research, pure and practical reason, facts and values, will render the traditional demarcations between the sciences an anachronism; it will lead to the merging of the natural and the social sciences into a Social Science of Nature. Classical concepts and distinctions will have to be revised in the light of the normatisation of the social politics of nature.

(2) We expect a gradual devaluation of the natural sciences and a corresponding revaluation of the social sciences. The scientific-technical disciplines

may well end up in a position analagous to that of medicine, exercising a kind of service function oriented towards the continuous re-establishment and maintenance of the "health" of our eco-system (which we would propose to define as a state of "ultra-stability", or the capacity of the eco-system to adapt to the impacts of its cultural environment without hitting back). The normative regulation of the human history of nature and the convergence of natural and social scientific research required by it, as well as the increasing number of problem-oriented interdisciplinary projects and the mutual opening of hitherto exclusive disciplinary world views might also have implications for academic education, for instance in the form of a "socio-ecologicum" (that is, a minor in the social sciences for natural scientists and in the natural sciences for social scientists).

(3) We expect a constant expansion of the social influence of political ecology. Scientific and environmental agencies in advanced industrialized societies are forced to interfere in a regulative capacity to shape external nature and the impending reconstruction of human nature. They will attempt to obtain a monopoly over the social regulation of human interaction with nature. The centre of social conflict will therefore shift away from the old industrial conflict between capital and labour and towards the post-industrial conflict between technocrats and those whose lives are affected by technocratic decision-making.[14] This new conflict will be mainly a struggle over the socialization of what Max Scheler has termed *Herrschaftswissen* – knowledge for the sake of domination. We therefore need to investigate the grass-roots opposition to ecological despoliation and ruin (critique of environmentalism) together with an investigation of the bureaucratic development of a technocratic politics of nature (critique of political ecology).

(4) We expect the societal reproduction of nature to be added to the customary – codified or non-codified – objectives of societies. How this can be most effectively carried out will undoubtedly become the subject of major debate (for example, should governmental responsibilities be amended or extended, should there be a widening of the candidates for legislation, a new ethics, or a transformation of the mode of production?). So far these questions have developed almost spontaneously through the development of legislation for the protection of nature, or the stipulation of technical norms and devices. It is a necessary task of research to investigate where these tendencies lead and whether they stimulate the political debate over the social integration of nature and allow it to remain open to the fundamental interests and demands of future generations.

(5) We expect that the construction of natural environments, which now

takes place in an unplanned and in this sense a "natural" way will be replaced by a conscious social construction of such environments. The neolithic revolution in agriculture – the first qualitative transformation of nature by humans – has led to the cultivation of the natural world which could be described as "naturally social". The second qualitative leap can, by contrast, only be a strategic and intentional one. It will be chracterized by the planned transformation of natural cycles and processes, by the consciously controlled design of desirable environments, and by the intentional construction of new natures and new human–nature systems. This is no small challenge: it means that "we can consciously begin to develop the human history of nature just as in the eighteenth century work began on the conscious development of the human history of society."[15] It also means that in our politics of nature we now bear the responsibility for setting rational goals for this human history of nature.

NOTES

[1] *Cf.* Dülmen (1978).
[2] See 'Normative Finalization' in this volume, pp. 223 f.
[3] See Council on Environmental Quality (1980).
[4] *Cf.* Böhme and Grebe (1980); Schäfer (1980).
[5] *Cf.* Noble (1979).
[6] *Cf.* Prigogine, Allen and Herman (1977); Prigogine and Stengers (1981).
[7] *Cf.* Moscovici (1968).
[8] See, for example, Carson (1962).
[9] Merchant (1980), p. 1. Carolyn Merchant narrates the not only metaphoric transformation of the old world into a storehouse of "resources".
[10] *Cf.* Bish (1977), pp. 217–228.
[11] See, for example, Singer (1977).
[12] United Nations Conference on the Environment (1972), 'Declaration of Principles', cited in Science Action Coalition (1980), p. 291.
[13] *Cf.* Böhme (1981).
[14] *Cf.* Touraine, Hegedus and Wieviorka (1980); Touraine (1981).
[15] Moscovici (1976), p. 111.

BIBLIOGRAPHY

Bish, R. L.: 1977, 'Environmental resource management – public or private?', in Hardin, G., and Baden, J. (eds.): 1977, *Managing the Commons*, Freeman, San Francisco.
Böhme, G.: 1981, 'Towards a reconstruction of Kant's epistemology and theory of science', *The Philosophical Forum* **VIII**, 75–102.
Böhme, G. and Grebe, J.: 1980, 'Soziale Naturwissenschaft. Über die wissenschaftliche

Bearbeitung der Stoffwechselbeziehung Mensch-Natur', in Böhme, G.: 1980, *Alternativen der Wissenschaft*, Suhrkamp, Frankfurt, pp. 245–270.

Carson, R.: 1962, *Silent Spring*, Houghton Mifflin Co., Boston.

Council on Environmental Quality (ed): 1980, *The Global 2000 Report to the President*, US Government Printing Office, Washington.

Dülmen, R. van: 1978, *Die Utopie einer christlichen Gesellschaft. Johann Valentin Andreae (1586–1654)*, Part I, Frommann-Holzboog, Stuttgart-Bad Canstatt.

Merchant, C.: 1980, *The Death of Nature. Women, Ecology, and the Scientific Revolution*, Harper & Row, San Francisco.

Moscovici, S.: 1968, *Essai sur l'histoire humaine de la nature*, Flammarion, Paris.

Moscovici, S.: 1976, 'Die Wiederverzauberung der Welt', in Touraine *et al.*: 1976, *Jenseits der Krise. Wider das politische Defizit der Ökologie*, Syndikat, Frankfurt.

Noble, D.: 1979, *America by Design. Science, Technology and the Rise of Corporate Capitalism*, Oxford University Press, Oxford, New York, Toronto and Melbourne.

Prigogine, I., Allen, P. M., and Herman, R.: 1977, 'Long term trends and the evolution of complexity', in Laszlo, E. and Bierman, J. (eds.): 1977, *Goals in a Global Community. A Report to the Club of Rome,* Vol. 1, Pergamon Press, New York, Toronto, Oxford, Sydney, Frankfurt and Paris.

Prigogine, I. and Stengers, I.: 1981, *Dialog mit der Natur. Neue Wege naturwissenschaftlichen Denkens*, Piper, Munich and Zurich.

Schäfer, W.: 1980, 'A note on the Social Natural Science Project (SNSP) – Science, Politics and Nature', *Social Science Information* **19**, 663–670.

Science Action Coalition: 1980, *Environmental Ethics*, Anchor Books, Garden City.

Singer, P.: 1977, *Animal Liberation. A New Ethics for our Treatment of Animals*, Avon Discus, New York.

Touraine, A.: 1981, *The Voice and the Eye. An Analysis of Social Movements*, Cambridge University Press, Cambridge.

Touraine, A., Hegedus, Z., and Wieviorka, M.: 1980, *La prophétie anti-nucléaire*, Editions du Seuil, Paris.

APPENDIX

INTRODUCTORY NOTE

The Finalization Debate refers to a political and theoretical dispute over the "Finalization in Science" thesis which took place in West Germany during 1976.

Regular listeners to the Bavarian radiostation *Bayerischer Rundfunk* who tuned in on the 8th March 1976 would have heard an ominous report: a group of social scientists in Germany were planning to provide a justification for a type of policy towards science which had been characteristic of the USSR under Stalin – the guidance of science in line with ideological pre-judgements (notorious as Lysenkoism). The warning also named the male-factors: researchers into the history and philosophy of science who were spreading the message that science could be directed in accordance with external objectives. However, a "Study Group" had been formed to combat the danger and was expected to meet for the first time in Munich shortly thereafter. Inquiries revealed that the Study Group was to hold a three-day meeting at the Hilton Hotel in Munich: the public were not to be admitted. No more precise details of the organizers or purpose of the meeting were forthcoming.

The conference took place between the 19th and 21st March 1976. Later reconstructions of events showed that the Study Group "Wissenschaftsforschung in der BRD" (Science Research in the Federal Republic of Germany) had been established in the summer of 1975 by Professor Kurt Hübner (Kiel), Professor Hermann Lübbe (Zürich) and Professor Gerard Radnitzky (Bochum). Finance for the conference was provided by the Fritz Thyssen Foundation – with which, at least, Professors Lübbe and Radnitzky were associated. The conference was chaired by Professor Nikolaus Lobkowicz, the Rector of the University of Munich. Other participants – that is those whose names featured in subsequent press reports – included Professors Hans Albert (Mannheim), Werner Becker (Frankfurt), Hans Poser (Berlin), Erwin K. Scheuch (Cologne) and Ernst Topitsch (Graz).

In the weeks following the conference it emerged that the Thyssen Study Group had provided the slogans for a political campaign against neo-marxists, dialectical ideologues and the like, and in particular against the Max Planck Institute for the Study of Conditions of Life in the Scientific-Technical

G. Böhme et al., Finalization in Science, ed. by W. Schäfer, 273–274.

World located in Starnberg, Bavaria, and more particularly still, the Institute's project group "Alternatives in Science". The subsequent polemic was initially propagated through the more conservative media – *Bayerischer Rundfunk, Süddeutsche Zeitung, Die Welt, Frankfurter Allgemeine Zeitung* and the *Münchner Merkur*. The various articles and broadcasts accused the Alternatives Group of seeking a "Final Solution" for science, of threatening science with a "1984 situation", of practising "Lysenkoism" – in short of endangering the freedom of science.

The scientists under attack – ourselves – were naturally eager to know more about the reasoning and the arguments behind this polemic. However, although it was clear that certain journalists – that is, those in agreement with the aims of the Thyssen Group – had been admitted, this invitation did not extend to ourselves. Neither were we given full access to the documents under discussion although Professor Lobkowicz subsequently passed on the same documents to *Der Spiegel*. Our opportunity to study the theses of our critics came later when the papers of the Munich conference were published, in "revised form" in a collection of – according to the blurb – "the most significant papers of the meeting of the Study Group 'Science Research in Federal Germany' which aroused unusually strong public interest" under the title *Die politische Herausforderung der Wissenschaft. Gegen eine ideologisch verplante Forschung*, edited by Hübner, Lobkowicz and Radnitzky (Hamburg, 1976).

The following critical discussion of the neo-conservative critique of the concept of finalization gives *both* sides a hearing. The anti-finalist position is confronted with the central working hypotheses of the finalists, with specific reference to (1) the orientation of basic research, (2) the partisanship of scientists, and (3) the convergence of theoretical and practical discourse. The conclusion to the Appendix consists of a bibliographical documentation of all the contributions known to us about the finalization discussion and debate.

WOLF SCHÄFER

THE FINALIZATION DEBATE: A REPLY TO OUR CRITICS

With a bibliography of the finalization discussion and debate

1. INTRODUCTION

The critics who resolved to put the finalization thesis on trial were clearly aware that "underhand insults and malicious imputations" can be most effectively deployed in the guise of "strictly objective argument".[1] Two examples will show, however, that the "arguments" of our critics did not require the efforts of the press to turn them into insults and imputations: in essence they were already such.

(a) Underhand insults, or Lobkowicz and the press:

In order to avoid any possible misunderstanding of what I am saying I would add that far from regarding Habermas as tilting towards being a "radical left ideologue" – something which is occasionally claimed – I consider that although such a political reproach may have some basis in his questionable effects, it overlooks the fact that as a philosopher Habermas is clearly above average, a fact which puts him several rungs higher than many of his critics.[2]

One of the central sources of this confusion was cited as being the Max-Planck-Institute, Starnberg, which under the direction of Jürgen Habermas and Carl Friedrich von Weizsäcker has concocted the theory of the "finalization of science" out of a mixture of fundamentalist Marxism, misunderstood American theories of science and extra-scientific "emancipatory" urges.[3]

(b) Malicious imputations, or Radnitzky and the press:

In addition, the "finalization theorists" of the Federal Republic would have to go much further (as in the Soviet Union) to obtain such a position of virtually unlimited power. Those disciplines which they consider to have attained the "finalization stage" will be subjected to the complete and total cessation of basic research by reference to this "fact" – not merely kept away from delicate subjects.[4]

What is more dangerous than what the finalists wish to encourage is, as a rule, their desire to obstruct certain types of research on political or ideological grounds. Stalin's Soviet Union not only defamed genetics as "bourgeois pseudo-philosophy" but occasionally included relativity-theory, cybernetics, symbolic logic and sociology as well. Whereas clear signs of a relaxation can be seen in this sphere in the East in recent years, the opposite tendency is visible in the West.[5]

The rosy view that the press "exaggerated" the arguments of the "Study Group on Science in the Federal Republic of Germany" is hard to swallow,

G. Böhme et al., Finalization in Science, ed. by W. Schäfer, 275–299.

as a comparison of the printed version of the papers given at the Study Group's first (and last) conference with the newspaper reports shows. The collection published by our honourable opponents is entitled *The Political Challenge to Science* and, according to the sub-title, is aimed *Against an Ideologically Planned Research.*[6] My response and reply is aimed at a different target: it is a political challenge to neo-conservative science research which advocated – and continues to advocate – an ideological *pro*scription of critical research into science. However, in the course of this reply I will strive *not* to be polemic.

In order to do full justice to the opponents of finalization and to ensure that the rational core of the anti-finalist position has a chance to show through I propose to distinguish between the reasoning and the emotion in the anti-finalists' case. Since there are some grounds for supposing that the power of the anti-finalists lies more in their emotional impact than in their arguments, this exercise may well result in much of what was distinctive about the Study Group on Science Research disappearing entirely. In particular, the separating out of their emotions will eliminate that strain of prejudice which seems to lead perfectly reasonable scientists to the bizarre view that finalization is a neo-Marxist "emancipatory doctrine of salvation".[7]

Extracting the emotional element from the anti-finalists' position reveals that its more or less rational core consists of three proscriptions. The first places a taboo around the *orientation of basic research*. The second places a taboo around *partisanship on the part of scientists*. And the third places a taboo around the *convergence of theoretical and practical discourse*. In the world of the anti-finalists these three taboos fulfil a magic function – they serve to banish finalization.

The critique of the finalization thesis argues as follows: because this century has already seen "numerous attempts to exercise political and economic tutelage over science or to conscript it"[8] and in view of the possibility that such a "strict conscription" of science could also come about in Federal Germany,[9] the finalist notion that external goals for science can become the guidelines for the development of theory is regarded as a stab in the back since, in the language of the "critique", this argument is preparing the ground for the "legitimation of the external steering of science, including the prohibition of certain areas of basic research."[10] The "road to totalitarianism" has therefore found its pathbeaters – the finalists,[11] who have presented the "demand for political control of science as something which is scientifically defensible" and – although this scientific defence is "scientifically insignificant, but not without political impact" – this has to be proscribed.[12]

This is a strong position. Even if one were to concede that the anti-finalists distort matters by overestimating the practical impact of the finalization thesis and underestimating its theoretical significance, the argument that the finalization thesis represents a form of ideological support for a totalitarian direction of science would still remain, and the three proscriptions would be justified. However, this strong position is not tenable. Finalization research – in disregard for proscriptions of any kind – has indeed dealt with the subject of how science is conscripted; nevertheless, it is a malicious imputation to suppose that research in finalization is responsible for the threat of a "totalitarian conflation" of evil social interests and good scientific knowledge.[13]

2. THE FIRST PROSCRIPTION: IT IS FORBIDDEN TO STUDY THE PROGRESSIVE GOAL ORIENTATION OF SCIENCE

2.1. *The Humus Theory of Basic Research – an Anti-finalist Position*

A finalization of research according to the Starnberg model would signify the end of free basic research. All that would remain would be research planned according to social objectives. Just as was the case with the depradation of the soil, such a depradation of science might produce short term successes but in the end would destroy the fertile humus for application-related research and technology.[14]

Radnitzky, Andersson and other anti-finalists predict the "end of free basic research".[15] Their prognosis runs as follows: if a "finalization according to the Starnberg model" were to succeed, the result would be the "depradation" and "exploitation of science", since finalization, "would destroy the fertile humus for application-related research and technology". This is a truly classical notion, reminiscent of a dispute in which Justus von Liebig became involved around the middle of the last century. At that time, to quote Liebig himself, people believed

that the power of the soil was borne by a particular agent, and that this agent was the humus; this term referred to a specific combustible material of organic origin which could not be more closely defined, a sort of manure; it was believed that the decrease and increase in the yields of fields stood in proportion to the content, or the decrease or increase of humus in the field . . . What was true about this supposition was that more plants grow in a fertile field than in an infertile field . . . [16]

Agricultural chemistry has long since become finalized and abandoned the pre-paradigmatic stage of humus theory. But this venerable thesis has now found a new application: the theory of basic research *à la* Andersson. Andersson believes that the "utility and social relevance of research"[17] is borne by a particular agent, and that this agent is basic research; this term refers to a particular institution of historical origin, a type of theoretical "manure" which cannot be more closely defined; he believes that the decrease and increase in the social utility of science stands in proportion to the content, or decrease or increase in free basic research. What is true about this supposition is that fertile basic research is more useful than infertile.

The fact that Andersson has nothing more to say than the notorious humus theorists of Liebig's time – that is, that one simply takes a lot of humus or a lot of free basic research – is an indication that he does not know what the real causes of the production of useful science are. Apart from that we can only agree to the maxim of anti-finalist husbandry that a "good quantity of free basic research be applied".

If we follow the humus-theory tradition back to the nineteenth century we find a trail leading from the metaphorical "depradation of science" to real depradation, "the robbing from the soil of the conditions of their fertility."[18] In the wake of the industrial and demographic revolution, the problem of the impossibility of feeding an exponentially growing population emerged, and was recognized for example by Malthus. A subsistence crisis of historic dimensions seemed unavoidable and threatened to become a major obstacle to European development. The humus theorists of that time – of which Thaer was the most prominent – recommended the English system (intensive crop rotation), and in particular, careful mixed cultivation; "once it was thought that the humus was the agent of fertility, naturally the infertility of a field was attributed to a lack of humus."[19] Nevertheless, yields continued to lag behind the growth in population. It was the "true" explanation of the "chemical process of the nutrition of plants"[20] – *en route* to a finalization of the theory of life-processes – which both revealed existing agricultural techniques to be an "economy of despoliation"[21] and facilitated a modern agriculture, manifested externally in a sustained increase in agricultural yields (through the application of artificial manures), but whose essential significance consisted in a new form of ecological integration with nature – abandoning exploitation and resembling "a wave movement in a cycle".[22]

Since Andersson accuses the finalists not merely of "depradation" of the resources of science, but also of ignorance of the history of science,[23] this is perhaps an appropriate place to point out that the original humus theory of soil fertility, which was disproved by Liebig's mineral theory of plant nutrition, neither discovered nor prevented the actual depradation of the soil involved in the practice of traditional husbandry. It is therefore interesting to ponder upon whether the humus theory of basic research bears a similar relation to the finalization of science, as Thaer's theory did to Liebig's.

The neo-classical humus theorists equate free research, or science, with autonomous basic research.[24] They expect useful science – that is science which serves external objectives – to emerge from autonomous basic research. Their creed prescribes that "research without a goal yields the greatest degree of goal-realization."[25] This expression of maximum usefulness through maximum goallessness may succeed as an aphorism, but fails, however, to provide any answers to the following questions:

– How does *scientific* goal-orientation come about?

– What is the *theoretical* structure of application-related research and technology?

– When (at what stage of scientific development) can one expect the *strategic* production of scientific knowledge to be possible?

The mental leap from maximum goallessness to maximum goal-realization therefore – and completely impermissibly – connects the fact that *within* the domain of autonomous basic research it is possible to carry out a type of research which is largely "goalless", with the causal assertion that a high degree of "goal-realization" will be guaranteed *through* autonomous basic research. Since the fetish of autonomous basic research holds open the promise of *all* conceivable uses, a number of interesting questions simply never get asked.

But this is not the whole story. The humus theory of basic research, which treats the social and cognitive act leading to the goal-orientation of science with great discretion, nevertheless adopts a passionate anti-finalist position precisely because of its silence on this issue. This appears consistent since the finalists' opinion that objectives external to science can be made into socially progressive guidelines for the development of research, clearly contradicts the postulate of goallessness advanced by the Study Group on Science Research. As far as the anti-finalists are concerned, the finalization thesis advocates a total perversion of "goalless research"; worse still, the unnatural goal-orientation of science can only lead to science's "depredation" through the agency of finalization.

Despite the ominous implications suggested by the humus theory of pure science[26] and the alleged hazards posed by finalization, it is nevertheless possible to construct a less dramatic reading of the finalization thesis. Lively competition between hypotheses and an evident lack of a capacity to steer science result from difficulties encountered both in theoretically capturing and strategically developing the social and cognitive process which leads to the goal-orientation of science. And since science studies are still in a pre-paradigmatic phase, and not a phase of relative maturity, such a state-of-affairs should not be seen as being especially abnormal. Although we suspect that the continuing self-reflection of science will eventually enable this situation to be overcome, we cannot yet fully describe the theoretical, and in particular the normative, structure of that fundamental change in the relation between science and its social environment which we have tried to grasp with the concept of finalization. Hence, it is highly dubious that the essence of our thesis will be reached simply by denouncing our analysis of the external goal-orientation of science as "ideologically planned research".

Since if a group G advances the theory T, and T is valid, then the fact that the correctness of T serves the interests of G is no grounds for suspicion; it would not be meaningful to subject G to an ideological critique. The suspicion of ideology, and hence an ideological critique, are only appropriate when it is clear beforehand that T is not valid, or that there are no recognised procedures by which to judge the correctness of T. Only under these circumstances one can raise the question of why G might have an interest in advancing T.[27]

Expressed more concretely: since the actual incorrectness of the finalization theory (T_{fin}) cannot be regarded as proven, and since the question of the validity of T_{fin}, or any other theory of science (e.g., the humus theory of basic research) cannot at present be resolved via consensus (because of the pre-paradigmatic stage of competition between hypotheses), the suspicion of ideology can in fact extend *in any direction*. However, there are limits to the criticism that the theory is really ideology: the critique may ask what the theoretical concepts of useful science really mean,[28] but this will not impinge on the real explanandum – the social relevance of science. The finalists (G_{fin}) cannot therefore be held responsible for the fact that the theoretical advance of science takes place in accordance with social needs and interests.[29] The phenomenon of goal-oriented scientific development was not called into existence by virtue of being noted, examined and given a name.

Modern science research, irrespective of whether it is finalist or anti-finalist, has to begin with the fact that the theoretical development of science in accordance with social needs and interests actually does exist. Given that this is the case, two questions arise which retain their validity irrespective of the scientific value of the finalization thesis:

(i) Why do the anti-finalists ignore the social reality of the "increasing capitalist functionalization"[30] of research policy?

(ii) Why does this actual finalization lead to a profit-oriented "conscription" of science for powerful particularist interests, rather than the actualization of the emancipatory content which originally characterized early modern science?

The answer to the first question is perhaps best left to the Thyssen Foundation-financed "Study Group on Science Research in Federal Germany". My own provisional answer to the second question runs as follows: this happens because scientists do not conceive their work in political terms and because society does not conceive its politics in rational terms.

2.2. *The Dissolution of the Coincidence of the Social and Cognitive Autonomy of Basic Research – a Finalist Hypothesis*

The examination of the structures of science dynamics and their compatibility with planning in accordance with goals is an open programme, whose results have to be awaited. One foreseeable result is, nevertheless, that "autonomy" and "the internal logic of development" are not simply properties of science, nor of basic research, but arise in the course of a discipline's development and are dismantled again with its success. Autonomy in the sense of a strict rejection of external goals as the guide to development is the sign of an immature science, not a mature one.[31]

The model of the finalist conception of science locates autonomous basic research midway between pre-paradigmatic and post-paradigmatic science dynamics. The dispute with the anti-finalists centres on the question of whether basic research *always* requires autonomy or not. The anti-finalists advance the view that this is the case; we have maintained that both the cognitive and social autonomy of basic research, can, and must be, terminated through an orientation to external goals. As far as we are concerned, the development of a discipline reaches its apogee through the finalization of basic research.

In order to clarify what is meant by this we need to make a short excursion into the questions of (a) the structuring of the dynamics of science in its three phases; (b) the historic boundedness of the value-neutrality of positive science; (c) the possibility of a reconstruction of the disciplinary community structure in science. The demands both for the emancipatory partisanship of scientists, and the evolutionary convergence of discourse rest upon these three points.

(a) *The three-phase model*. The finalization thesis contends that scientific development can be broken down into three stages (a pre-paradigmatic or explorative phase, a paradigmatic or autonomous phase, and a post-paradigmatic or finalized phase). A Kuhnian view of science would tend to define the preparadigmatic phase in negative terms, through the absence of theory. In contrast to this, we have emphasised the positive elements of the early modern extension of experience.[32]

The subsequent phase of development, the paradigmatic stage, is dominated by the search for the fundamental theory of an object-field previously structured in experiential terms. We have seen that science customarily develops and should develop autonomously during this phase; an internally defined and determined research front emerges from the conjunction of a number of theoretical approaches, its subject-matter constituted by their internal problems. The dominance of a theoretical dynamic which is internally regulated by science therefore acts to ward off external goal-setting during the paradigmatic phase. The anxiety of the anti-finalists about the autonomy of basic research is therefore misplaced here, since we regard "free basic research" as the precondition for scientific progress in the paradigmatic phase.

If the fundamental theory of a research field can be formulated in terms of the cognitive advances achieved during the autonomous phase, it should then be possible to link the further development of the field to social objectives. Thus, the disputed orientation of basic research becomes an option in the post-paradigmatic phase. Only when the theoretical problems of the respective research field have been solved can the structural resistance of autonomous basic research towards external goal-setting be regarded as obsolete.

(b) *The origin of pure science*. The founding of the Royal Society in 1662 and the Académie Royale des Sciences in 1666 mark a decisive turning-point in the social and cognitive history of the "New Science". Experimental natural science acquired a political constitution through a sovereign act of the Crown mainly by divesting itself of all normative claims.[33] This is not without relevance to the finalization debate, since the historic compromise shows that the normative neutrality of modern natural knowledge was not a cognitive necessity when it was established, but rather represented a political condition for the social institutionalization of science. The alleged illegitimacy of any normative "conscription" of science is therefore rooted in the fact that "Moralls and Politicks" dropped out of those sciences which were selected in the seventeenth century and which became stabilized in the nineteenth century.

Science was intimately bound up and identified with religious, moral, political, pedagogic and social considerations in the sociocultural "humus" of the early seventeenth century which provided the initial environment for modern natural science. Chemical philosophy (iatro-chemistry), for example, owed a major debt to the hermetic-magic tradition and generated a vision of a mystic-religious knowledge of nature. The Baconian reform programme of the Puritan period established a bridge between empirical natural knowledge and the promotion of the public good, in particular the lot of the poor. Moreover, this period was still undecided as to which concept of science was to count as the "true" one.[34]

Now, however, the contingent circumstance that only "value-neutral" science has been regarded as serious science for the last three hundred years has been provided with a rationalistic "explanation". This rationalization, which bears many features of an "internalist" reading of the history of science, does not regard the normative regression of the seventeenth century as an adaptation to changed political circumstances after 1660, but rather as a cognitive transformation into an endogenous scientific history of intellectual structures.[35] Since the internalist programme not only attempts a historical reconstruction of the victory of modern science, but also seeks to establish its developmental logic, the genesis of the forms of knowledge is construed as being an independent variable of sociocultural evolution. As a consequence, the expulsion of "Moralls and Politicks" from science acquires the appearance of a cognitively grounded necessity of the logic of scientific development.

This oversystematisation of the actual course of history can accommodate neither the real historical process nor the empirical demands which are placed on any reconstruction of history. Even an internalist historiography, which does not reconstruct the selection of the normatively neutral variant of science by simply burying *actual* history as a deficient variant of *true* history in the footnotes,[36] requires the existence of a historical context of justification. Such a context cannot merely be constituted by a logical principle of posthumous evaluation, but requires a real context of debate and communication between scientists, in which the validity and relevance of concepts, the truth-value of hypotheses and the acceptability of theories can be resolved through discursive means. Only if such a context exists is it possible for the historical and the rational "explanation" to coincide. However, no such context of justification can be identified for the transition from the ambitious programmes of the early modern scientific movement to "experimental philosophy". It was in fact the institutionalization of normatively neutral science which created the social structures out of which such a context was able to develop.

I therefore continue to hold that the politically opportune separation of knowledge and emancipation in the second half of the seventeenth century favoured the social and cognitive development of that scientific alternative which promised progress solely on the side of *pure knowledge*. Whether this science, cleansed of all its emancipatory aspirations, was the only *true* form of natural knowledge is, in the final analysis, an open question. What is clear is merely that pure knowledge as such no longer automatically carries with it emancipatory consequences.[37] For this reason I question normatively neutral science.

(c) *The task community*. The programmatic demand for the social reconstruction of normatively neutral science is not confined to the finalists. Initial steps towards a new social organization of science can be detected in the fact that science and society are drawing closer together – the re-entry of norms into modern forms of knowledge constitutes a real contemporary process. This is evidenced in the emergence of multidisciplinary "task communities". Cancer research represents one example of such a non-disciplinary organisation of science. Although cancer researchers in theory can be pigeon-holed into the classical "scientific communities" – biology, genetics, cell-biology, radiology and so on – the reference of these scientists to the problem they are engaged with cannot be deduced solely from the logic of one individual disciplinary matrix. The disciplinary logic of the traditional "scientific communities" is broken through – decentred – and becomes adapted to finding a solution to the cancer problem.

One basis for distinguishing between a disciplinary "scientific community" and a multidisciplinary "task community" is the contrast between *extensive problem generation* and *intensive problem orientation*. The "scientific community" is a type of problem generator, which constantly produces new puzzles from within itself (extensive problem generation). The "task community" leaps over these disciplinary puzzles; it locates itself ahead of the theoretical front and seizes on a social problem which is marked by its complexity. The "task community" then initiates and selects appropriate scientific developments (intensive problem orientation). The solution as far as such a community is concerned is not that of the scientific problems which are implicit in the task, but of the social problem, which is the complex unity of all the puzzles involved. The weakness of a task community is therefore more likely to be related to its scientificity than to its social relevance – a situation which presents itself in reverse for a scientific community. Although the disciplinary logic of extensive problem generation guarantees the maintenance of professional scientific standards, it often cannot prevent the loss or neglect of orientation to important social problems.

Since the multidisciplinary approach represented by cancer research is not the product of a disciplinary research programme, but rather follows from the intention to inaugurate certain social actions (prevention, diagnosis, therapy and the remission of the disease), cancer research can be conceptualized as a social challenge to disciplinary fronts. This challenge to specialist communities leads, if accepted by the specialists, to a *decentred focusing* of disciplinary logics. The terms "decentring" and "focusing" provide a shorthand expression for the cognitive orientation to external problems. From the standpoint of the scientific community, the disciplinary appropriation of a non-disciplinary research perspective will appear to constitute an impermissible problem shift. However, a less discipline-biased standpoint will reveal this process to be a *decentring* of a disciplinary matrix. Conversely, from the standpoint of a "task community" the disciplinary enforcement of a non-disciplinary problem orientation will appear as the *focusing* of disciplinary logics. "Decentred focusing" accordingly links professional scientificity and social relevance.

Conclusion. Our contention that the linkage of social and cognitive autonomy of basic research is a sign of an immature rather than a mature science can now be made more precise. We consider that the historic compromise which facilitated both the normative neutrality of modern natural science as well as its social institutionalization has been overtaken by recent developments in science. Advances in the elaboration of fundamental theories for more and more object-fields have – as in the early instance of agricultural chemistry – opened the further scientific development of these theories to social goal-setting and revealed cognitive autonomy to be a transitory phenomenon of the paradigmatic phase. Task communities – such as cancer research – which are problem-oriented and structured in a non-disciplinary manner, will throw open the social autonomy of traditional specialist communities and initiate a reorganization of science in relation to problems of social action. It follows from this, firstly, that basic research in the paradigmatic stage is linked to cognitive autonomy – but does not necessarily require a socially autonomous research institution[38] and secondly, that the resocialisation of science in task communities can be linked with a cognitive deautonomization of basic research – if the development of theory formation in the post-paradigmatic stage takes place through the structures of finalization.[39]

3. THE SECOND PROSCRIPTION: IT IS FORBIDDEN TO ADVOCATE THE EMANCIPATORY PARTISANSHIP OF SCIENTISTS

The *separation of political validity and truth is at the heart of the liberal political order, of the Open Society*. If this is denied, or revoked, this order, and with it science, is placed in jeopardy.[40]

According to Hans Albert, one of the contributors to the anti-finalist collection, the "modern cult of partisanship"[41] could dangerously blur the distinction between social interests and scientific truth, were this cult to be practiced in our society. However, in view of the fact that this "cult" has few adherents in advanced industrial societies, what is the aim behind the anti-finalists' preventive verbal onslaught? They allude to "intolerance"[42] and to the "thoroughgoing critique of that 'critical' philosophy of science"[43] which seeks to throw a bridge from the world of facts across to the world of values; they want to "take the claim to progressiveness out of the hands of the representatives of Marxism."[44] They put themselves forwards as the ideological bodyguard of the Open Society, and defenders of our liberal political order. It seems that the neo-conservative vanguard are very much engaged in political activity themselves, that they are highly adept at conflating descriptive and evaluative propositions, and that by no stretch of the imagination can they be said to be practising non-partisan science. In fact, the anti-finalists seem more aggressively partisan than ourselves. However, their partisanship points in a different political direction. Whereas the anti-finalists are more oriented to the status-quo, and even backward-looking, we insist on a social reconstruction of science; we pursue the idea of an emancipatory alternative in science. And this idea is linked with the belief that scientific work should be considered as a social contribution, not least by scientists themselves.

The separation of theoretical knowledge from practical interests has traditionally had an analytical significance – values and facts have been distinguished as separate categories, which does not rule out the fact that one and the same person might have both an evaluative and a descriptive view about one and the same thing. The demand for the emancipatory partisanship of scientists lends a normative aspect to the possibility of this double reflection. *Scientists ought to conceptualise their theoretical work politically*. The justification for this norm follows from our analysis of the current structural change in modern science.

When scientists who have been socialized within traditional scientific disciplines find themselves working together in task communities in order to develop a practical solution to social problems, and when the scientification of social problems takes place on the basis of mature theories, then the development of science itself has undermined the ideal sphere of pure knowledge. Science is then well on its way to becoming a central instance in social development: Daniel Bell's contention that theoretical knowledge will become the strategic resource of post-industrial society may well then become a reality. This is an empirically meaningful and practically fruitful perspective. It not only implies the conscious societation of science, but also a change in scientists' conception of their own role. Scientists will no longer regard themselves exclusively as students of natural laws, but rather as social actors; they will be *political* scientists. Of course, this does not necessarily mean that they will represent emancipatory interests.

The transformation of autonomous science into a means for serving the entrenched interests of power and privilege, whether private capitalist or bureaucratic collectivist, is an evident consideration. Finalized science can serve both particular as well as general interests. And since the general interest has no clearly identifiable bearer, it is in fact more likely that science will be used in conformity with dominant interests. Confronted with this possibility one could conclude that the preservation of the classical autonomy of science would be worth advocating – not as a means for furthering "good" science, but for making "bad" science more difficult. In my view such a defence is not adequate to the challenge; it will prevent neither the reorganisation of the social organisation of science nor the deautonomization of postparadigmatic theory formation. The politicization of science is inevitable.

It is highly unlikely that any society will designate a free enterprise zone for emancipatory science, and such a niche would not in fact be desirable. Emancipatory science can only succeed if scientists break through their privileged social isolation. It would also be naive to think that one fine day society will at last be "mature" enough to use the opportunities presented by finalization "correctly". The assertion of a normative interest in emancipatory science requires that scientists begin to work for a fundamental change in the conventional relation between science and its social environment here and now. This naturally carries with it the implication that partisan scientists will have to represent, at least for the time being, particular interests.[45]

The partisan scientist who will reflect both the social and scientific relevance of his or her work, and consequently will be able to distinguish values and facts, will no longer cling to the separation between science and society, which although permitting a theoretical discourse with scientific colleagues, imposes a ban on the consideration of normative questions. He or she will violate the convention which delegates practical discussion of the aims of finalized research to the "appropriately empowered bodies."[46] He or she will exercise the practical orientation of theoretical discourse. Of course, the danger remains that our mental anticipation of a social reconstruction of science may be proved illusionary by the actions of scientists themselves. As Marx noted: "it is not enough that thought should strive for realisation, reality must also strive towards thought."[47] Is there in fact a reality which is striving towards the ideal of a social reconstruction of science? I will consider this question in the context of the third of the anti-finalists' proscriptions.

4. THE THIRD PROSCRIPTION: IT IS FORBIDDEN TO BELIEVE IN THE EVOLUTIONARY CONVERGENCE OF THEORETICAL AND PRACTICAL DISCOURSE

4.1. *The Disjunction of Theory and Practice – an Anti-finalist Position*

> . . . would not a state of fusion of theory and practice be the appropriate transcendental orientation for our striving for a state of universally guaranteed peace based on a freedom which has, at last, become rational? The answer is, not peace and not rational freedom, but rather intellectual and perhaps worse forms of terror will be the consequences of a political orientation which regards such a state as being one which is realisable, even if only as an ideal.[48]

The anti-finalists refuse to regard as legitimate either the politicization of science – through progressive goal-orientation and emancipatory partisanship – or a corresponding scientification of politics. The idea that people not only make their history but might also shape it rationally, not only with hindsight, but as a strategic prospect in the course of events lies at the core of the disagreement. The thought that binding arguments and compelling discussion might allow a rational development of our history encounters vehement opposition from the anti-finalists who predict ideological dictatorship and terror were such a philosophy ever to be implemented. In order to demonstrate why an anti-finalist like Hermann Lübbe resists the "demand for politicization"[49] as far as science is concerned, and especially, the analogous "demand for scientification" as far as politics is concerned, I have to return to the categorical distinction between theoretical and practical reason.

Following the decisionist school of Carl Schmitt, Lübbe distinguishes between the "logical and the social validity of norms", "between discursive and political compulsion"[50] and insists that the logical validity of norms which may be demonstrated in theoretical discourse[51] does not carry with it any implications for political behaviour. "There is nothing automatic about the transfer from the cognitive level to the social level. Norms do not obtain social validity through 'compelling discussion', by virtue of 'binding argument', that is, not through the procedures which justify norms but solely through the procedures of those appropriately empowered bodies which enact norms – the norm thus decided is only valid by virtue of a decision reached through the appropriate voting procedure."[52]

This decisionistic observation rests, in my opinion, on the underlying assumption that a fundamental dichotomy exists between descriptive and evaluative judgements: evaluative judgements guide action, descriptive judgements do not. As a consequence, the compelling aspects of social values cannot be ascertained discursively, since discursive necessity corresponds with logical but not teleological validity. If one shares the view that this disjuncture is valid, then

– one is obliged to differentiate strictly, but not symmetrically, between theory and practice, fact and value, truth and interest, description and evaluation, logical and teleological validity, discursive and political necessity, causal and intentional explanation, theoretical and practical discourse, science and politics;

– the traditional *primacy of politics* – that is, the legitimately installed bodies of decision-making – is retained;

– and any scientification of politics – in other words, the theoretical orientation of practical discourse – is rejected, but not the *official* politicization of science.

This is an interesting conclusion, because Lübbe is willing to allow a certain amount of politicization of science, or at least he does not "deny the dependency of scientific practice on external objectives which are given validity through planning by the legitimized bodies."[53] However, he is not willing to entertain a practical orientation of theoretical discourse via progressive goal orientation and the emancipatory partisanship of scientists. The overall setting of goals by the state, or the planning of science, is accepted as a "trivial necessity" within the context of "institutional procedures";[54] but a "political obligation in the name of an entity which lacks an identifiable institutional subject, where science has to interpret this obligation itself"[55] is rejected. Why?

Lübbe does not depart from the view that goal-orientations which have a bearing on action should be subject to the control solely of official decision-making agencies and by enactment, and that responsibility for such orientations should never be transferred to the "academic-scientific intelligentsia".[56]

His reasoning is that if one were to accede to "demands which have been induced through a 'critical' philosophy of science" and "which in their exclusivity and obtrusiveness have penetrated the academic milieu in the last few years in such an irritating fashion" then one would grant critical science the status of a "fourth 'critical' power in the state"[57] – which would not only be unjust, but would pave the way for a number of unacceptable consequences.

The unconditional rejection of the view that "scientific theory might incidentally constitute an authority in the establishment of normative obligations"[58] is the product, on the one hand, of the radical disjunction between theory and practice, which logically leads on to the conclusion that any idea of bridging this dichotomy by discursive means appears as a cultural revolutionary fusion of theory and practice, and on the other, of the primacy of political decision, which in turn allows the step towards the scientification of politics to be denounced as a "totalitarian confusion of logical and social validity."[59]

Nevertheless, the mere fact that the primacy of political decision has been asserted with such an (anti-) intellectual vigour will not prevent the idea of a discursive convergence of scientific theory and social practice from becoming a subject of discussions which have no immediate bearing on action. The necessity of decisions may remain, and this may well be because two factors stand in the way of conclusive discourses: firstly, values may be irrational, and secondly, final justifications may succumb to new knowledge. However, even if one were to concede that decisions through the political mechanisms advocated by Lübbe were necessary, this would not justify prohibiting consideration of the issue of the scientification of politics and the politicization of science unless the area of irrationality is to be extended to cover the domain of rational discourse as well. I cannot believe that this is what the anti-finalists intend.

4.2. *The Convergence of Theoretical and Practical Discourse – a Finalist Hypothesis*

Natural science will one day incorporate the science of man, just as the science of man will incorporate natural science: there will be a single science.[60]

This sentence from the *Economic and Philosophical Manuscripts* in which Marx predicts the fusion of natural science and social science into one science is both obscure and suggestive. The sentence from the first essay on

"Finalization in Science", that "the reference point of the generality of science will be social universality under the conditions of normatization, and no longer that of universality in nature"[61] is no less unclear. Both propositions point to a future present of which it is not known whether it will fulfil what the hopes placed on it promise.[62] I therefore regard it as more fruitful to take a closer look at the path of historical development which led to the first signs of a social reconstruction of science.

The aim of a social reconstruction of science is the strategic coordination of the laws of nature and the needs of society – that convergence of theoretical and practical discourse which allows the rational development of nature *and* society, both in terms of objectives and values, and with the assistance of science. The path of historical development which I consider here is conceptualized by the finalization thesis as movement towards the practical orientation of theoretical discourse: social goals which are made into the guiding thread for post-paradigmatic theoretical formation break open the inherited structure of pure science and constitute the new structures of a *social natural science*. In order to justify and illustrate this view of modern scientific development I propose to make a short re-interpretation of the case of agricultural chemistry (dealt with in detail in Part I of this volume).

In 1840 Liebig came to the view that chemistry was *ripe* for "useful applications".[63] "At this time chemistry had attained sufficient autonomy within the natural sciences to take part in the development of other fields, and inasmuch as the works of chemists turned to the investigation of the life of plants and animals, this impinged on agriculture."[64] The success of the "impinging" of chemistry onto agriculture has already been noted – it prevented a secular subsistence crisis.

The *social goal* which the formation of theory in agricultural chemistry internalized in the nineteenth century was formulated by Liebig in the following terms: "The task of the farmer does not consist in obtaining high harvests at the expense of the field, which simply means that it will be impoverished all the sooner, but that it is in his interest and in the interests of human society to strive for high and ever increasing harvests in perpetuity."[65] It is evident that the desired value for agricultural production fitted closely with the development of the population of Europe over the previous century: Large, Rising and Permanent – this was the norm.[66]

The paradigm of agricultural chemistry is oriented to this external goal: one half consists of a theory of natural cycles (the nitrogen, hydrogen and the interdependent oxygen and carbon cycles), and the other of a theory of the *variation* of these cycles. The paradigm of agricultural chemistry is therefore suitable for uniting both knowledge of the objective interrelationships in nature, which had been obtained in the well-tried scientific manner, and in addition a particular norm which is the product of certain social factors. Finalization referred to the strategic linking of both parts of the paradigm.

The basic solution of this demanding achievement – "high and ever increasing harvests in perpetuity" – broke through the theory of general life processes, which was confined to the first part of the paradigm, and extended it through a theory of the cultural variation of these life processes. Agricultural chemistry was thus constituted as a *social natural science*. Its characteristic was not simply in the discovery of the laws of the reproduction of vegetable nature, but primarily in facilitating the purposive construction of natural cycles in the acknowledged interest of human society. What it means to state that a social interest is "acknowledged" can be illustrated by this case of a practical orientation of a theoretical discourse.

When Liebig stated, "the agricultural institutes teach the art of cutting as many shoes as possible from the inexhaustible reserves of the soil, and the best teacher is said to be the one who has advanced this intensive husbandry the furthest"[67] he did not make this remark in order to oppose the generally agreed aim of an intensification of agriculture. His critique was directed at the ideology of the practical farmer, who although, as Liebig noted, was an "inveterate opponent of all 'theory' ", nevertheless subscribed to the then prevalent theory that "his soil's fertility was inexhaustible."[68] This theory – or more accurately, this false interpretation of practice and of nature – had to be delegitimised by Liebig.

The theoretical discourse over the Chemical Process of the Nutrition of Plants had led to the mineral theory of plant-nutrition, and allowed the practical orientation of this discourse to the social objective of "high and ever increasing harvests in perpetuity". The seemingly reliable and progressive technique of the practical farmer was consequently rendered obsolete: it was established once and for all that this method only allowed rising harvests for a *short* duration.

The scientification of the social interest in an increase in agricultural yields through agricultural chemistry shows that social natural science not only offers the instrumental potential for developing efficient means for attaining pre-given social goals, but also for allowing existing means and objectives to be subjected to a rational critique.[69] Practical goal-orientation and theoretical goal-critique are interrelated features of social natural science. The practically oriented theoretical discourse produces knowledge both *for* and *about* the objectives which we set. What may initially appear as legitimate objectives can therefore be questioned and ultimately completely revised in the light of new critical knowledge. Such revisions are made via discursive reasoning based on binding argument. It is distressing indeed that the anti-finalists were only able to denounce this potential of critical science as the "historicist self-privileging of scientific knowledge." [70]

5. CONCLUSION

I would not regard myself as fully entitled to offer a final assessment of the finalization debate; were I to attempt such an evaluation I would not try, nor would I want, to do so *sine ira et studio*. My conclusion, therefore, is simply to stress that the problems of a socially desirable orientation of the further progress of science, which were raised so vigorously in this dispute, have not disappeared from the sphere of the public discussion of science since 1976. The reproach made by the critics of finalization, that the treatment of the problems we had taken up was either irrelevant or dangerous, has not succeeded in convincing those who can still distinguish between a real and an ideological challenge. Ilya Prigogine writes, for example, and with this I propose to conclude,

> What we once discussed under the catchword 'finalization' proved to be a question of institutionalization: how should communication between science and society be organised so that the resources of research are fully used in the service of the general good?
>
> Such a problematic had had no place within the problem-sphere of academic science as it has been constituted in universities and related institutions of higher learning of the last two hundred years. What was, and is, relevant, as a criterion is whatever happened to correspond to the personal interests of the researcher. Quality control was in the hands of a small group of specialists. What various social groups could expect from this research remained a background issue. It's usefulness for society was, so to speak, self-evident.
>
> This model is now being questioned for a number of different reasons . . . For one, there is the enormous financial cost of certain projects in basic research, which makes the decision about it political in the narrowest sense of the term. In addition, the conception of a fundamentally beneficent use has been thrown off course. Science is no longer being held responsible for progress, but rather for the breakdown of progress. This in turn leads to the conclusion that the distinctions between pure and applied science have lost their validity: research can currently only be carried out functionally, towards particular social objectives. This challenge has been met almost solely with resistance from within the academic world. [71]

NOTES AND REFERENCES

[1] See Lobkowicz, N., 'Preface', in Hübner, K. *et al.* (eds.): 1976, *Die politische Herausforderung der Wissenschaft*, Hoffmann and Campe, Hamburg, p. 9.

[2] Lobkowicz, N., 'Erkenntnisleitende Interessen', in Hübner *et al.* (eds.), 1976, *op. cit.*, p. 65, Note 10.

[3] See Zehm, BFD 35. – Titles listed in the following *Bibliography of the Finalization Discussion and Debate* are referred to by BFD and the number of the entry.

[4] Radnitzky, BFD 80, p. 38.

[5] Zehm, BFD 35.

[6] *Cf.* BFD 65. – The Study Group on Science Research in the Federal Republic of Germany financed by the Thyssen Foundation (which exercised "no influence at all" on the work of the Group, as Lobkowicz stressed, nor, as is evident, needed to) was not of course against "the" but rather against "a" planning of research – namely the fancy "radical left" variant. This preference even failed to convince the *Frankfurter Allgemeine Zeitung*: that the political "steering of science can come just as well from the reactionary right as from the radical left is a fact which should not be overlooked especially in Federal Germany, where the threat of restoration is greater than that of revolution" (Schostack, BFD 31).

[7] Besides, I don't think that I should be expected to reply to questions such as the following: "What can be opposed to the emancipatory doctrine of salvation which the self-appointed boyscouts in the ideological philosophy of history of the 'clergy of the modern religion of emancipation' (Schelsky) preach at us?" (Radnitzky, BFD 80, p. 44).

[8] *Cf.* Topitsch, BFD 92, p. 159. – The horror stories taken as examples by the anti-finalists mostly originate in fascist or Stalinist science policy. This creates the impression that totalitarian science policy and finalization are two sides of the same coin (one side represents an evil practice, the other the theory of this practice). The finalists, as intended, are then revealed as the "advocates of a 'final solution' to the science problem" (*cf.* Hochkeppel, BFD 15).

[9] According to some generous anti-finalist insinuations, the "apostles of relevance" (H. Albert) who are seeking this "strict conscription" (see Note 10) have already begun the "ruthless practice of their politics in the domains of institutionalized science" (E. W. Orth, 'Externe Bestimmung der Geisteswissenschaften?', in Hübner *et al.* (eds.), 1976, *op. cit.*, p. 122).

[10] This originates from Radnitzky. In reply to his own question, "What impact does the finalization thesis have?" he gives the answer, "It is completely sufficient to offer a legitimation of the external steering of science including the prohibition of certain areas of basic research, since whenever it is claimed that the discipline involved has reached the stage of 'maturity', the strict conscription of science is legitimated *eo ipso*, as dreamt of by the Bernalists with Bukharin and the other apologists of Stalin's science policy, and demanded at that time by the National Socialists" (Radnitzky, BFD 80, p. 30).

[11] The finalists, "may have the good will to improve society; in relation to the trend towards totalitarianism they not only have no chance, but without wanting to be are in fact its pathbeaters" (Radnitzky, BFD 80, p. 45). Radnitzky's final sentence reveals what it really means to be a "finalist" or a "representative of the sceptical position":

the finalists "undermine, without willing it, the foundations of the Open Society. They are a part of that force which always wants the Good, but always creates Evil" (*ibid.*).

[12] *Cf.* Radnitzky, BFD 80, p. 30.

[13] *Cf.* Lübbe, BFD 72, p. 20.

[14] Andersson, BFD 3. *Cf.* too BFD 44, p. 67: "In opposition to the radical demands for the planning of science, Liberals in England have objected that the planning of basic research could have devastating consequences for the growth of knowledge. Just as with depradation of the soil and exploitation of nature . . . ".

[15] *Cf.* Note 10; or H. Albert, 'Die Idee der Wahrheit und der Primat der Politik. Über die Konsequenzen der deutschen Ideologie für die Entwicklung der Wissenschaft', in Hübner *et al.* (eds.), 1976, *op. cit.*, p. 156: The goal-orientation of science, "suggests a tying of knowledge to practical needs, which makes it then seem plausible to additionally subject basic research to steering from outside, which would be fatal for its development." Or Lübbe, BFD 72, p. 22: "The obligation of science to the general aim of emancipation generates an academic atmosphere in which the human interest in pure theory appears frivolous, and even cynical . . . the feeling for the human and practical use of pure theory, including basic research, withers away in a manner inimical to science." And so on.

[16] J. Liebig, *Die Chemie in ihrer Anwendung auf Agricultur und Physiologie*, Vol. 1, Introduction, Vieweg, Brunswick, 7th edition, 1862, pp. 2 f.

[17] Andersson, BFD 3.

[18] Liebig, 1862, *op. cit.*, p. 111.

[19] Liebig, 1862, *op. cit.*, p. 3.

[20] The title of the first volume of the 7th edition of Liebig's 1862 *Agricultural Chemistry*.

[21] Liebig, 1862, *op. cit.*, p. XV.

[22] Liebig, 1862, *op. cit.*, p. 87.

[23] *Cf.* Andersson, BFD 3 and BFD 44, p. 67: "Nothing learnt from the attempts and earlier errors in the East, or no desire to learn". In reply to this I would say: we might learn from the history of learning or not-learning from history; we do not learn "from" history but rather *in* history; and we are clearly a long way from understanding how this learning-process can be learnt.

[24] *Cf.* Andersson's motto for this section, according to which the finalist "end of free basic research" is equated with the "depradation of science". Although the conflation of the concepts of freedom and autonomy on the one hand, and research, science and basic research on the other allows a series of language games, they all end up by stating one thing: that freedom of research = autonomy of basic research.

[25] Tenbruck, F. H., 'Wissenschaft und Verwissenschaftlichung', in Hübner *et al.* (eds.), 1976, *op. cit.*, p. 215. – Poser, BFD 78, p. 87, claims that the postulate of goallessness is meaningful at least for mathematics: "Irrespective of how the position is as far as exclusive goal-orientation is concerned for the other sciences, for the sake of application it is to be rejected for mathematics." (Poser's article shows, by the way, that the controversial question of the instrumentalization of mathematics can be dealt with in both a directed and rational way.)

[26] Such talk will of course continue as long as a certain kind of scientific ignorance – giving the old answers without posing new questions – continues to prevail. The often repeated view that "basic research is necessary in order to secure the technical

potential in the long-run, and that basic research can be compared with a fertile humus for applications and technology" (Andersson, BFD 45, p. 69) has nevertheless a good chance to wither away when it is recognized that the old separation of basic and applied research is no longer valid for an increasing number of scientific fields.

[27] Lobkowicz, 'Erkenntnisleitende Interessen', in Hübner *et al.* (eds.), 1976, *op. cit.* pp. 57–58.

[28] For example, from our standpoint the humus-model of basic research represents an expression of a "conservative ideology of science" fixated on the ideal-type of pure science, which "formulates the issue of socially-determined alternatives in science into a question leading outside of science" (*cf.* Böhme, BFD 50, p. 44).

[29] The possibility conjured up by Radnitzky that the finalization thesis, "could serve as a subsequent legitimation for measures planned or already undertaken" (BFD 80, p. 30) and the spirit of the age invoked by Hübner – "one currently finds academics and educational policy makers in universities and ministeries who express the view that science has to be finalized" (BFD 66, p. 89) – consequently only refer to the fact of an interest in the planning of science. Beyond that, it makes a difference whether the term "finalization" is being used as a theoretical concept or a slogan in science policy; in contrast to slogans, theoretical concepts are clearly defined and cannot be used simply at will. The utility of the finalization thesis in legitimizing planning is not, however, available to all planning interests since this thesis also delegitimizes certain interests; for example, it delegitimizes the planning of basic research in the paradigmatic phase.

[30] *Cf.* Ronge, BFD 84, p. 271. Ronge notes that, "in practice more social determination of research takes place than is supposed by the finalization approach and conceded by some opponents of this approach" (p. 271). He specifies this observation by stating that, "social functionalization basically means private-industrial determination, or depoliticization" (p. 261). The sober comment that the finalists have linked "clearly something 'emancipatory'" (p. 271) with social determination, but that the "structure of social reality allows the normative aspect of the finalization thesis to become problematic" (*ibid.*) seems highly appropriate from my own view of current developments. What follows from this? In Ronge's view, "the finalization thesis must be 'transcended' in a discussion of the 'politicization' of (capitalist) society" (p. 272). I believe that the "vacancy in social theory" identified by Ronge (see p. 254) as lying within the finalization thesis must be filled. I would therefore support the demand for normative finalization, not in the sense of capitalist functionalization, but in the sense of a social politicization of the finalization theory.

[31] W. van den Daele: 1975, 'Autonomie contra Planung: Scheingefecht um die Grundlagenforschung?', *Wirtschaft und Wissenschaft* **XXIII/2**, 32.

[32] *Cf.* Böhme, G. and van den Daele, W.: 1977, 'Erfahrung als Programm. Über Strukturen vorparadigmatischer Wissenschaft', in Böhme, G., van den Daele, W., and Krohn, W. (eds.): 1977, *Experimentelle Philosophie. Ursprünge autonomer Wissenschaftsentwicklung*, Suhrkamp, Frankfurt, pp. 183–236. Böhme and van den Daele contend that currently pre-paradigmatic structures can be found, for instance, in the research fields of high energy physics, behavioural biology, embryology, neuro-physiology, and add, "Even in the explorative phase any new discipline today begins within science, science does not begin itself – which could be said to apply to the pre-paradigmatic sciences" (p. 225). In addition, the non-simultaneity of development in the social and natural sciences should also be taken into account: "A scientification of experience comparable

to the processes of the pre-paradigmatic sciences of the seventeenth and eighteenth centuries is now only likely within the field of the social sciences" (*ibid.*).

[33] *Cf.* Mendelsohn, E.: 1977, 'The social construction of scientific knowledge', in Mendelsohn, E., Weingart, P., and Whitley, R., (eds.) *The Social Production of Scientific Knowledge*, Reidel, Dordrecht/Boston, pp. 3–26; van den Daele, W., 'The social construction of science: institutionalization and definition of positive science in the latter half of the seventeenth century', in Mendelsohn *et al.* (eds): 1977, *op. cit.*, pp. 27–54.

[34] *Cf.* Debus, A.: 1970, *Science and Education in the Seventeenth Century: The Webster-Ward Debate*, MacDonald & American Elsevier, London and New York, p. 62: "There was still room in the third quarter of the seventeenth century for serious disagreement over the choice of the new science which nearly all agreed must supersede the Peripatetic philosophy of the past."

[35] One representative of the internalist reading of the seventeenth century revolution in science is Dijksterhuis who seeks to reconstruct the emergence of modern science as "The Mechanisation of the World-View": *cf.* Dijksterhuis, E.: 1956, *Die Mechanisierung des Weltbildes*, Springer, Berlin, Göttingen and Heidelberg.

[36] *Cf.* Lakatos, I.: 1971, 'History of science and its rational reconstructions', in *Boston Studies in the Philosophy of Science*, Vol. VIII, Reidel, Dordrecht, p. 107: "One way to indicate discrepancies between history and its rational reconstruction is to relate the internal history *in the text* and indicate *in the footnotes* how actual history 'misbehaved' in the light of its rational reconstruction". In his following 'Notes on Lakatos', Kuhn replied sarcastically: " . . . what Lakatos conceives as history is not history at all but philosophy fabricating examples" (*Boston Studies in the Philosophy of Science, op. cit.* p. 143).

[37] The original assumption that the disjuncture between knowledge and emancipation in the end is not damaging, since the advance in knowledge as such has an emancipatory effect, has lost its power of conviction in our century and realized a warning already given three hundred years previously: "In 1668 Comenius warned the virtuosi of the Royal Society that their work should it not pursue aims transcending the mere knowledge of nature might well turn out to be a 'Babylon turned upside down, building not toward heaven, but towards earth'" (van den Daele, 'The social construction of science', in Mendelsohn, 1977, *op. cit.*, p. 47). The disturbing thought that the consequences of an (apparently) normatively neutral scientific development could strike the earth like a bomb, "has recently been revived. The shock of atomic weapons, the threats to the ecological balance through scientifically produced technology, the role of scientists in the Vietnam War, the visions of genetic manipulation have all given rise to severe criticism of the existing form of science" (*ibid.*, pp. 47–48).

[38] We have so far assumed that the dissolution of the coincidence of the social and cognitive autonomy of basic research would only be possible in the third phase of scientific development, once the cognitive autonomy of basic research has ceased to be necessary. However, I do not see that the *social* autonomy of institutionalized research is necessary in any stage of scientific development – cognitively autonomous basic research could be carried out within new scientific organisations which are, in principle, open to social goal-setting.

[39] The conditional formulation is deliberate since our perspective encountered resistance. From hindsight, it is almost to be welcomed that the neo-conservative vanguard

of 1976 struggled against finalization mainly through prohibitions on what might or might not be thought. However, those who polemicised like Hübner built up fronts which were no longer intended to be the subject of argument alone. "Existing organizational forms of science are not the product of some conceited professors, or the power hunger of the bourgeoisie, but are the necessary and rational product of the essence of science itself. There is no reason to change anything fundamental in them, and to become seduced by fashionable slogans. It was an outstanding achievement of a time past, of which nowadays little good is spoken, that it produced this graded system of channels in which mutual control, stimulation and competition could be exercised. Thus, science in Germany reached a leading position in many fields, and obtained a high regard throughout the world. It was not the failure of this system, but the interventions into this system by the National Socialists which brought it down. A good deal of lost territory was able to be won back after the war. Do we want to lose it again?" (Hübner, BFD 66, pp. 95–95). I might add that it was not the failure of our work, but the political interventions into it and related work by the neo-conservatives, such as Hübner, which helped to bring the Max-Planck-Institute in Starnberg down.

[40] Radnitzky, BFD 80, pp. 25–26.

[41] Albert, H.: 1976, 'Die Idee der Wahrheit und der Primat der Politik', in Hübner *et al.* (eds.), 1976, *op. cit.*, p. 152.

[42] Radnitzky, BFD 80, p. 26.

[43] Lübbe, BFD 72, p. 23.

[44] Becker, W.: 1976, 'Der Konservativismus des marxistischen Theorieverständnisses und seine gesellschaftspolitischen Folgen', in Hübner *et al.* (eds.), 1976, *op. cit.* p. 173.

[45] This would be in line with the historical fact that the rhetoric of social movements changed significantly in the time of the transition to the modern period (between 1750 and 1850). Pre-modern movements spoke a universalist language (up until the famous slogan: Liberty, Equality, Fraternity!); in contrast, modern social movements tend towards a much more particularist rhetoric. For example, in *Travail*, a journal edited by workers in Lyon in 1841, we read: "Only workers can speak the truth". Some of the political watchwords of the current women's movement are also a long way removed from universal "brotherhood". Nevertheless, the emancipation of women does not appear conceivable without the emancipation of men. In short: I suspect that we will only be able to bring about our universalist heaven, which has been occupied by sweeping generalizations for too long, through the consistent assertion of particular interests on earth.

[46] Lübbe, BFD 72, p. 19.

[47] Marx, K.: 1844, 'Contribution to the Critique of Hegel's Philosophy of Law', in Marx, K. and Engels, F.: 1975, *Collected Works*, Vol. III, Lawrence and Wishart, London, p. 183.

[48] Lübbe, H., BFD 72, p. 19.

[49] *ibid.* p. 23.

[50] *ibid.* p. 19.

[51] In order to dispel any misunderstandings, I would point out that Lübbe does not follow the distinction between theoretical discourse (concerning the truth of statements-of-fact) and practical discourse (concerning the rightness of norms) introduced by Jürgen Habermas – otherwise, the appropriate term here would be practical discourse. However, the fact that Lübbe distinguishes between the logical and social validity of

norms (in his decisionist way) means that the logical validity of norms is conceived as the cognitive product of a theoretical discourse, and, in contrast, their social validity is seen as the outcome of a political decision (*Dezision*). Lübbe does not acknowledge a practical discourse in Habermas' sense.

[52] Lübbe, H., BFD 72, p. 19.

[53] *ibid.* p. 22.

[54] *ibid.* p. 13.

[55] *ibid.* p. 22.

[56] *ibid.* p. 20.

[57] *ibid.*

[58] *ibid.*

[59] *ibid.*

[60] Marx, K.: 1844, 'Economic and Philosophical Manuscripts', *Collected Works*, 1975, Vol. III, Lawrence and Wishart, London, p. 304.

[61] Böhme, G., van den Daele, W., and Krohn, W.: 1973, 'Die Finalisierung der Wissenschaft', *Zeitschrift für Soziologie* **II**, 144. (*Cf.* 'Finalization in Science', *Social Science Information* **XV**, 325.)

[62] The young Marx and the early finalists, mediated through Herbert Marcuse who inspired the concept of finalization (see Marcuse, 1966, *On Dimensional Man*, Beacon Press, Boston, pp. 232–234), left behind a considerable volume of conceptual fragments which are difficult to place within our positivist academic tradition, such as the idea of a "true resurrection of nature" (Marx) or the idea of a normatively finalized science. Up until now these fragments can be seen as the broken egg-shells of particular intellectual developments, but also as fragments of the idea of a New Programme in Science, Technology, and Society.

[63] Carrière, J.: 1893, *Berzelius und Liebig. Ihre Briefe von 1831–1845*, J. F. Lehmann, Munich and Stuttgart, p. 211. In a letter to Berzelius written on 26th April 1840 Liebig expressed his "insuperable disgust and opposition to the practice of chemistry at this time" (*op. cit.*, p. 210), and asked what purpose all this discussion has served; neither in medicine, physiology nor industry do useful applications result" (*ibid.*, p. 211).

[64] Liebig (1862), *op. cit.*, pp. 9–10.

[65] *ibid.*, pp. 11–12.

[66] The objectives "high", "ever increasing" and "in perpetuity" should not be seen as implying that "in perpetuity" refers to "ever increasing", since permanent increase was not the dominant aim: rather, the issue was to guarantee that harvests at any level of production attained could be reproduced *in perpetuity.*

[67] Liebig (1862), *op. cit.*, p. 5.

[68] *ibid.* p. 9

[69] Agricultural chemistry facilitated a critique of existing *means*, since it recognized and named *Raubbau* agriculture. In other instances, the elaboration of goal-oriented knowledge may produce a thorough criticism of pre-given *objectives.*

[70] Lübbe, H., BFD 72, p. 21.

[71] Prigogine, I.: 1982, 'Wissenschaft und Gesellschaft im Wandel Europas', *Merkur. Deutsche Zeitschrift für europäisches Denken* XXXVI, p. 486.

BIBLIOGRAPHY OF THE FINALIZATION DISCUSSION AND DEBATE

Compiled by Wolf Schäfer

The following bibliographical documentation is divided into three sections, which cover the public debate of Spring 1976 (I), the reviews of the German edition of Parts I and II of this volume (II), and the contributions to the finalization discussion and debate mainly in scholarly journals (III).

I. THE ANTI-FINALIZATION CAMPAIGN AND PUBLIC DEBATE IN THE MEDIA

1. Albert, Hans: May 7, 1976, 'Wissenschaftsforschung in der Bundesrepublik: Finalisierung der Wissenschaft', Letter to the Editor in *Die Zeit*, No. 20.
2. Andersson, Gunnar: May 7, 1976, 'Freiheit oder Finalisierung der Forschung?', *Hochschulpolitische Informationen*, VII/9 pp. 3–5 and 7.
3. Andersson, Gunnar: Novemeber/December 1976, 'Nichts von der Geschichte gelernt', *Information Philosophie*, No. 5, p. 2.
4. Andersson, Gunnar and Radnitzky, Gerard: August 21/22, 1978, 'Finalisierung der Wissenschaft im doppelten Sinn', *Neue Zürcher Zeitung*, No. 191, p. 21.
5. Anonym: April 5, 1976, 'Etwas unfair', *Der Spiegel*, No. 15, pp. 207–208.
6. Behrmann, J., Böhme, G., van den Daele, W., Hohlfeld, R., Krohn, W., Schäfer, W., and Spengler, T.: April 7, 1976, 'Die Starnberger Forscher fühlen sich ausgetrickst, Letter to the Editor in *Die Welt*, No. 83, p. 6.
7. Behrmann, J., Böhme, G., van den Deale, W., Hohlfeld, R., Krohn, W., Schäfer, W., and Spengler, T.: April 10/11, 1976, 'Statt Auseinandersetzung eine Kampagne gegen Forscher', Letter to the Editor in *Süddeutsche Zeitung*, No. 85.
8. Behrmann, J., Böhme, G., van den Deale, W., Hohlfeld, R., Krohn, W., Schäfer, W., and Spengler, T.: April 16, 1976, 'Was ist das: Finalisierung?', *Die Zeit*, No. 17, pp. 33–34.
9. Behrmann, J., Böhme, G., van den Deale, W., Hohlfeld, R., Krohn, W., Schäfer, W., and Spengler, T.: April 17/18/19, 1976, 'Wer bedroht die Wissenschaft?' Letter to the Editor in *Münchner Merkur*, No. 90.
10. Böhme, Gernot: November/December 1976, 'Nichts zur Wissenschaftstheorie beigetragen', *Information Philosophie*, No. 5, p. 3.
11. Dahrendorf, Ralf: April 23, 1976, 'Bitte um Klarstellung', *Die Zeit*, No. 18.
12. Dahrendorf, Ralf: May 21, 1976, 'Die Unabhängigkeit der Wissenschaft. Vorläufiges Schlußwort in einer wichtigen Debatte', *Die Zeit*, No. 22, p. 53–54.
13. von der Haar, Friedrich: May 7, 1976, 'Wissenschaftsforschung in der Bundesrepublik: Finalisierung der Wissenschaft', Letter to the Editor in *Die Zeit*, No. 20.

G. Böhme et al., Finalization in Science, ed. by W. Schäfer, 301–306.

14. Hochkeppel, Willy: March 22, 1976, 'Der wissenschaftliche Bericht', *Bayerischer Rundfunk*, 2. Programm (broadcast), non-authorised MS, 8 pp.
15. Hochkeppel, Willy: March 24, 1976, 'Gefährdete Wissenschaft? Eine Tagung des Arbeitskreises Wissenschaftsforschung', *Süddeutsche Zeitung*, No. 70, p. 12.
16. Hübner, Kurt: May 7, 1976, 'Wissenschaftsforschung in der Bundesrepublik: Finalisierung der Wissenschaft', Letter to the Editor in *Die Zeit*, No. 20.
17. Küppers, Günter: May 6, 1976, 'Steuerung der Wissenschaft? Unwissenheit über die Bedingungen von Forschungserfolgen', *Süddeutsche Zeitung*, No. 104, p. 33.
18. Liese, W.: May 14, 1976, 'Thema: Die politische Wissenschaftsforschung', Letter to the Editor in *Neue Zürcher Zeitung*, No. 112.
19. Lobkowicz, Nikolaus: May 7, 1976, 'Freie Wissenschaftsforschung', Interview in *Hochschulpolitische Informationen*, VII/6, p. 7.
20. Lobkowicz, Nikolaus: May 21, 1976, 'Die erbetene Klarstellung', *Die Zeit*, No. 22, p. 53.
21. Mey(er, Christoph?): April 23, 1976, 'Politisierung der Wissenschaft. Streit um ein Finalisierungskonzept', *Neue Zürcher Zeitung*, No. 94.
22. Meyer, Christoph: March 24, 1976, 'Wie Wissenschaft in der Bundesrepublik bedroht wird. Bei einer Fachtagung in München: die Öffentlichkeit vor Ansprüchen der Neomarxisten warnen', *Münchner Merkur*, No. 70, p. 6.
23. Marko, Kurt: May 14, 1976, 'Thema: Die politisierte Wissenschaftsforschung', Letter to the Editor in *Neue Zürcher Zeitung*, No. 112.
24. Perger, Werner A.: May 29/30, 1976, 'Freiheit der Wissenschaft im Ausgedinge? Der Streit um die Autonomie der Forschung und die Wünsche des Geldgebers Staat', *Die Presse*.
25. Poser, Hans: April 1976, 'Wider die planende Destruktion', *Mitteilungen des Hochschulverbandes*, XXIV/2, pp. 69–72.
26. Radnitzky, Gerard: April 15, 1976, 'Die Münchner Tagung war kein Finalisierungsspektakel', Letter to the Editor in *Die Welt*, No. 90, p. 6.
27. Radnitzky, G. and Andersson, G.: April 23, 1976, 'Was wahr ist – was frei macht', *Die Zeit*, No. 18.
28. Radnitzky, Gerard: May 7, 1976, 'Wissenschaftsforschung in der Bundesrepublik: Finalisierung der Wissenschaft', Letter to the Editor in *Die Zeit*, No. 20 (also in *Hochschulpolitische Informationen*, VII/6, May 7, 1976, p. 6).
29. Radnitzky, Gerard: May 14, 1976, 'Thema: Die politisierte Wissenschaftsforschung', Letter to the Editor in *Neue Zürcher Zeitung*, No. 112.
30. Richter, Herbert: May 14, 1976, 'Thema: Die politisierte Wissenschaftsforschung', Letter to the Editor in *Neue Zürcher Zeitung*, No. 112.
31. Schostack, Renate: March 25, 1976, 'Widerspruch aus der Wissenschaft. Gegen Ideologie und Bürokratie', *Frankfurter Allgemeine Zeitung*, No. 72.
32. Schuster, Hans: May 29/30, 1976, 'Freiheit oder Lenkbarkeit der Forschung – ist das die Frage? Verwirrende Fronten in der Gelehrtenrepublik', *Süddeutsche Zeitung*, No. 123.
33. Vossenkuhl, Wilhelm: April 2, 1976, 'Neue Glaubwürdigkeit. Wissenschaften im Widerstand gegen die Ideologisierung', *Rheinischer Merkur*, No. 14.
34. von Weizsäcker, Carl Friedrich: May 5, 1976, 'Ueber Wissenschaftspolitik und die begreifliche Angst des Bürgertums', *Neue Zürcher Zeitung*, No. 104.

35. Zehm, Günter: March 24, 1976, 'Droht der deutschen Wissenschaft ein 1984?', *Die Welt*, No. 71.
36. Zimmer, Dieter E.: April 16, 1976, 'Verschwörung gegen die Wissenschaft? Das Starnberger Max-Plank-Institut plant angeblich das Ende freier Forschung – Ein Vorwurf und eine Antwort', *Die Zeit*, No. 17.
37. Zimmer, Dieter E.: April 23, 1976, 'Finalisierung – und kein Ende', *Die Zeit*, No. 18.

II. REVIEWS OF STARNBERGER STUDIEN 1

38. Freitag-Poppe, Claus-Peter: 1979, 'Starnberger Studien 1', *Das Argument* **116**, pp. 584–585.
39. Klima, Rolf: 1979, 'Wissenschaft', *Soziologische Revue*, II/4, pp. 407–411.
40. Mocek, Reinhard: 1980, 'Starnberger Studien 1', *Wissenschaftliche Zeitschrift Halle-Wittenberg*, XXIX/5, pp. 147–152.
41. Setzer, Hans: 1980, 'Starnberger Studien 1', *PVS-Literatur*, Heft 1, p. 71.
42. Teich, Mikuláš: 1980, 'Essay Review Starnberger Studien 1', *Annals of Science*, XXXVII, pp. 101–106.
43. Topitsch, Ernst: 1980, 'The End of Finalization: The Retreat of a Spurious Ideal', *Minerva. A Review of Science, Learning and Policy*, XVIII/1, pp. 179–195.

III. CONTRIBUTIONS TO THE ACADEMIC FINALIZATION DISCUSSION AND DEBATE

44. Andersson, Gunnar, 1976, 'Freiheit oder Finalisierung der Forschung?', in Hübner, K. *et al.* (eds.), *Die politische Herausforderung der Wissenschaft*, pp. 66–76.
45. Andersson, Gunnar: 1978, 'Praxisbezug und Erkenntnisfortschritt', in Hubig/von Rahden (eds.), *Konsequenzen kritischer Wissenschaftstheorie*, pp. 58–80.
46. Andersson, Gunnar: 1979, 'Zu einer Apologie der Finalisierungsthese', *Conceptus*, XIII/32, pp. 90–92.
47. Andersson, Gunnar: 1979, 'Kritik der Finalisierungsthese', in Bärmark, J. (ed.), *Perspectives in Metascience*, Kungl. Vetenskaps- och Vitterhets-Samhället, Göteborg, pp. 27–37.
48. Bodewitz, Henk and Stemerding, Dirk: 1979, 'Wetenschap en Samenleving en het werk van de Starnberg-groep', *Kennis en methode. Tijdschrift voor wetenschapsfilosofie en methodologie*, III, pp. 158–172.
49. Böhme, G., van den Daele, W., and Krohn, W.: 1976, 'Gesellschaftliche Konstitution und gesellschaftliche Steuerung von Wissenschaft. Antwort auf Lutz Hiebers Kritik am Konzept der Finalisierung', *Leviathan. Zeitschrift für Sozialwissenschaft*, IV/2, pp. 277–283.
50. Böhme, Gernot: 1978, 'Alternativen in der Wissenschaft – Alternativen zur Wissenschaft', in Hubig/von Rahden (eds.), *Konsequenzen kritischer Wissenschaftstheorie*, pp. 40–57; English: 'Alternatives in Science – Alternatives to Science', in Helga Nowotny and Hilary Rose (eds.), *Counter-Movements in the Sciences*, Reidel, Dordrecht/Boston 1979, pp. 105–125.

51. Böhme, Gernot: 1979, 'Wie kann es abgeschlossene Theorien geben?', *Zeitschrift für allgemeine Wissenschaftstheorie*, X/2, pp. 343–351; English: 'On the Possibility of Closed Theories', *Stud. Hist. Phil. Sci.* 11, 1980, 2, pp. 163–172.
52. Brunner, Walter Beat: 1981, *Bakterieller Abbau von Dichlormethan und 1, 2 Dichlorethan*, ADAG Administration & Druck AG, Zürich, pp. 85–90.
53. Büchel, Wolfgang: 1972, 'Finalisierung der Wissenschaft und Lyssenkoismus', *Zeitschrift für allgemeine Wissenschaftstheorie*, X/2, pp. 352–357.
54. Dudek, Peter: 1979, *Naturwissenschaften und Gesellschaftsformation. Zum Problem der historischen und begrifflichen Konstitution der klassischen bürgerlichen Naturwissenschaften*. Campus, Frankfurt/New York, pp. 15–22.
55. Düsberg, Klaus-Jürgen: 1979, 'Bemerkung zum Begriff der "abgeschlossenen" Theorie', *Zeitschrift für allgemeine Wissenschaftstheorie*, X/2, pp. 358–362.
56. Düsberg, Klaus-Jürgen: 'Finality or Finalization of Science?', (unpublished MS), 18 pp.
57. Ewers, Michael, 1981: *Menschliche Natur und Umwelterziehung*, päd. extra buchverlag, Bensheim, pp. 79–81.
58. Halfmann, Jost: 1980, *Innenansichten der Wissenschaft. Die Vergesellschaftung der abstrakten Geistesarbeit und die Krise der Metatheorien*, Campus, Frankfurt/New York, pp. 90–92; 157–159.
59. Hieber, Lutz: 1975, 'Zum Konzept: Finalisierung der Wissenschaft', *Leviathan. Zeitschrift für Sozialwissenschaft*, III/4, pp. 449–472.
60. Hieberg, Lutz: 1976, 'Gesellschaftliche Steuerung der Wissenschaft und das Konzept der Finalisierung. Anmerkungen zur Gegenkritik von Böhme/van den Deale/Krohn', *Leviathan. Zeitschrift für Sozialwissenschaft*, pp. 284–289.
61. Hronszky, Imre: 1979, 'Die Finalisationsthese' (unpublished Lecture, April 10, Interuniversity Dubrovnik), 36 pp.
62. Hubig, Christoph, von Rahden, Wolfert (eds.): 1978, *Konsequenzen kritischer Wissenschaftstheorie*, Walter de Gruyter, Berlin/New York, 398 pp.
63. Hubig, Christoph: 1978, 'Das Defizit der Finalisierungsdebatte und eine pragmatische Alternative', in Hubig/von Rahden (eds.), *Konsequenzen kritischer Wissenschaftstheorie*, pp. 111–138.
64. Hucklenbroich, Peter: 1978, *Theorie des Erkenntnisfortschritts. Zum Verhältnis von Erfahrung und Methoden in den Naturwissenschaften*, in Monographien zur Philosophischen Forschung, Vol. 165, Anton Hain, Meisenheim am Glan, pp. 440–447.
65. Hübner, K., Lobkowicz, N., Lübbe, H., and Radnitzky, G. (eds.): 1976, *Die politische Herausforderung der Wissenschaft. Gegen eine ideologisch verplante Forschung*, Hoffmann und Campe, Hamburg, 226 pp.
66. Hübner, Kurt: 1976, 'Die Finalisierung der Wissenschaft als allgemeine Parole und was sich dahinter verbirgt', in Hübner *et al.* (eds.), Die *politische Herausforderung der Wissenschaft*, pp. 89–96.
67. Johnston, Ron: 1976, 'Finalization: A new start for science policy?', *Social Science Information*, XV, pp. 331–336.
68. de Jong, Teus: 1979, 'Het internalisme-externalisme-debat en het werk van de Starnberg-groep', *Kennis en methode. Tijdschrift voor wetenschapsfilosofie en methodologie*, III, pp. 142–157.

69. Koller, Peter: 1979, 'Ein Gespenst von Freiheit', *Conceptus*, XIII/32, pp. 79–88.
70. Koller, Peter: 1979, 'Antwort auf die Erwiderungen von Andersson und Radnitzky', *Conceptus*, XIII/32, pp. 97–100.
71. Kutschmann, Werner: 1981, 'Wissenschaftsforschung und Wissenschaftskritik. Am Beispiel des Max-Planck-Instituts Starnberg', *Merkur. Deutsche Zeitschrift für europäisches Denken*, XXV/5, pp. 529–534.
72. Lübbe, Hermann: 1976, 'Planung oder Politisierung der Wissenschaft. Zur Kritik einer kritischen Wissenschaftsphilosophie', in Hübner, K. *et al.* (eds.), *Die politische Herausforderung der Wissenschaft*, pp. 13–23.
73. Mocek, Reinhard: 1981, 'Wissenschaftstheorie im Kontext bürgerlicher Weltanschauung', *Deutsche Zeitschrift für Philosophie*, XXIX/2, pp. 183–198.
74. Molitor, Bruno: 1977, 'Finalisierung der Wissenschaft?', *Institut für Wirtschaftspolitik an der Universität Köln. Wirtschaftspolitische Chronik*, No. 1, pp. 13–22.
75. Nauta, L. W.: 1979, 'Wetenschapssociologie en wetenschapsfilosofie', *Kennis en methode. Tijdschrift voor wetenschapsfilosofie en methodologie*, III/1, pp. 51–71.
76. Nowotny, Helga: 1973, 'On the Feasibility of a Cognitive Approach to the Study of Science', *Zeitschrift für Soziologie*, II/3, pp. 282–296.
77. Pfetsch, Frank R.: 1979, 'The Finalization Debate in Germany: Some Comments and Explanations', *Social Studies of Science*, IX, pp. 115–124.
78. Poser, Hans: 1976, 'Mathesis instrumentalis? Zur Problematik einer ausschließlich anwendungsorientierten Mathematik', in Hübner, K. *et al.* (eds.), *Die politische Herausforderung der Wissenschaft*, pp. 77–88.
79. Prüß, Karsten: 1974, *Kernforschungspolitik in der Bundesrepublik Deutschland*, Suhrkamp, Frankfurt, pp. 284–290.
80. Radnitzky, Gerard: 1976, 'Dogmatik und Skepsis: Folgen der Aufgabe der Wahrheitsidee für Wissenschaft und Politik', in Hübner, K. *et al.* (eds.), *Die politische Herausforderung der Wissenschaft*, pp. 24–51.
81. Radnitzky, Gerard: 1979, 'Gibt es Irrlehren, die die Freiheit der Wissenschaft bedrohen?', *Conceptus*, XIII/ 32, pp. 93–96.
82. Radnitzky, Gerard: 1980, 'Der politische Wert der nicht-politisierten Wissenschaft', in M. Zöller (ed.), *Aufklärung heute. Bedingungen unserer Freiheit*, Edition Interfrom, Zürich, pp. 99–115.
83. Rip, Arie: 1981, 'A Cognitive Approach to Science Policy, *Research Policy*, X/4, pp. 294–311.
84. Ronge, Volker: 1978, 'Möglichkeiten und Restriktionen der Forschungspolitik', in Hubig/von Rahden (eds.), *Konsequenzen kritischer Wissenschaftstheorie*, pp. 252–272.
85. Sander, Hans-Dietrich: 1978, 'Der Finalismus-Streit', *Criticón*, IL, pp. 241–242.
86. Schäfer, Wolf: 1978, 'Zur Frage der praktischen Orientierung des theoretischen Diskurses. Ein Plädoyer gegen drei Denkverbote der antifinalistischen Wissenschaftsforschung', in Hubig/von Rahden (eds.), *Konsequenzen kritischer Wissenschaftstheorie*, pp. 81–110; English: see 'The Finalization Debate' in this volume.
87. Schanz, Günther: 1974, 'Funktionalisierung der Wissenschaft? Marginalien zum Systemdenken in der Betriebswirtschaftslehre', *Schmalenbachs Zeitschrift für betriebswirtschaftliche Forschung*, XXVI/8, pp. 544–560.

88. Schopman, J.: 1980, 'Finalization and Functionalization', *Zeitschrift für allgemeine Wissenschaftstheorie*, XI/2, pp. 347–353.
89. de Sousa Santos, Bonaventura: 1978, 'Da Sociologia da Ciência à Política Científica', *Revista Crítica de Ciencias Sociais*, I, pp. 11–56.
90. Symes, John M. D.: 1976, 'Policy and Maturity in Science', *Social Science Information*, XV, pp. 337–347.
91. Tietzel, Manfred: 1978, 'Die Finalisierungsdebatte oder: Viel Lärm um nichts', *Zeitschrift für allgemeine Wissenschaftstheorie*, IX/2, pp. 348–360.
92. Topitsch, Ernst: 1976, 'Politisierte Wissenschaft – Ansprüche ohne Legitimation', in Hübner *et al.* (eds.), *Die politische Herausforderung der Wissenschaft*, pp. 159–172.

Addendum to Section II

93. Lenhardt, Christian: 1981, 'Review', *Philosophy of the Social Sciences* **11**, 509–513.

NOTES ON AUTHORS

Dr. GERNOT BÖHME (b. 1937) is Professor of Philosophy and Philosophy of Science at the Technische Hochschule Darmstadt. He studied mathematics, physics and philosophy at the Universities of Göttingen and Hamburg, and was a staff member of the Max Planck Institute in Starnberg (from 1969 to 1977), Visiting Professor at the Institute for Advanced Studies in Vienna (Spring 1974), and Visiting Scholar at the Harvard University Science Center (Spring 1981). His areas of interest include classical philosophy as well as history and sociology of science; he is author of many papers and some books and has currently co-authored a book on *Das Andere der Vernunft. Zur Entwicklung von Rationalitätsstrukturen am Beispiel I. Kants*, Frankfurt, 1983 (together with his brother Hartmut Böhme).

Dr. WOLFGANG VAN DEN DAELE (b. 1939) is Research Associate at the Bielefeld University Science Studies Unit. He studied law and philosophy at the Universities of Hamburg, Tübingen and Munich, and was a staff member of the Max Planck Institute in Starnberg (from 1970 to 1980), and Visiting Professor at the University of Amsterdam (1981). His current areas of interest are the political direction of scientific development and bioethics.

Dr. RAINER HOHLFELD (b. 1942) is Research Associate at the Institute for Science and Society in Erlangen, where he is currently comparing the biomedical research policies in East and West Germany. He studied biology and philosophy in Berlin, Freiburg, Tübingen, and Cologne and did his thesis on the replication pattern of the bacterial chromosome. He was a staff member of the Max Planck Institute in Starnberg (from 1974 to 1980). His areas of interest are the social and political assessment of genetic engineering and the shortcomings of the "biomedical model" for people's health maintenance. Recent contributions are: 'Two Scientific Establishments which Shape the Pattern of Cancer Research in Germany: Basic Science and Medicine', in N. Elias, H. Martins and R. Whitley (eds.), *Scientific Establishments and Hierarchies*, D. Reidel, 1982, and 'The Political Direction of GDR Health Research and Technology', in *Studies in GDR Culture and Society*, Vol. 3, Washington, D.C., 1983.

Dr. WOLFGANG KROHN (b. 1941) is Research Associate at the Bielefeld University Science Studies Unit and Lecturer at the University of Bremen. He studied philosophy and social sciences at the universities of Hamburg, Göttingen, and Marburg, and was a staff member of the Max Planck Institute in Starnberg (from 1970 to 1981) and Visiting Scholar at the Harvard University Science Center (1977). His research focuses on the social origins of modern science and the interrelation between epistemology and sociology of science.

Dr. WOLF SCHÄFER, M.A. (b. 1942) is currently conducting a research project on the concept of Social Natural Science at Frankfurt University. He was a freelance artist (painter) in the early sixties and studied history, philosophy and political science at the universities of Marburg, Bonn, London, and Munich. He was a staff member of the Max Planck Institute in Starnberg (from 1973 to 1981), Visiting Scholar at the Boston University Center for Philosophy and History of Science (Spring 1981, 1982, and 1983), the Harvard University Science Center (Spring 1982), and the M.I.T. Program in Science, Technology, and Society (Fall 1983). His areas of interest are history and theory of the social and natural sciences as well as history of social movements. He has edited classical texts of Wilhelm Weitling (1971) and Ferdinand Lassalle (1972). Recent works include the editing and co-authorship of a small volume of more immediate interest entitled *Neue Soziale Bewegungen – Konservativer Aufbruch im bunten Gewand?*, Frankfurt, 1983, and a collection of his papers dealing with the historical reconstruction and theoretical synthesis of natural and social history: *Historische Umrisse einer anderen Natur- und Sozialgeschichte* (forthcoming).

INDEX OF NAMES

Compiled by Carolyn R. Fawcett

Ackerman, J. S. 199, 200, 203
Adorno, T. 172
Agassi, J. 140, 168, 169, 170
Aitken, H. G. J. 202, 203
Albert, H. 228, 229, 273, 286, 294, 295, 298, 301
Alembert, J. L. R. d' 76, 77, 86
Allen, P. M. 268, 269
Andersson, G. 167, 168, 170, 228, 230, 277, 278, 295, 296, 301, 302, 303
Andreae, J. V. 251
Apian, P. [Petrus Apianus] 177
Aristotle 223
Artz, F. B. 203
Averdunk, H. 200, 203

Bacon, F. 201, 225, 238, 249, 250, 252, 255
Baden, J. 268
Bärmark, J. 303
Baltimore, D. 121, 122, 123
Bauer, K. H. 118, 120, 123
Becker, W. 273, 298
Behrmann, J. 301
Bell, A. G. 189
Bell, D. 9, 11, 287
Bellmann, R. 89
Benhabib, S. xviii
Berenblum, I. 118, 119, 120, 123
Bernal, J. D. 202, 203
Bernard, C. 66
Bernoulli, James 78
Berthollet, C. L. 23, 66
Berzelius, J. J. 23, 24, 30, 39, 48, 49, 50, 58, 66, 67, 86, 88, 299
Bessemer, H. 191, 202
Betz, A. 86, 88, 90
Bieri, P. 172
Bierman, J. 269
Bish, R. L. 268
Boas-Hall, M. 201, 203
Bodewitz, H. 303
Boecker, W. 122, 123
Böhme, G. xvii, xviii, 47, 50, 86, 87, 88, 166, 167, 168, 169, 170, 171, 203, 204, 226, 227, 228, 229, 230, 268, 269, 296, 299, 301, 303, 304
Bohr, N. 134, 167, 171
Borelli, G. A. 79
Borscheid, P. 50, 51
Bossuet, Abbé Jacques 86
Boussingault, J. B. 35, 49, 50
Boutron *see* Boutron-Charland, A. F.
Boutron-Charland, A. F. 66, 67, 68
Boveri, T. 102, 120, 123
Brooks, H. 168, 171
Browne, C. A. 49, 50, 51
Brunner, W. B. 304
Bruno, G. 222
Bucher, O. 122, 123
Buchholz, K. 169, 170, 171, 203
Buchner, E. 60, 67, 70, 86, 88
Büchel, W. 304
Bukarin, N. I. 294
Bunge, G. 48, 51
Bunsen, R. W. von 48
Burgess, P. xviii
Burnet, M. F. 120, 123

Cagniard de la Tour, C. 56, 58, 67
Camus, A. 137
Carnot, S. 190, 202
Carrière, J. 50, 51, 299
Carson, R. 268, 269
Casperson, T. 120, 123
Cassirer, E. 179, 201, 203
Celsius, A. 184
Charles II 223
Clark, G. 201, 203
Cohen, R. S. xviii

Comenius, J. A. 222, 252, 297
Conant, J. B. 67, 89
Condorcet, M. J. A. N. 86
Cooke, W. F. 189
Cope, J. 231
Crick, F. 7, 107, 119, 121, 123
Cruz, O. 124

Dahrendorf, R. 301
Dalton, J. 23
Darwin, C. 3, 10, 11, 17, 225
Davy, H. 26–28, 29, 43, 48, 51, 58
de Jong, T. 304
de Sousa Santos, B. 306
Debus, A. 297
Dee, J. 222
Deffossez, L. 203
Descartes, R. 177, 178, 187, 200, 201, 203, 225, 229, 230, 238, 255
Dijksterhuis, E. J. 202, 203, 297
Döbereiner, J. W. 58
Drebbel, C. 185
Dudek, P. 304
Dülmen, R. van 268, 269
Düsberg, K.-J. 304
Dulbecco, R. 121, 122, 124
Dumas, J. B. A. 25, 48
Durand, W. F. 90

Einstein, A. 138
Elliot, P. 170, 171
Ellis, R. E. 250
Engels, F. 225, 298
Epstein, S. 123, 124
Ertel, H. 50, 52
Eucken, A. T. 203
Euler, L. 56, 61, 73, 74, 76, 77, 78, 79
Ewers, M. 304

Fabroni, A. 58
Fahrenheit, G. D. 184
Faraday, M. 189
Farrington, B. 200, 203
Fehling, H. von 184
Feyerabend, P. K. 248, 250
Finsterwalder, S. 81, 88, 90
Flachsbart, O. 86, 90
Fliedner, T. 118, 124
Fludd, R. 185, 222
Flügge, S. 90
Ford, E. 122, 124
Foulds, L. 121, 124
Frankl, P. 199, 203
Franklin, B. 6
Freitag-Poppe, C.-P. 303
Frémy, E. 66, 67, 68
Frentzel-Beyme, R. 123, 124
Frick, P. 122, 124
Fruton, J. S. 86, 89

Gahrton, G. 123
Galen 223
Galileo 185, 187
Gallmeier, W. M. 122, 123, 124
Gatlin, L. L. 119, 124
Gauss, C. F. 189
Gay-Lussac, J. L. 23, 25, 58
Giacomelli, R. 86, 88, 90
Gilbert, J. H. 45
Gilbert, W. 177, 200
Gilchrist, P. 191
Gille, B. 200, 204
Grebe, J. 268
Greenstein, J. P. 104, 118, 120, 124
Griesbach, L. 124
Gross, L. 121, 122, 124
Grunberg, E. 124

Haar, F. von der 301
Habermas, J. xvii, 201, 204, 226, 227, 229, 230, 250, 264, 275, 298, 299
Hagstrom, W. O. 119, 124
Haldane, E. S. 203, 230
Halfmann, J. 304
Hall, A. R. 204
Hamilton, W. R. 138
Hammond, A. L. 11
Handler, P. 120, 121, 124, 169, 171
Hansen, E. C. 87
Hardin, G. 268
Hartlib, S. 222
Hauksbee, F. 186
Hausen, H. 122, 124
Heath, D. D. 250

Hecker, E. 121, 124
Heckscher, E. 201, 204
Hegedus, Z. 268, 269
Hegel, G. W. F. 239
Heidelberger, C. 120, 124
Heisenberg, W. 7, 73, 74, 131, 133–135, 167, 171
Helmholtz, H. von 78, 80, 90, 189
Hempel, C. G. 168
Henry, J. 189
Henschen, R. 119, 124
Herman, R. 268, 269
Hertz, H. R. 189
Hieber, L. 304
Hiebert, E. N. 48, 51
Hill, C. 222, 226, 228, 229, 230
Hjelt, E. von 49, 51
Hochkeppel, W. 294, 302
Hohlfeld, R. xvii, 118, 124, 166, 169, 170, 301
Holorenshaw, H. 229, 230
Hooke, R. 201, 228
Horstmann, R. 172
Hossfeld, D. 123
Hronszky, I. 304
Hubig, C. 303, 304, 305
Hübner, K. 170, 228, 229, 230, 231, 273, 274, 294, 295, 296, 298, 302, 303, 304, 305, 306
Hucklenbroich, P. 304
Humboldt, A. von 23, 49
Hund, F. 131, 166, 171

Immendorff, H. 50, 51
Ince, S. 86, 90
Ingen-Housz, J. 49
Ichikawa, K. 100, 119, 126

Jacobi, C. G. J. 138
Jacobi, M. H. von 189
Janich, P. 201, 204
Johnston, R. 304
Jonas, H. 248, 250
Jones, H. W. 231
Joukowski, N. 79, 80
Jollow, E. D. 125
Jungk, R. 242, 250

Kant, I. 134, 179, 201, 239, 264
Kärcher, K. H. 123, 124
Kármán, T. von 62, 85, 86, 88, 90
Kellner, O. 50, 51
Kelly, W. 202
Kelvin, W. T. 184
Kennaway, E. 119, 124
Kepler, J. 132, 138
Kinzel, V. 125
Kirchhof, G. S. C. (Gottlieb Sigismund Constantin) *see* Kirchhof, K. S.
Kirchhof, K. S. (Konstantin Sigizmundovich) 58
Kirchhoff, G. R. 78
Kirschbaum 203
Klein, F. 61
Klima, R. 303
Köllmann, W. 48, 51
König, R. 88
Kohler, N. 81, 86, 88, 90
Kohler, R. E. 120, 124
Koller, P. 305
Kopp, H. 48, 51
Koselleck, R. 229, 230
Koyré, A. 201, 204
Kräger, L. 172
Krohn, W. xvii, 11, 50, 87, 123, 168, 169, 170, 171, 172, 200, 201, 203, 204, 230, 296, 299, 301, 303
Kühne, W. 60
Küppers, G. 168, 169, 170, 171, 229, 230, 302
Kuhn, T. S. 6, 7–8, 97, 119, 124, 132–133, 137, 138, 167, 169, 171, 282, 297
Kutschmann, W. 305
Kutta, M. 60, 79, 80, 81, 90, 197

Lakatos, I. 10, 11, 171, 297
Lamb, H. 86, 87, 88, 90
Lanchester, F. W. 79
Laszlo, E. 269
Lavoisier, A. L. 6, 23, 24, 57, 58, 202
Lawes, J. B. 40, 45
Layton, E. 171, 172, 202, 203, 204
Le Châtelier, H. L. 193
Leeuwenhoek, A. van 58

Lehninger, A. L. 120
Lenz, H. F. E. 189
Leonardo da Vinci 79, 176, 178, 201, 204
Lepenies, W. 229, 230
Liebig, J. von 3, 19–20, 21–22, 23, 24–26, 27, 29–31, 33, 34, 35–36, 37, 39, 40–42, 43, 44–45, 46, 47, 48, 49, 50, 51, 56, 57, 58, 65–71, 86, 87, 89, 149, 184, 194, 224, 226, 229, 256, 277, 278–279, 291, 292, 295, 299
Liese, W. 302
Lilienthal, O. 79, 81
Lindsten, J. 123
Lobkowicz, N. 230, 273, 274, 275, 294, 296, 302, 304
Long, F. 11
Louis XIV 223
Lübbe, H. 228, 230, 273, 288, 289, 290, 295, 298, 299, 304, 305
Luhmann, N. 170, 171, 248, 250
Lundgren, P. 230
Luria, S. E. 121, 124

MacCurdy, E. 201, 204
Mach, E. 10, 11
Malthus, T. R. 3, 20–21, 48, 51, 278
Manegold, K.-H. 170, 171
Marconi, G. 189
Marcuse, H. 216, 299
Marey, E. 79
Marko, K. 302
Marx, J. L. 118, 121, 123, 125
Marx, K. 216, 224, 225, 226, 229, 230, 288, 290, 298, 299
Mason, S. F. 201, 204
Maugh, T. H. 118, 121, 122, 123, 124, 125
Maxwell, J. C. 134, 167, 187, 189
Mayr, O. 49, 51, 202–203, 204
Mendel, G. 106
Mendelsohn, E. xviii, 171, 203, 220, 228, 230, 297
Mercator, G. 177
Merchant, C. 258, 268, 269
Mey (*see* Meyer, C.)
Meyer, C. 302
Meyer, E. von 51
Middleton, W. E. K. 201, 204
Mignot, J. 175
Miller, E. C. 120
Miller, J. A. 120
Mitscherlich, E. 56, 58, 67, 68, 69, 86, 87, 89
Mittelstrass, J. 204
Mocek, R. 303, 305
Molitor, B. 305
Moscovici, S. 10, 11, 221, 228, 230, 264, 268, 269
Müller, J. 99, 119, 125
Müller-Reinhard, J. 200, 203
Muller, H. J. 120
Multhauf, R. 202, 204
Musgrave, A. 171
Musson, A. E. 202, 204

Nails, D. xviii
Nauta, L. W. 305
Navier, C. L. M. H. 73, 74, 75, 76, 77, 148, 197
Needham, J. 230
Nelkin, D. 248, 250
Némenyi, P. F. 86, 90
Newton, I. 132, 138, 182, 186, 187, 201
Nitschke, A. 10, 11
Nixon, R. M. 93
Noble, D. 268, 269
Nowotny, H. 303, 305

Oberling, C. 118, 119, 125
Oersted, H. C. 188, 189
Oettgen, H. F. 123, 125
Olby 121, 125
Olschki, L. 200, 204
Oppenheim, P. 168
Orth, E. W. 294
Oswatitsche, K. 87, 90
Ott, G. 124

Pagel, W. 200, 204
Panofsky, E. 199, 204
Paolini, C. 89

Paracelsus [Theophrastus Bombast von Hohenheim] 177, 200
Paré, A. 177, 200
Paris, J. A. 48, 52
Partington, J. R. 52
Pascal, B. 181
Pasteur, L. 56, 59, 66, 67, 68–70, 71, 86, 87, 89, 193–194, 202, 204
Payen, A. 67
Péligot, E. M. 48
Perger, W. A. 302
Persoz, J.-F. 67
Pfetsch, F. R. 230, 305
Piaget, J. 179, 201, 204
Pilgrim, I. 118, 125
Pistolesi, E. 86, 88, 90
Poggendorff, J. C. 89
Poiseuille, J. L. M. 87
Polanyi, M. 119, 125, 171
Pollack, R. 122, 125
Popper, K. 135–137, 138, 167, 171, 172
Poser, H. 273, 295, 302, 305
Potter, R. 104, 118, 120, 125
Prager, F. 200, 204
Prandtl, L. 60, 61, 73, 74, 77, 78, 79, 80, 81, 82, 85, 88, 90, 132, 197
Price, D. J. de Solla 62, 87, 88
Priestley, J. 48
Prigogine, I. 268, 269, 293, 299
Proust, J. L. 23
Prüß, K. 305
Ptolemy 223

Radnitzky, G. 140, 167, 168, 172, 227, 228, 230, 273, 274, 275, 277, 294, 295, 296, 298, 301, 302, 304, 305
Rahden, W. von 303, 304, 305
Rayleigh, J. W. 78
Redtenbacher, F. 192–193, 204
Reichenbach, H. 10, 11
Reiner, N. 75, 90
Reynolds, O. 72, 87
Richter, H. 302
Rip, A. 305
Robinson, E. 202, 204
Ronge, V. 229, 231, 296, 305
Rose, H. 303
Ross, G. R. T. 203, 230
Roszak, T. 248, 250
Rous, P. 110, 121, 125
Rouse, H. 86, 90
Rubin, H. 121, 125
Rumford, B. T. 27

Sander, H.-D. 305
Santorio [Sanctorius], S. 185
Saussure, N. T. de 43, 58
Scaglia, G. 200, 204
Schäfer, W. xvii, 168, 268, 269, 301, 305
Schanz, G. 305
Scheler, M. 204, 264, 267
Schelsky, H. 294
Scheuch, E. K. 273
Schilling, P. 189
Schlichting, H. 62, 63, 80, 86, 90
Schmidt, C. 123
Schmidtgall, H. 49, 52
Schmitt, C. 288
Schmitt, L. 50, 52
Schnitzer, R. J. 124
Schofield, R. S. 202, 204
Schopman, J. 306
Schostack, R. 294, 302
Schuster, H. 302
Schwann, T. 67
Scribner, J. D. 125
Senebier, J. 49
Setzer, H. 303
Sherman, M. J. 122, 125
Sherman, P. 75, 90
Shimkin, M. B. 118, 125
Siegenthaler, W. 124, 125
Singer, P. 268, 269
Skinner, B. F. 245
Smith, Adam 33, 49
Smith, C. S. 204
Sneed, J. D. 88, 138, 140–141, 142, 143, 166, 168, 172
Snell, J. B. 202, 204
Sohn-Rethel, A. 264
Solter, D. 122, 125
Spedding, J. 250

Spengler, T. xvii, 301
Spratt, T. 229, 231
Sprengel, C. 41, 52
Stalin, J. 294
Stanbridge, E. 122, 125
Stegmüller, W. 88, 138, 166, 167, 168, 172
Stehr, N. 88
Stemerding, D. 303
Stengers, I. 268, 269
Stent, G. S. 10, 11, 119, 121, 125, 131, 166, 172
Stephenson, R. 192, 202
Stokes, G. G. 73, 74, 75, 76, 77, 148, 197
Storer, N. 170, 172
Stornaloco, G. 199–200
Sträuli, P. 122, 125
Stürner, W. 10, 11
Sturtevant, A. H. 120, 125
Süss, R. 118, 119, 125
Symes, J. M. D. 168, 172, 306
Szabadváry, F. 48, 52

Tartaglia, N. 176, 177, 200
Taylor, G. I. 132
Taylor, H. M. 86
Teich, M. 303
Temin, H. M. 122, 125
Tenbruck, F. H. 228, 231, 295
Thaer, A. D. 18, 28, 29, 40, 43, 49, 52, 278, 279
Thénard, L. J. 58
Thomas, C. L. 123
Thomas, L. 158, 169, 172
Thomas, S. G. 191, 202
Thünen, J. H. 29, 43, 49, 52
Tietzel, M. 306
Tilden, W. A. 48, 52
Tollmien, W. 86, 90
Tooze, J. 118, 120, 121, 122, 123, 125
Topitsch, E. 273, 294, 303, 306
Torricelli, E. 181
Touraine, A. 268, 269
Traube, M. 66, 67, 69, 87, 89
Triolo, V. A. 118, 125
Truckenbrodt, E. 86, 91
Truesdell, C. 72, 87, 91, 132, 167, 172

Utke, A. R. 249, 250

van den Daele, W. xvii, 47, 50, 51, 86, 87, 123, 124, 168, 169, 170, 171, 201, 203, 216–217, 222, 228, 229, 230, 296, 297, 299, 301, 303
Vesalius, A. 177, 200
Viviani, V. 181
Vogelpohl, G. 86, 90
Volhard, J. 48, 49, 50, 52, 86, 89
Volta, A. 186
Vossenkuhl, W. 302

Wallace, A. R. 10, 225
Wallerius, J. G. 18, 52
Warburg, O. 103–104, 120, 125
Wartofsky, M. xviii
Watson, J. D. 7, 107, 118, 121, 122, 126
Weber, M. 218, 237
Weber, W. 189
Weil, E. 249, 250
Weinfurter, F. 87, 89
Weingart, P. 47, 51, 123, 169, 170, 171, 203, 230, 297
Weiss, R. 121, 126
Weisskopf, V. 138, 166, 167, 168, 172
Weizsäcker, C. F. von xvii, 134–135, 167, 172, 205, 227, 231, 249, 250, 275, 302
Wendel, G. 86, 88
Wheatstone, C. 189
Whitley, R. D. 171, 203, 297
Wieviorka, M. 268, 269
Winstanley, G. 222, 224, 229, 231, 252
Wöhler, F. 25, 31, 48, 67, 86, 89
Wright, O. 60, 81, 197
Wright, W. 60, 81, 197

Yamagiwa, K. 100, 119, 126
Yuan, L. 172

Zamboni, G. 200, 201, 204

Zehm, G. 294, 303
Zierep, J. 91
Zilsel, E. 10, 11, 200, 201, 205
Zimmer, D. E. 303
Znaniecki, F. 167, 172
Zöller, M. 305

BOSTON STUDIES IN THE PHILOSOPHY OF SCIENCE

Editors:
ROBERT S. COHEN and MARX W. WARTOFSKY
(Boston University)

1. Marx W. Wartofsky (ed.), *Proceedings of the Boston Colloquium for the Philosophy of Science 1961-1962.* 1963.
2. Robert S. Cohen and Marx W. Wartofsky (eds.), *In Honor of Philipp Frank.* 1965.
3. Robert S. Cohen and Marx W. Wartofsky (eds.), *Proceedings of the Boston Colloquium for the Philosophy of Science 1964-1966. In Memory of Norwood Russell Hanson.* 1967.
4. Robert S. Cohen and Marx W. Wartofsky (eds.), *Proceedings of the Boston Colloquium for the Philosophy of Science 1966-1968.* 1969.
5. Robert S. Cohen and Marx W. Wartofsky (eds.), *Proceedings of the Boston Colloquium for the Philosophy of Science 1966-1968.* 1969.
6. Robert S. Cohen and Raymond J. Seeger (eds.), *Ernst Mach: Physicist and Philosopher.* 1970.
7. Milic Capek, *Bergson and Modern Physics.* 1971.
8. Roger C. Buck and Robert S. Cohen (eds.), *PSA 1970. In Memory of Rudolf Carnap.* 1971.
9. A. A. Zinov'ev, *Foundations of the Logical Theory of Scientific Knowledge (Complex Logic).* (Revised and enlarged English edition with an appendix by G. A. Smirnov, E. A. Sidorenka, A. M. Fedina, and L. A. Bobrova.) 1973.
10. Ladislav Tondl, *Scientific Procedures.* 1973.
11. R. J. Seeger and Robert S. Cohen (eds.). *Philosophical Foundations of Science.* 1974.
12. Adolf Grünbaum, *Philosophical Problems of Space and Time.* (Second, enlarged edition.) 1973.
13. Robert S. Cohen and Marx W. Wartofsky (eds.), *Logical and Epistemological Studies in Contemporary Physics.* 1973.
14. Robert S. Cohen and Marx W. Wartofsky (eds.), *Methodological and Historical Essays in the Natural and Social Sciences. Proceedings of the Boston Colloquium for the Philosophy of Science 1969-1972.* 1974.
15. Robert S. Cohen, J. J. Stachel and Marx W. Wartofsky (eds.), *For Dirk Struik. Scientific, Historical and Political Essays in Honor of Dirk Struik.* 1974.
16. Norman Geschwind, *Selected Papers on Language and the Brain.* 1974.
18. Peter Mittelstaedt, *Philosophical Problems of Modern Physics.* 1976.
19. Henry Mehlberg, *Time, Causality, and the Quantum Theory* (2 vols.). 1980.
20. Kenneth F. Schaffner and Robert S. Cohen (eds.), *Proceedings of the 1972 Biennial Meeting, Philosophy of Science Association.* 1974.
21. R. S. Cohen and J. J. Stachel (eds.), *Selected Papers of Léon Rosenfeld.* 1978.
22. Milic Capek (ed.), *The Concepts of Space and Time. Their Structure and Their Development.* 1976.
23. Marjorie Grene, *The Understanding of Nature. Essays in the Philosophy of Biology.* 1974.

24. Don Ihde, *Technics and Praxis. A Philosophy of Technology.* 1978.
25. Jaakko Hintikka and Unto Remes, *The Method of Analysis. Its Geometrical Origin and Its General Significance.* 1974.
26. John Emery Murdoch and Edith Dudley Sylla, *The Cultural Context of Medieval Learning.* 1975.
27. Marjorie Grene and Everett Mendelsohn (eds.), *Topics in the Philosophy of Biology.* 1976.
28. Joseph Agassi, *Science in Flux.* 1975.
29. Jerzy J. Wiatr (ed.), *Polish Essays in the Methodology of the Social Sciences.* 1979.
31. Robert S. Cohen and Marx W. Wartofsky (eds.), *Language, Logic, and Method.* 1983.
32. R. S. Cohen, C. A. Hooker, A. C. Michalos, and J. W. van Evra (eds.), *PSA 1974: Proceedings of the 1974 Biennial Meeting of the Philosophy of Science Association.* 1976.
33. Gerald Holton and William Blanpied (eds.), *Science and Its Public: The Changing Relationship.* 1976.
34. Mirko D. Grmek (ed.), *On Scientific Discovery.* 1980.
35. Stefan Amsterdamski, *Between Experience and Metaphysics. Philosophical Problems of the Evolution of Science.* 1975.
36. Mihailo Marković and Gajo Petrović (eds.), *Praxis. Yugoslav Essays in the Philosophy and Methodology of the Social Sciences.* 1979.
37. Hermann von Helmholtz: *Epistemological Writings. The Paul Hertz/Moritz Schlick Centenary Edition of 1921 with Notes and Commentary by the Editors.* (Newly translated by Malcolm F. Lowe. Edited, with an Introduction and Bibliography, by Robert S. Cohen and Yehuda Elkana.) 1977.
38. R. M. Martin, *Pragmatics, Truth, and Language.* 1979.
39. R. S. Cohen, P. K. Feyerabend, and M. W. Wartofsky (eds.), *Essays in Memory of Imre Lakatos.* 1976.
42. Humberto R. Maturana and Francisco J. Varela, *Autopoiesis and Cognition. The Realization of the Living.* 1980.
43. A. Kasher (ed.), *Language in Focus: Foundations, Methods and Systems. Essays Dedicated to Yehoshua Bar-Hillel.* 1976.
46. Peter L. Kapitza, *Experiment, Theory, Practice.* 1980.
47. Maria L. Dalla Chiara (ed.), *Italian Studies in the Philosophy of Science.* 1980.
48. Marx W. Wartofsky, *Models: Representation and the Scientific Understanding.* 1979.
50. Yehuda Fried and Joseph Agassi, *Paranoia: A Study in Diagnosis.* 1976.
51. Kurt H. Wolff, *Surrender and Catch: Experience and Inquiry Today.* 1976.
52. Karel Kosík, *Dialectics of the Concrete.* 1976.
53. Nelson Goodman, *The Structure of Appearance.* (Third edition.) 1977.
54. Herbert A. Simon, *Models of Discovery and Other Topics in the Methods of Science.* 1977.
55. Morris Lazerowitz, *The Language of Philosophy. Freud and Wittgenstein.* 1977.
56. Thomas Nickles (ed.), *Scientific Discovery, Logic, and Rationality.* 1980.
57. Joseph Margolis, *Persons and Minds. The Prospects of Nonreductive Materialism.* 1977.

58. Gerard Radnitzky and Gunnar Andersson (eds.), *Progress and Rationality in Science.* 1978.
59. Gerard Radnitzky and Gunnar Andersson (eds.), *The Structure and Development of Science.* 1979.
60. Thomas Nickles (ed.), *Scientific Discovery: Case Studies.* 1980.
61. Maurice A. Finocchiaro, *Galileo and the Art of Reasoning.* 1980.
62. William A. Wallace, *Prelude to Galileo.* 1981.
63. Friedrich Rapp, *Analytical Philosophy of Technology.* 1981.
64. Robert S. Cohen and Marx W. Wartofsky (eds.), *Hegel and the Sciences.* (Forthcoming).
65. Joseph Agassi, *Science and Society.* 1981.
66. Ladislav Tondl, *Problems of Semantics.* 1981.
67. Joseph Agassi and Robert S. Cohen (eds.), *Scientific Philosophy Today.* 1982.
68. Władysław Krajewski (ed.), *Polish Essays in the Philosophy of the Natural Sciences.* 1982.
69. James H. Fetzer, *Scientific Knowledge.* 1981.
70. Stephen Grossberg, *Studies of Mind and Brain.* 1982.
71. Robert S. Cohen and Marx W. Wartofsky (eds.), *Epistemology, Methodology, and the Social Sciences.* 1983.
72. Karel Berka, *Measurement.* 1983.
73. G. L. Pandit, *The Structure and Growth of Scientific Knowledge.* 1983.
74. A. A. Zinov'ev, *Logical Physics.* (Forthcoming).
75. Gilles-Gaston Granger, *Formal Thought and the Sciences of Man.* 1983.
76. R. S. Cohen and L. Laudan (eds.), *Physics, Philosophy and Psychoanalysis.* 1983.
77. G. Böhme et al., *Finalization in Science*, ed. by W. Schäfer. 1983.